SOURCES OF POWER

SOURCES OF POWER

How Energy Forges Human History

VOLUME 1
BEFORE OIL: THE AGES OF
FORAGING, AGRICULTURE, AND COAL

Manfred Weissenbacher

Praeger Perspectives

PRAEGER
An Imprint of ABC-CLIO, LLC

A B C C L I O

Santa Barbara, California • Denver, Colorado • Oxford, England

Library of Congress Cataloging-in-Publication Data

Weissenbacher, Manfred.
    Sources of power : how energy forges human history / Manfred Weissenbacher.
        p. cm.
    Contents: v. 1. Before oil : the ages of foraging, agriculture, and coal—v. 2. The oil age
and beyond.
    Includes bibliographical references and index.
    ISBN 978-0-313-35626-1 (set)—ISBN 978-0-313-35627-8 (set : ebook)—
ISBN 978-0-313-35628-5 (v. 1 : hardcopy : alk. paper)—ISBN 978-0-313-35629-2
(v. 1 : ebook)—ISBN 978-0-313-35630-8 (v. 2 : hardcopy : alk. paper)—
ISBN 978-0-313-35631-5 (v. 2 : ebook)
1. Agriculture and energy—History.   2. Coal—Social aspects.
3. Petroleum—Social aspects.   4. Civilization.   I. Title.
    S494.5.E5W33 2009
    333.79—dc22   2009023895

13   12   11   10   09   1   2   3   4   5

This book is also available on the World Wide Web as an eBook.
Visit www.abc-clio.com for details.

ABC-CLIO, LLC
130 Cremona Drive, P.O. Box 1911
Santa Barbara, California 93116-1911

This book is printed on acid-free paper (∞)

Manufactured in the United States of America

To my parents

Please visit the author's Web site at
www.manfredweissenbacher.com

CONTENTS

PART III. COAL AGE

VOLUME 2. THE OIL AGE AND BEYOND

INTRODUCTION

We have all asked these questions. Why are some of the world's regions so rich and others so poor? Why are European languages spoken dominantly on four of the world's six populated continents? Why is Africa's largest lake named for a British queen? Why were the Russians first to put a man into space? And how come the United States emerged as the world's sole remaining superpower at the end of the 20th century? However, answers that explain the big picture and the overall course of world history are rare. They are often designed to flatter the culture, religion, or intellect of those who are better off. Or they focus on environmental factors such as pronounced seasons, and on resistance to disease. And many historians reject ideas of orderly patterns in world history altogether: they consider the course of history chaotic, or at least emphasize the unpredictability of human behavior.

There is, however, a wide consensus on one issue: most people would agree that the emergence of agriculture has been the most critical milestone in human history. Agriculture triggered the rise of complex civilizations, as farmers were able to produce surplus food that sustained not just themselves and their families, but many other members of society who in turn specialized in newly emerging occupations to become full-time potters, weavers, miners, soldiers, priests, bankers, and so on. As it has long been known that agriculture emerged much earlier in Eurasia than in Africa and the Americas, it should thus not surprise us that Eurasians eventually wiped out or enslaved native Africans and Americans, and not vice versa. Eurasians simply capitalized on their agricultural head start, which provided them with higher population densities, more advanced technology, and better weapons.

This leaves us with the question why agriculture actually *did* emerge earlier one place than another. Charles Darwin, among many others, pointed at climatic conditions to provide an answer. In *The Descent of Man* (1871), he explained that agricultural production is necessary to provide a basis for civilization, and speculated about the role of a temperate climate, though he concluded that "the problem, however, of the first advance of savages towards civilisation is at present much too difficult to be solved."[1] A more differentiated theory about how climatic conditions somewhat indirectly facilitated the emergence of agriculture has been popularized by Jared Diamond's 1997 book *Guns, Germs, and Steel*.[2] Diamond argues that the world's asymmetric endowment with species that lend themselves well to domestication is behind the delays in the onset of agriculture on different continents, and thus the course of world history. Research of the past decades has indeed shown that some regions are very rich, and others very poor, in terms of natural occurrence of species that agricultural people tend to eat. Most important were large-seeded grass species such as wheat, rice, and maize, which still now account for the bulk of the nutrition sustaining the global human population. As many as 32 of the world's 56 species of large-seeded grasses are native to the Mediterranean. The situation is similarly lopsided with respect to the world's large terrestrial mammal species. (These were not all that important in terms of food supply, but became indispensable for draft and portage, and decisive in warfare.) Worldwide, only 14 such mammal species proved suitable for domestication, with 13 of them descending from wild species found exclusively in Eurasia and North Africa.

The main problem associated with this theory is that it does not explain modern history. True, native Africans, Americans, and Australians may have been disadvantaged by the delayed onset of their farming, but there was not much time difference in the emergence of agriculture in western Eurasia compared to eastern Eurasia. Why isn't Chinese the main language spoken in the Americas today? Why did India become a colony governed by Europeans? Why did Britain rise to superpowerdom? And why did British global power decline at the expense of the United States of America and the Soviet Union? If we attempt to understand history beyond the 16th century, we certainly have to find answers that go beyond agriculture and its effects on societal development.

My personal quest for a general theory of world history began with a set of questions concerning the 20th century. Most importantly, I was curious how it was possible for the Soviet Union, a nation still languishing in technological backwaters in the early 20th century, to emerge as a superpower that controlled all of Eastern Europe and challenged the United States internationally. After all, the Soviet Union enjoyed little internal cohesion and practiced an economic system that did not allocate resources very well.

My approach to this question may have been quite conservative at first. My education was straightforwardly humanistic, and I studied Latin for six years, translating such classics as Caesar's *De Bello Gallico* (on the Roman conquest of what is now France) and Ovid's *Metamorphoses*. History had been presented to me as a long list of empires and battles, and each country in the world seemed to cherish the supposed brilliancy of its own respective generals and statesmen.

However, as I moved on to study first chemistry and then economics, my views and methods changed. After developing energy storage systems during my PhD years, I eventually took a position at the Stanford Research Institute, where one of my early assignments concerned the advantages and disadvantages of central governments to control energy prices. Around this time I began formulating a theory of world history that would apply to all epochs. Most notably, my consternation regarding Soviet superpowerdom disappeared like the Cheshire cat as soon as I saw the statistical figures of international oil production. Which two nations pumped the most oil by far cumulatively during the 20th century? The Soviet Union and the United States of America, the two principal superpowers of the 20th century. Coincidence? Hardly. In the century before, when coal was the most critical fuel, there was just one single country mining far more of it cumulatively than any other nation: It was Britain, the sole superpower of the 19th century and creator of the largest empire ever to exist on the face of the Earth. These observations in themselves prompted me to postulate that command of energy has probably always been the most important factor in the fate of societies and nations. What else could have been? After all, energy is formally defined as the ability or capacity to do work. But before I could claim this hypothesis to be universally correct, I had to check whether it applied to preindustrial times as well.

Let's go to the very beginning. Whenever I ask people what they think was the first major energy source commanded by humans, they tend to say: fire! This is a tempting choice indeed, but it is actually wrong. The first kind of energy that humans had access to, as trivial as it may sound at first, was the food that they found in nature. Accordingly, the human body itself was the first prime mover that people had at their disposal. And what a prime mover it was! Strong, endurable, and incredibly flexible! Useful to build shelters, to produce tools, and to find nutritional energy to fuel itself. And yet the notion of the human body being a prime mover was pushed to a whole new level as soon as slave-based agrarian societies emerged.

We have already heard that the emergence of agriculture is considered the most important milestone in human history. But why did agriculture make such a big difference and trigger the rise of complex civilizations? Because it was a source of energy! Think about it for a moment. What differentiates plants from animals is that plants can directly utilize sunlight to fuel their life. Animals (including humans), on the other hand, depend on the chemical energy that plants store in their tissue during growth. As humans happen to be unable to digest most of the cellulose-based plant material that tends to fill natural landscapes, original foraging societies had quite limited amounts of nutritional energy at their disposal. They had to collect edible fruits, nuts, seeds, and tubers from wide landscapes that were dominated by biomass people could not eat. Farmers, in contrast, removed natural vegetation with all its leaves, branches, bushes, and trunks, and directed sunlight towards the growth of edible plant material. Most importantly, they cultivated thin-stemmed grass species that store a lot of chemical energy in large digestible seeds.

By this stage, we have already identified the three principal fuels that have critically influenced the fate of human societies up until this day: grain, coal, and oil. Accordingly, we can divide human history into four main Energy Eras:

Foraging Age
Agricultural Age
Coal Age
Oil Age

Each of these eras is defined by its principal pair of fuel and prime mover. The work done by grain-fueled muscles dominated the Agricultural Age. Human workers (often slaves) were the principal engines of this era, and animal power was utilized where suitable beasts were available and fields (or natural landscapes) provided enough nutritional energy to feed them. In first approximation, the pairs of the succeeding eras were coal and the steam engine, followed by oil and the internal combustion engine. Coal-fired steam engines were initially used to pump water from mines, but soon powered all sorts of factory equipment. Once steam engines were fitted to boats and railroad cars, they also revolutionized human mobility and the transport of goods. However, oil-fueled internal combustion engines had an even bigger impact, as they powered motor vehicles and airplanes.

As societies took the three consecutive *energetic steps*—towards grain, coal, and oil—the amount of energy controlled by humans increased radically. Pick up a bag of rice and it will tell you that 100 grams of pealed rice contain about 350 kcal (kilocalorie) worth of energy. A man of average size needs to eat some 2,400 kcal worth of food energy a day to be able to function and to do work. In comparison, the amount of technical (inanimate) energy consumed daily per person living on the planet averaged around 2,100 kcal in 1860 and 18,000 kcal in 1950. By the end of the 20th century the figure was 41,000 kcal. In terms of food energy, 41,000 kcal per day would be sufficient to nourish 17 working men. Hence the average world citizen by now has a whole horde of personal slaves at her or his disposal in terms of technical energy.

But in reality there is no such thing as the average world citizen. At the beginning of the 21st century Americans consumed some 220,000 kcal, Latin Americans 35,000 kcal, and South Asians 11,000 kcal per day and person in technical energy. It is such differences in the command and consumption of energy that point at a fundamental understanding of world history: Those who commanded more energy have always been better off. They used energy in productive ways to increase their wealth, and they used energy in destructive ways to defeat their neighbors. Throughout history those who commanded more energy have killed, expelled, enslaved, or, at the minimum, politically controlled those who commanded less energy.

Although a one-sentence summary of this book would read "Command of energy is the most important determinant in world history," I am not going to claim that energy has fully determined every aspect of the fate of societies or nations. Rather, it laid out the broad patterns: energy resources set the outer limits as to what could be done within or by a society. Within their energetic boundaries people were free to make choices. There was thus space for all the curiosity, greed, love, hate, jealousy, folie de grandeur, and all the other human elements that may have influenced world history at one point or another. Energy realities can explain why the Aztecs did not conquer Spain, and why Germany lost World War II, but it would have been impossible to predict that Hitler would provoke a large-scale war (despite Germany's obvious lack of oil resources), and that the Spanish rather than the Portuguese would be first to sail to America. By the time the Soviet Union collapsed, it was the world's largest oil producer: Something had obviously gone wrong despite the availability of vast energy resources. (But the Siberian oil deposits are still there, and we can only guess what role the Russian Federation is going to play in the 21st century.)

Much energy technology served both peaceful and belligerent purposes. Horses, trains, ships, trucks, and airplanes—by increasing human mobility—improved productivity and trade as well as military operations. However, the principal imperative of warfare was the ability to kill from a distance: It increased from javelin to bow-and-arrow, to firearms, to bombs dropped from airplanes, and finally to intercontinental ballistic missiles. Societies seem to have invaded other societies throughout history simply because they had the means to do so. But the disposition or aptness to act aggressively was generally enhanced when the balance between command of destructive versus productive energy was out of proportion within a society. The mounted peoples of central Asia harassed settled Eurasia for centuries. They commanded lots of destructive energy as they had plenty of horses and superior compound bows, but their productive energy was minimal. Hence they kept being attracted by the riches of settled agrarian societies and attacked them regularly up until the time sedentary civilizations developed firearms.

An imbalance between productive and destructive energy command may also arise when destructive energy resources prevail while productive energy resources diminish. For various reasons some agricultural systems sustained empires literally for millennia, while other farming schemes collapsed after just a few centuries. (And what is happening in America right now? The United States rose to superpowerdom based on its enormous domestic oil resources, which are now rapidly declining.) A more obvious path towards such energetic imbalance is the spread of arms technology. The recent proliferation of weapons of mass destruction to relatively poor countries created an immense gap between destructive and productive energy commanded in these societies. Generally, the existence of weapons of such enormous destructive energy has decreased predictability and widened the bandwidth of possible futures: with enough nuclear warheads around to blow every person off the face of the Earth, who is to say what the future of humankind is going to be?

This set of books is divided into two volumes and five parts. The first four parts demonstrate how people proceeded towards more energy command; how their life changed in the process; and how political history unfolded as societies took the three energetic steps in different regions at different times. The fifth part is a look into the future.

Part I, titled "Foraging Age," very briefly presents what was by far the longest era in human history. Globally, it lasted from the time various proto-humans roamed the Earth until about 10,000 years ago, when agriculture first emerged. This was the period when people were still spreading to all the continents; when there were hardly any differences between and within human societies; and when no artificial borders existed. The Foraging Age represents something of an energy baseline for human societies, an era in which people were sustained by natural energy flows. Gathering-hunting, the original way of life, provided no consistent oversupply in food energy, and the principal energy innovation of this era, the command of open fire, had somewhat limited applications. Mobile lifestyle and nutritional limitations kept population growth in check, and humans lived a lot like less intelligent animals do. Thus our distant ancestors seemed to represent a fairly inconspicuous species: it would have been hard to guess that these creatures would eventually reshape landscapes, overfish the oceans, and fly to the moon.

Part II, titled "Agricultural Age," covers the long time period from 10,000 years ago until the end of the 18th century C.E.,[3] when coal technology emerged. The start of this era marks an extraordinary turning point in human evolutionary development: Humans stopped adapting to natural environments (as animals normally do), and instead began to consciously shape the environment to their own needs on an ever vaster scale. This lacks comparison with any other species[4] and marks the point when humans truly began conquering the world. They became the dominant species on Earth and henceforth adapted to the artificial environment they had themselves created, rather than to nature as it once was.

The Agricultural Age proceeded in three periods. The first period concerned developments within continents only. Agriculture emerged independently in western Eurasia, eastern Eurasia, Mesoamerica, South America, and sub-Saharan Africa. People used quite different sets of domesticated species to capture sun energy in these regions, but the consequences were the same everywhere: as more nutritional energy was available, population growth accelerated, more human prime movers were around, and a lot more work could be done. Professional specialization promoted technological progress, and soon farming societies produced materials (most importantly hard metal) that do not occur in nature and were thus inaccessible to gatherer-hunters. Settled lifestyle and professional specialization triggered the emergence of social classes, while centralized governance and organized religions arose to manage agrarian communities.

Soon after the emergence of agriculture we see the first *energy campaigns*. Accelerated population growth prompted farming societies to expand further and further into gatherer-hunter regions in search for more nutritional energy and arable land. These expansions were bad news for indigenous foragers, as farmers commanded a lot more energy (more food and more people) and had better weapons. In this first wave of energy invasions agrarians committed something of a global genocide of gatherer-hunters, which irreversibly changed the genetic make-up of the human population. The only option for foragers to escape the fate of being killed, enslaved, or expelled was to rapidly adopt agriculture, or else to retreat to marginal lands.

Some of the early agrarian expansions are surprisingly well documented. At first it was rather loose groups of farmers migrating to new areas, but when empires emerged the campaigns into gatherer-hunter land became more organized. Empires kept expanding until they bordered seashores, areas unsuitable to agriculture, or other empires. (Sometimes they disintegrated into smaller entities, but the total area under agrarian administration then tended to remain the same.) And for a quite simple reason territorial expansion remained the name of the game for the entire Agricultural Age: The energy commanded by an empire was directly related to the field area under its authority. What is more, campaigns yielded slaves who served as additional prime movers. (The world's first empires appropriately emerged around the original centers of agriculture. Our whirlwind tour through political history will therefore start in Mesopotamia. From there we will quickly proceed to Egypt, Greece, Rome, and to the more northern regions of Europe.)

The Agricultural Age's second period was initiated by Europeans crossing the world's oceans and opening the world. In this period differences between continents became relevant. This era saw the near or full extinction of peoples native to the Americas and the beginnings of large-scale enslavements of West Africans. Wherever Europeans encountered people with inferior weapons and no resistance to Eurasian disease they slaughtered, infected, and enslaved them much the same way as was previously done by technologically advanced people on an intra-continental scale. And with natives disappearing from overseas fertile land, Europeans proliferated into many the world's temperate regions by introduction of Fertile Crescent domesticates. (The Fertile Crescent is the original center of agriculture in western Eurasia.) Hence, this was really a classic agrarian expansion, just on a global scale.

Compared to the Americas, the situation was entirely different in South and East Asia. In India and China European explorers encountered civilizations that were richer and more advanced than their own. Europeans nevertheless had two main advantages in terms of energy command, but these were limited to the open seas and did not enable them to actually conquer or occupy Asia's densely populated areas. First, Europeans had superior sailing vessels, which allowed them to harness wind energy for mobility as nobody else could. Second, they had the world's best naval cannons, thus commanding vast amounts of destructive energy. This combination enabled Europeans to become pirates of the Indian ocean, to terrorize Asian coastlines, and to control much of the intra-Asian maritime trade.

The third and last period of the Agricultural Age was what I call the "Super-Agricultural Era." It was characterized by the effects of New World domesticates spreading to the Old World and vice versa. As maize, potato, and sweet potato was grown in Eurasia and Africa, and wheat in the Americas, the world's arable land was better matched with domesticates suited for various local climates. (This was especially relevant as the global climate was changing around that time.) Hence, the worldwide flow of agricultural energy soared to unprecedented levels, and all the effects associated with the initial emergence of agriculture were accelerated. From around 1650 C.E. the global human population grew about ten times faster than before, and by 1750 C.E. some 750 million people populated the planet (compared to 300 million in the year 1000 C.E., and 500 million in 1650 C.E.). As more people will work, experiment, and think more, this was a period of fast technological progress. Europe experienced a scientific revolution that involved agricultural reforms and eventually climaxed into an industrial revolution in which waterpower drove complex machinery, transforming first the textile and then other industries.

Part III, titled "Coal Age," presents the time period that began in the late 18th century and lasted until the early 20th century. In this period global energy consumption skyrocketed as the development of steam engines allowed the chemical energy contained in coal to be translated into mechanical work. The ascent of coal began with its adoption as a fuel by certain industries in timber-lacking England. Most importantly coal was soon mined in large quantities to fuel the expansion of the iron industry, which had earlier exclusively relied on charcoal (that is, a biofuel). The coal-fueled steam engine was an entirely new type of prime mover, based on the expansion and condensation of water vapor. It remained crude and inefficient for a long time, but eventually matured to propel anything from production machinery to railroad trains and steam ships. An even more powerful prime mover, the coal-fired steam turbine, further increased the speed of steam ships and allowed for coal energy to be efficiently turned into electricity.

These developments redefined the energy mix that influenced political history. World energy (and thus political power) was no longer dominated by agricultural energy. The previous motivation for empires to expand diminished as coal deposits delivered energy that was unrelated to land area exposed to sunlight. Once nations had secured access to coal, the aim was to control large international (colonial) markets. These served as outlets for products delivered by efficient coal-fueled domestic industries. Britain (where coal technology first emerged) and other western European countries (where coal technology was rapidly adopted) competed fiercely in the international arena. And with nearly no delay the United States of America, while still expanding towards the Pacific coast, joined the contest.

Coal energy and the availability of unprecedented quantities of iron and steel revolutionized warfare and military power during the Coal Age. Coal-fired steamers were capable of penetrating rivers, the hitherto secure inland lifelines of agricultural societies. The Western powers could thus overcome the main shortcoming of their wind-powered oceangoing vessels, and finally made a break into Asia's interiors. India was made a colony, and China was firmly controlled. Africa was partitioned between all western European Coal Powers, and Britain snatched remote Australia and New Zealand. The United States began to engage in coal-fueled imperialism in Latin America as well as the North Pacific, where Hawaii was absorbed as a state and the Philippines acquired as a colony. What is more, U.S. cannon steamers forced Japan out of seclusion: Looking for fuel, they threatened Japanese coastal cities, demanding the construction of coaling stations and access to Japanese markets.

On land, coal-fired railroad locomotives redefined transport (and therefore military strategy). Most people had traditionally been living along shorelines and rivers. These regions were now being connected to the continents' interiors by railroad tracks. Much of the radical territorial expansion of the United States and Russia during the Coal Age, for instance, would have been impossible without steam trains. Meanwhile short-distance travel and agriculture was still dominated by animate power. The Coal Age saw a resurgence of slavery (predominantly in the U.S. South), and the world's horse population soared to unprecedented levels. (The cities were full of horse-drawn carts, and Californian harvester-thresher combines were being pulled by teams of up to 40 horses.)

As we would expect, the enhanced energy flows in Coal Age societies accelerated the effects observed following the emergence of agriculture. More work was taken over by machines, more people were released towards new tasks, and the rate at which knowledge accumulated gained more speed. But the shift towards coal-energy involved great hardships: Many of the traditional jobs disappeared, and scores of laborers and their families were pushed into urban poverty. In the longer run, though, the Coal Age saw major improvements for workers, a prolonged life expectancy for all, and an increase in the global rate of population growth. Chemicals extracted from coal tar were used to make both the first mass-produced pharmaceuticals and a new class of very powerful explosives (nitroglycerin, dynamite, TNT). The latter helped to increase the range of firearms, which were developed into more precisely engineered and deadly weapons (repeating rifles, machine guns).

Part IV, titled "Oil Age," covers the time period that began around World War I and continues until the present. Crude oil prospecting and refining techniques were developed from the mid-19th century to serve the lighting market. Only kerosene was sold as lamp fuel, while the other crude oil components (including gasoline, diesel, and asphalt) were initially discarded. Eventually gasoline and diesel were used to fuel internal combustion engines, whose pistons are moved directly by the expanding gases generated by the combustion of a fuel inside a cylinder. (In contrast, steam engines burn coal externally and direct water vapor into the cylinder.) These new lightweight engines revolutionized mobility as they allowed for the construction of off-rail motor vehicles and airplanes.

Global energy flows in the Oil Age reached levels ten times those of the Coal Age. Hence the world experienced an unparalleled push towards further technological progress and increase in wealth. Electrical appliances took over many household tasks; automobiles provided for unmatched individual mobility; and large passenger jets eventually opened extreme long-distance mobility for the masses. Urbanization proceeded even further, while oil-fueled tractors and harvesting machines increased agricultural output in combination with artificial fertilizers and pesticides. Pharmaceuticals improved the life-expectancy of Oil Age people, and the global population soared from two billion in 1930 to six billion in 1999.

Oil technology emerged mainly in western-central Europe and soon dominated military strategy in its manifestation as tanks and bomber planes. But western Europe lacked oil resources. European nations therefore used their powerful international position (read: their Coal Age might) to initiate policies whose consequences are still now influencing world politics. Most notably Britain, the principal superpower of the Coal Age, had no oil at all, none domestically (the North Sea reservoirs were discovered only after World War II), and curiously almost none in its huge global colonial empire. (The oil resources of Burma, then part of British India, were the one exception.) Nevertheless, Britain in 1912 decided to switch its navy from British coal to foreign oil. This risky step, taken to increase the speed of the British battle fleet, involved dependence on fuel shipped to Europe half way around the globe from Burma and Persia (Iran). World War I was fought right at the interface of the Coal Age and the Oil Age. Primitive airplanes and tanks entered the scene, but the conflict was generally dominated by coal energy. Germany, rich in coal but entirely without oil, eventually capitulated without losing a single decisive battle.

World War II, on the other hand, was a full-fledged Oil Age war. Germany lost it after failing to reach the oil fields of Azerbaijan, and Japan entered it for the sake of gaining control over the (then Dutch-owned) oil provinces of Indonesia. In the end, World War II was won through American oil. However, right after Germany's defeat the Russians developed the rich oil fields of the Volga and Ural basins. Consequently the world's two principal oil producers, the United States and the Soviet Union, emerged as superpowers from the global conflict and divided the world into two opposing zones of influence. Outside the United States and Russia the richest oil deposits were discovered in the Middle East (southwestern Asia). Western companies developed the oil fields of this region, but the area turned out difficult to control politically. The United States soon showed a special interest in the Middle East, as domestic oil discoveries wound down: American oil production started its long, slow decline in 1970.

Protected by large oceans to its east and west, the United States during World War II had taken over the safe-island position that Britain had enjoyed for centuries until it came into the reach of German zeppelins, airplanes, and missiles. However, the United States and the Soviet Union both perfected unmanned German missiles into, on the one hand, rockets that carried satellites and people into space, and on the other hand, intercontinental ballistic missiles that could reach every corner of every continent on the planet. These missiles soon carried nuclear warheads to deliver destructive energy of a magnitude that evades imagination.

Part V, titled "Beyond the Oil Age," takes a look into the future. When global oil production is starting to decline, what alternatives do we have in terms of liquid fuels or other sources of energy for transportation? And how should we produce electricity for those large parts of the world that are still waiting for their energy liberation: the adoption of a more energy-rich lifestyle with all the consequent benefits as experienced in the industrialized world? As a review of the most recent population projections shows, this is not going to be easy, especially because the issue is complicated by the perceived or real threat of global climate change. Futurists tend to say,

"The best way to foresee the future, is to create it." But in terms of climate change, this is a bit tricky. The climate record recovered from nature clearly indicates that we must expect to soon enter a new Ice Age, not a very attractive thought. On the other hand, fossil fuel burning has elevated atmospheric concentrations of carbon dioxide, a greenhouse gas that keeps the planet's surface warm, to unprecedented levels. If these amounts of carbon dioxide are not absorbed by the system (additional dissolution in the oceans; or increased plant growth as indicated by recent global greening[5]), or countered by means of technology (releasing dust in the atmosphere to cool it, adding iron to the oceans to promote algae growth, etc.), we need to expect a trend of global warming that, in extent and speed, outstretches the warming periods observed in recent history. If we look exclusively at the Coal Age and Oil Age, then global temperatures reached a record peak in 1998, while regional temperatures did not. (NASA recently had to revise its temperature record for the United States. The warmest year since the late Coal Age was not 1998, as previously claimed, but rather 1934, the year the infamous Dust Bowl devastated the Midwest. Five of the 10 warmest years were between 1920 and 1939, only one of the 10 was in the 21st century.) On global average, 1998 remained the warmest year in the period 1998–2008, with temperatures now being at about the same level as they were during the Middle Ages, when vineyards flourished in England, and Viking farmers settled in Greenland. But it remains unclear what is going to happen with the Earth's climate in the future, and just about everyone would agree that it is a bad idea for humankind to keep conducting an uncontrolled experiment on its own habitat. Unfortunately, the computer models currently employed remain weak in terms of predicting what the effects of global warming might be on a regional or local scale. This question is not to be taken lightly, as the consequences of global climate change have to be weighed against the benefits that the unhampered use of inexpensive fossil energy would provide for the world, and especially for the developing world. The only carbon-neutral energy alternatives that are currently competitive on a large enough scale are nuclear energy and waterpower. However, much of the world's waterpower potential has already been tapped, and nuclear (fission) energy remains controversial due to questions of security and final waste storage. The other nuclear energy technology, nuclear fusion, holds the potential of leading the world into a great new Energy Era, but currently remains illusive. Billions of dollars are being pumped into the development of a nuclear fusion reactor, but severe material limitations have yet to be overcome. Thus, the world seems to be heading into a Nuclear Age based on established nuclear fusion technology, or into a Second Coal Age, in which much of the energy needs is met by cheap, abundant, and fairly widely distributed coal, supplemented by various forms of renewable energy. Technology to capture carbon dioxide directly at the coal-fired plant already exists, but currently costs about a third of the energy gained. Similarly, coal liquefaction technology is in place (as it had been developed by Nazi Germany during World War II, and in South Africa during the international anti-apartheid embargo), but it is not yet economical. Thus, oil will in the near future remain the critical liquid fuel for transportation, and in terms of economic and military power. But how will this translate into global political developments?

We are now in the period that began with the collapse of the Soviet Union, but is dominated by the fact that the sole remaining superpower, the United States of America, is running out of oil. The traditional U.S. approach to this problem was to strive for control of the oil-rich Middle East, but the disastrous developments during the Iraq occupation for a moment seemed to change the attitude. At least in part under the disguise of a sudden interest in climate protection, an issue long marginalized by America to the despair of the rest of the world, the U.S. government seemed to aim for a new kind of energy self-sufficiency in terms of transportation fuel, notably by promoting the production of bioethanol. However, this was set in perspective when the United States announced in July 2007 that arms sales and military aid to U.S. allies Saudi Arabia, Israel, and Egypt were stepped up to arm the Middle East with more sophisticated weapons systems than ever before. It is questionable whether this will help the United States cling to global power. There are currently strong indications that relative or absolute U.S. decline will be accompanied by the rise of such populous Asian countries as China and India. But these countries do not have oil either. Energy-rich Russia will likely be able to improve its global strategic position, and has already reversed the trend of liberalizing and privatizing its oil and gas sectors. The Middle East will be in possession of increasingly larger shares of the remaining oil, and the whole picture has to be viewed in terms of the continuing global proliferation of weapons of mass destruction. To get a clear view of what is most likely to happen in the near future, we need to review the lessons learned from the previous Energy Eras and their transitions, and apply them to the current situation. And we need to assess how strong the energy theory of human history is in comparison to other theories explaining world history.

NOTES

1. Charles Darwin, *The Descent of Man, and Selection in Relation to Sex* (New York: Barnes & Noble Publishing, 2004), 113.

2. Jared Diamond, *Guns, Germs, and Steel: The Fates of Human Societies* (New York: W. W. Norton & Company, 1997).

3. C.E., or Common Era, replaces the religiously and culturally (more) biased term A.D. Accordingly, the now outdated term B.C. has been replaced by B.C.E., Before the Common Era. However, the new system indicates the same points in time, or time periods, as the old system according to the Gregorian calendar.

4. There are a few exceptions at a lower level. Semi-aquatic beavers, for instance, quite dramatically shape their environment through the construction of dams.

5. United Nations Environment Programme, *Global Environment Outlook, GEO Year Book 2003, International Environmental Agenda*, Box 4: Greening of the biosphere (Nairobi: UNEP, 2003), http://new.unep.org/geo/yearbook/yb2003/box7a.htm.

BIBLIOGRAPHY

Darwin, Charles. *The Descent of Man, and Selection in Relation to Sex*. New York: Barnes & Noble Publishing, 2004. (Original edition, London: John Murray, 1871, available online at Project Gutenberg, http://www.gutenberg.org/etext/2300).

Diamond, Jared. *Guns, Germs, and Steel: The Fates of Human Societies*. New York: W. W. Norton & Company, 1997.

United Nations Environment Programme. *Global Environment Outlook, GEO Year Book 2003, International Environmental Agenda*, Box 4: Greening of the biosphere. Nairobi: UNEP, 2003. http://new.unep.org/geo/yearbook/yb2003/box7a.htm.

PART I

Foraging Age

Of the few remaining gatherer-hunter groups, the San of southern Africa's Kalahari desert are my favorites. They seem to have an especially friendly mindset, they have no real warrior tradition, and they speak Khoisan languages that are intriguingly unique in their use of various click sounds as consonants.[1] What the San share with all other present-day foragers is that they survive in areas unsuited for agriculture. All the planet's hospitable and fertile places are now in the hands of farmers who expelled gatherer-hunters from these regions a long time ago. The gatherer-hunters we observe today are therefore somewhat of foraging superstars, able to survive where no one else could. They may thus not represent the lifestyle of the Foraging Age perfectly, but they certainly come quite close.

People of the Foraging Age acquired all food as it occurs in nature. They hunted, trapped and fished animals, and they gathered fruits, nuts, seeds, and tubers. As they did not engage in food production at all, there was no consistent oversupply in nutritional energy, which restrained population size and limited the amount of work potentially done by any foraging society. On the other hand, gatherer-hunters did not invest much work into acquiring nutritional energy either. Even though present-day foraging societies exclusively inhabit marginal areas, they enjoy amounts of leisure time that would embarrass any farmer. The !Kung, a San group of the Kalahari desert, spend only about two and a half hours a day searching for food, which adds up to less than 20 hours a week. (The "!" in !Kung denotes a click sound.) Similarly, the Hadza nomads of Tanzania, who are Khoisan speakers as well, devote around 14 hours each week to obtain food,[2] though food processing has to be added to these work hours. Canadian Inuits hunt aggressively for but one month in spring and another month in fall during the seasonal caribou migrations. Meat is then stored (in permafrost) to be consumed during the off-season.

NUTRITIONAL ENERGY

Like that of any other species, the survival of humans depended on a simple imperative: People had to extract more energy out of their environment than they expended in the process of doing so. Only under this condition would individuals be able to survive and produce offspring, and populations to maintain their size. The energy balance sheet explains why the human diet was initially overwhelmingly vegetarian in most regions. Humans are omnivores and can digest both plant and animal food, but animals attempt to escape while plants are fixed in location. Chasing an animal can easily turn the energy balance into the negative, while gathering plant food tends to be very rewarding. Chimpanzees, genetically our closest surviving relatives, eat up to 40 different kinds of fruit a day and supplement their diet with ants and termites. The large mountain gorillas do not eat flesh at all and even consume insects only if they happen to be present on ingested leaves.

Similarly, most present-day gatherer-hunter groups acquire over three-quarters of their food calories by gathering. (That's why it is appropriate to change the traditional term hunter-gatherer into gatherer-hunter.) The highest net energy return, amounting to 30 to 40 units of food energy acquired for every unit expended, is achieved by gathering roots. More typically, gathering tends to return 10 to 20 times the energy expended in the process. Hunting small- to medium-sized (or arboreal) mammals, on the other hand, may easily result in a net energy loss.[3] San hunters save energy by developing expert track-reading skills: When they hit an animal with a poisoned arrow, they do not immediately rush after it, but slowly follow its track until it finally falls. However, for our distant ancestors, the most energy-efficient way to secure meat may have been by scavenging, which is speculated to have been the first mechanism to include significant shares of animal protein in the human diet. Perhaps people began to scavenge animal fat and meat seasonally during the African dry period that leaves many animals (such as antelope) dead. Once people had learned how to open bones with stone tools, they quite possibly began to eat relatively large amounts of bone marrow. (This was a resource for which faster scavengers such as vultures and hyenas could not compete.) The only instance when actual hunting would approach the energy returns of food gathering was when large fatty mammals were concerned. Such animals were found right in those cold climates where a lack of adequate plant food did not leave people much choice other than to base their diet on meat. (Energy-rich fat serves as insulator for animals adapted to cold temperatures.) Among the surviving gatherer-hunter societies it is the Inuit of the Canadian Arctic who depend most heavily on meat. To avoid

Gatherer-hunters While hunting scenes tend to look more spectacular, foraging societies in most regions secured the bulk of their nutritional energy by gathering, which was typically the task of women. It is therefore more meaningful to refer to such societies as gatherer-hunters rather than hunter-gatherers. The top picture was taken in 1907 and is titled "A Negrito huntsman with his aboriginal war weapon, the long bow and arrow, Philippine Islands." The photograph on the bottom was taken in North Queensland, Australia, in ca. 1920. It shows a family group of aboriginal natives posing with a dog in front of a low hut. The men are holding spears and present their prey. (Library of Congress images LC-USZ62-108516, edited, and LC-USZ62-46880, edited.)

nutritional deficiencies due to the lack of vegetables and fruit in their environment, they supplement their diet by eating the partially digested stomach contents of caribou.

Tools for acquiring and processing food were made of wood, stone, and bone, all materials readily found in nature. Simple wooden tools were almost certainly used first, but already 2.5 million years ago hominids produced crude stone tools by hitting one stone onto another, with minimal shaping of the sharp parts that split off. Increasingly symmetrical stone tools appeared from about 1.7 million years ago and may have served as hand axes and cleavers. Around 200,000 years ago a major step forward in stone tool-making technology was taken, and by 40,000 years ago people had learned to produce strong, standardized stone blades of triangular cross-section. These were used as spear- and arrowheads, and for other composite devices, while bone was used to make fish hooks (and needles).

MOBILE WAY OF LIFE

Foraging communities lived mobile lives out of seasonally shifting campsites by following ripening plants or migrating game. (The San usually shift their homesites once a month, as soon as the local food supply is exhausted.) Housing took the form of simple huts, tents or lean-tos made of plant materials or animal skin. If available, caves were incorporated into the yearly cycle of seasonal camps.

The mobile lifestyle imposed several limitations on gatherer-hunter societies. We observe that mothers in foraging societies keep their children spaced, at something like four-year intervals. They tend to bear several young during their lifetime, but a mother must carry her baby until it is old enough to keep up with the adults. (In addition, the available food is often not easily digested by the very young, and breast-feeding thus may have to continue for over three years, which delays the return of fertility after childbirth.) Spacing children contributed to the low population density of about one to several hundred foragers per hundred square kilometers. Both annual birth and death rates were very high in comparison to Oil Age standards, but they equaled one another, which kept the size of foraging populations constant. Active ways of limiting population size (and removing weaker parts of society) have been observed among present-day gatherer-hunters as well. These include infanticide (killing of infants), invalidicide (killing of invalids), geronticide (killing of the old), and delayed marriage.[4]

Another feature of the nomadic lifestyle is that people have little private property. All possessions are limited to what people can carry with them, which helps to maintain a classless society. No individual accumulates more wealth, knowledge, or property than another, as the degree of professional specialization is extremely low, and each member of society engages in almost all activities as defined by gender roles. In those current foraging groups that indeed practice both hunting and gathering, men hunt while women gather plant food and do most domestic chores. San women, for instance, provide for most of the society's nutritional energy by gathering readily available plant and animal foods. San men hunt with light bows and poisoned arrows, but this is often the less successful activity.

In some present-day foraging groups extreme discrimination of women has been observed, with men often retaining a sense of ownership over their family.[5] There are usually rules that require to marry outside of the immediate family or band, and it is the brides who have to leave their families and are exchanged between bands. Arguably the fact that females rather than males have to leave their families may have practical reasons. Inuit couples take residence with the husband's band, as this allows him to keep on hunting in his familiar environment and with his familiar partners, while the wife's activities are more focused on the camp itself.

Within a band, which is effectively an extended family or a group of several akin families, all individuals are closely related either by marriage or birth or both. Early humans presumably wandered and foraged in small groups of perhaps a few dozen individuals, and later gatherer-hunter bands probably did not exceed 200 individuals. In such mobile egalitarian societies leadership is often nonhereditary, and increased prestige or contribution to decision making is only attained by proving special talent or skill. There is usually no political organization except for the web of kinship that links individuals. However, some groups of North American Inuits establish an additional network to ensure survival in the harsh Arctic environment. Every hunter forms partnerships with people (outside of his designated kinsmen) with whom he shares certain parts of his prey. These partnerships are extremely close and used to involve wife-sharing as well as avenging each other's death.

SEDENTARY GATHERER-HUNTERS

To be sure, there were also a few regions globally that offered such extraordinary natural food supply that gatherer-hunters were able to adopt a quasi or fully sedentary lifestyle featuring many aspects of the later agricultural way of life. Among these regions were the handful of areas where agriculture was eventually going to emerge based on large-seeded grasses. But there are also examples for gatherer-hunter societies that became sedentary based on nutritional energy derived from animals. Mammoth hunters living in the Moravian loess region of the present-day Czech Republic built stone houses, fired clays, and produced a variety of excellent tools. (Even a small single mammoth provided as much edible energy as fifty reindeer.)[6] In some coastal regions gatherer-hunter populations expanded on the basis of gathering shellfish, trapping fish, or hunting fatty marine mammals such as seals. (Killing migrating baleen whales in northwestern Alaska may provide net energy returns of more than 2000-fold.[7]) In the Pacific Northwest (North America's northwest coast) massive yearly runs of fatty salmon supported settlements of several hundred inhabitants. This, too, is an example of sedentary hunters expending little energy by trapping seasonally migrating prey. Lacking Arctic permafrost for storage, these foragers smoke-dried their salmon for consumption during the off-season and gathered acorns to make flour.

COMMAND OF FIRE

Gatherer-hunters did not have draft animals, and they did not utilize wind or waterpower. They relied fully on their own muscle power, fueled by nutritional energy obtained from food as it was found in nature. Gradually more sophisticated tools helped foragers to augment their muscle power, but the distance from which humans could kill remained directly proportional to their muscle strength: It limited how far a spear could be thrown, and how hard a bow be drawn.

Eventually people learned to control fire, which provided for an entirely new kind of energy flow in form of visible and heat radiation. Open fires allowed people to cook food, warm themselves, light up campsites, drive game animals, harden wooden spear heads, and to clear landscapes. Ambiguous evidence, such as burnt bones and reddened sediments, suggests that humans may have controlled fires in Africa 1.5 million years ago. Hard evidence, in form of the oldest known fire hearths discovered in China, Germany, and Hungary, only dates to some 400,000 years ago, though a more recent find in northern Israel seems to push back previously accepted dates for people's fire-making ability to 790,000 years.[8] Fire hearths indicate that hominids of the time had entirely diverged from the usual primate eat-on-the-go strategy of feeding. Cooking did not just warm food. It gradually broadened the human diet as it made many plants more easily digestible or even detoxified them.[9] Grilled meat spoils less easily because it contains less water than raw meat, and heating food before consumption generally kills harmful bacteria.

The warmth and light provided by open flames may have had an even more profound impact on people's life than cooking, especially in areas of cold climate and long winter nights. Not even rain would extinguish torches made of resinous wood such as pine. (These burn for a very long time due to their turpentine content.) Some 35,000 years ago people in Europe also started to use oil lamps that burned animal fat. On the other hand, the undirected, untamed energy released by open fires had relatively limited productive applications. Most notably, open fires could not be used to do any constructive mechanical work. All the energy used to collect food, make clothing, build shelters, and produce tools derived from nutrition that fueled human muscles.

TOWARD A NEW ERA

Due to their low energy command people remained largely inconspicuous during the Foraging Age. Nevertheless this era has special significance: It saw the emergence of anatomically modern humans and their spread to all continents. The earliest creatures classified as truly hominid according to their fossilized remains appeared in Africa some five million years ago. The oldest skeletons of anatomically fully modern people were also unearthed in Africa: they date to nearly 200,000 years ago.[10] Some evidence of cultural traits such as religious practices and the creation of art and music showed up in the archaeological record from about 80,000 years ago, but is not extensively documented until some 50,000 years ago. Around this time something of a Great Leap Forward towards modern human behavior seems to have occurred.[11]

Hominids had begun migrating out of Africa about one and a half million years ago,[12] spreading over wide parts of Eurasia but not yet to the Americas and Australia. Anatomically fully modern people followed them from just before 50,000 years ago. Soon thereafter all earlier types of humans disappeared in Eurasia just as they did in Africa. Neanderthals actually survived in western Eurasia until (at least) a short 30,000 years ago, when they finally lost out against the new human species that had better ways to extract nutritional energy from natural environments. This extinction ended the wonderful five-million-year-or-so epoch during which different human species and subspecies had populated the Earth side-by-side: all people now living on the planet belong to the same species of anatomically fully modern humans. (Individuals are formally defined as members of the same species if they are capable of producing fertile offspring together.)

The anatomically fully modern people of our kind soon spread beyond Eurasia as well. Already some 40,000 years ago they reached Australia, and at least 14,000 years ago some had crossed the Bering land bridge into North America. A short millennium later, around 13,000 years ago, the first people had reached the southernmost tip of South America.[13] Soon thereafter, about 10,000 years ago, people in especially rich environments began to cultivate plants and raise animals. The Foraging Age came to a close. Earth was never going to be such equal place again. Gatherer-hunter societies were internally egalitarian, but they were also quite equal in comparison to one another: people all over the world commanded similar energy flows and developed technology at a similar pace. But this epoch, after thousands of centuries of gatherer-hunter history, which was free of empires and artificial borders, ended when people in some regions began to use new technology to extract unprecedented amounts of energy from their environment.

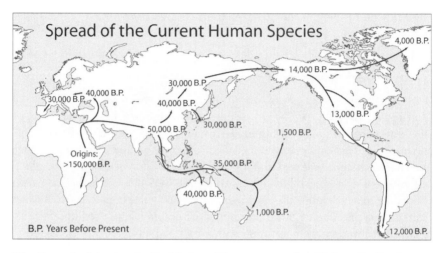

The Spread of Anatomically Modern Humans over the Globe   Earlier human species, which are now extinct, lived millions of years ago confined in Africa, while *Homo erectus* had spread quite widely over southern parts of Eurasia by one million years B.P. (before the present). Anatomically fully modern humans, belonging to the only human species that exists today, emerged in Africa nearly 200,000 years ago, and had begun migrating out of Africa just before 50,000 years ago. Such humans in turn even spread beyond Eurasia, into the Americas and Australia.

NOTES

1.   Listen to the distinctive sounds of a Bushman click language, !Ora (pronounced kora), at the National Geographic Magazine's Web site: "Bushmen: Last Stand for Southern Africa's First People," http://ngm.nationalgeographic.com/ngm/0102/feature6/index.html. Possibly, low population numbers and density have contributed to the peaceful mindset now observed in the !Kung San of the Kalahari desert. David Adams, "There Is No Instinct for War," translated from a Russian article published in the Moscow-based Academy of Sciences of the USSR's *Psychological Journal* 5 (1984): 140–44, http://www.culture-of-peace.info/instinct/title-page.html. To be sure, as peaceful as they might be now, even the !Kung and similar cultures did engage in some warfare in earlier times. R. B. Lee, *The !Kung San* (Cambridge: Cambridge University Press, 1979).

2.   Jared Diamond, "The Worst Mistake in the History of the Human Race," *Discover*, May 1987, 64–66.

3.   Vaclav Smil, *Essays in World History: Energy in World History* (Boulder, CO: Westview Press, 1994), 21.

4.   Joel E. Cohen, *How Many People Can the Earth Support?* (New York: W.W. Norton & Company, 1995).

5.   Find an example of discrimination of women among current foraging groups in Robert M. Glasse and Shirley Lindenbaum, "The Highlands of Neuguinea—

A Review of the Ethnographic and Related Problems," in *Essays on Kuru*, ed. R. W. Hornabrook (Faringdon, England: EW Classey, 1976).

6. Vaclav Smil, *Essays in World History: Energy in World History* (Boulder, CO: Westview Press, 1994).

7. Find more on this topic in G. W. Sheehan, "Whaling as an Organizing Focus in Northwestern Alaskan Eskimo Societies," in *Prehistoric Hunter-Gatherers: The Emergence of Cultural Complexity*, ed. T. D. Price and J. A. Brown (San Diego: Academic Press, 1985), 123–54.

8. The oldest evidence of fire control used to date from around 500,000 years ago in Europe. Compelling, although not conclusive, evidence that fire was used 790,000 years ago (likely by homo erectus people) was then found in Israel in 2004. A 2008 study presented hints in the form of the spatial distribution of burned and unburned flint microartifacts that hominids at the Gesher Benot Ya'aqov archaeological site actually made fire at will rather than just preserving fire from natural conflagrations 790,000 years ago. This gave rise to speculations that the control of fire actually helped such hominids to migrate out of Africa.

James Randerson, "Charred Remains May Be Earliest Human Fires", *NewScientist. com*, April 29, 2004. The article cites a journal reference: *Science* 304 (2004): 725, http://www.newscientist.com/channel/being-human/human-evolution/dn4944. Nira Alperson-Afil, "Continual Fire-making by Hominins at Gesher Benot Ya'aqov, Israel," *Quaternary Science Reviews* 27 (2008): 1733–39. http://www.sciencedirect.com/science/article/B6VBC-4TC2S1S-1/2/cc0a30ae30e55da73541e977a5f5 caea.

9. For instance, many kinds of dry beans that we are now familiar with contain quite large quantities of a toxic protein that is destroyed by boiling them in water—but this was still out of reach for gatherer-hunters. (The young pods of many bean species, on the other hand, may be eaten raw.)

10. Those who prefer a story of early humans based on genetic markers can find information here: H. Kaessmann, and S. Pääbo, "The Genetical History of Humans and the Great Apes," *Journal of Internal Medicine* 251 (2002): 1–18. http://www.ncbi.nlm.nih.gov/entrez/query.fcgi?cmd=Retrieve&db=PubMed&list_uids=11851860&dopt=Abstract; Bryan Sykes, *The Seven Daughters of Eve: The Science That Reveals Our Genetic Ancestry* (New York: W.W. Norton & Company, 2001); L. L. Cavalli-Sforza, P. Menozzi, and A. Piazza, *The History and Geography of Human Genes* (Princeton: Princeton University Press, 1994).

11. T. D. Price and J. A. Brown, eds., *Prehistoric Hunter-Gatherers: The Emergence of Cultural Complexity* (San Diego: Academic Press, 1985), 123–54. It is speculated that this Great Leap Forward may have been associated with development of speech. Wolfgang Enard et al., "Molecular Evolution of FOXP2, a Gene Involved in Speech and Language," *Nature* 418 (2002): 869–72. http://www.ncbi.nlm.nih.gov/entrez/query.fcgi?cmd=Retrieve&db=PubMed&list_uids=12192408&dopt=Abstract.

12. Possibly, the initial out-of-Africa migration began a bit earlier. Scientists have dated hominid remains found in Israel and Georgia, for instance, to up to 1.8 million years ago.

13. National Geographic, "Atlas of the Human Journey," https://www3.national geographic.com/genographic/atlas.html.

BIBLIOGRAPHY TO PART I

Adams, David. "There Is No Instinct for War." Translated from a Russian article published in the Moscow-based Academy of Sciences of USSR's *Psychological Journal* 5 (1984): 140–44. http://www.culture-of-peace.info/instinct/title-page.html.

Alperson-Afil, Nira. "Continual Fire-making by Hominins at Gesher Benot Ya'aqov, Israel." *Quaternary Science Reviews* 27 (2008): 1733–39. http://www.sciencedirect.com/science/article/B6VBC-4TC2S1S-1/2/cc0a30ae30e55da73541e977a5f5caea.

Cavalli-Sforza, L. L., P. Menozzi, and A. Piazza. *The History and Geography of Human Genes*. Princeton, NJ: Princeton University Press, 1994.

Cohen, Joel E. *How Many People Can the Earth Support?* New York: W.W. Norton & Company, 1995.

Cook, Earl. *Man, Energy, Society*. San Francisco: W.H. Freeman, 1976.

Cottrell, Fred. *Energy and Society—The Relation between Energy, Social Change, and Economic Development*. New York: McGraw-Hill Book Company, 1955.

Diamond, Jared. "The Worst Mistake in the History of the Human Race." *Discover*, May 1987.

Diamond, Jared. *Guns, Germs, and Steel: The Fates of Human Societies*. New York: W.W. Norton & Company, 1997.

Enard, Wolfgang, M. Przeworski, S. E. Fisher, C. S. Lai, V. Wiebe, T. Kitano, A. P. Monaco, and S. Pääbo. "Molecular Evolution of FOXP2, a Gene Involved in Speech and Language." *Nature* 418 (2002), 869–72. http://www.ncbi.nlm.nih.gov/entrez/query.fcgi?cmd=Retrieve&db=PubMed&list_uids=12192408&dopt=Abstract.

Forde, C. Daryll. *Habitat, Economy and Society: A Geographical Introduction to Ethnology*. New York: E.P. Dutton & Co., Inc., 1946.

Glasse, Robert M., and Shirley Lindenbaum. "The Highlands of Neuguinea—A Review of the Ethnographic and Related Problems." In *Essays on Kuru*, edited by R. W. Hornabrook. Faringdon, England: EW Classey, 1976.

Kaessmann, H. and S. Pääbo. "The Genetical History of Humans and the Great Apes." *Journal of Internal Medicine* 251 (2002): 1–18. http://www.ncbi.nlm.nih.gov/entrez/query.fcgi?cmd=Retrieve&db=PubMed&list_uids=11851860&dopt=Abstract.

Lee, R. B.. *The !Kung San*. Cambridge: Cambridge University Press, 1979.

National Geographic Magazine. "Bushmen—Last Stand for Southern Africa's First People." http://ngm.nationalgeographic.com/ngm/0102/feature6/index.html.

National Geographic Magazine. "Atlas of the Human Journey." https://www3.nationalgeographic.com/genographic/atlas.html.

Price, T. D., and J. A. Brown, eds. *Prehistoric Hunter-Gatherers: The Emergence of Cultural Complexity*. San Diego: Academic Press, 1985.

Randerson, James. "Charred Remains May Be Earliest Human Fires." *NewScientist.com*, April 29, 2004. http://www.newscientist.com/channel/being-human/human-evolution/dn4944.

Sahlins, Marshall. *Stone Age Economies*. Chicago: Aldine-Atherton, 1972.

Sheehan, G. W. "Whaling as an Organizing Focus in Northwestern Alaskan Eskimo Societies." In *Prehistoric Hunter-Gatherers: The Emergence of Cultural Complexity*, edited by T. D. Price and J. A. Brown, 123–54. San Diego: Academic Press, 1985.

Smil, Vaclav. *Essays in World History—Energy in World History*. Boulder, CO: Westview Press, 1994.

Sykes, Bryan. *The Seven Daughters of Eve: The Science That Reveals Our Genetic Ancestry*. New York: W.W. Norton & Company, 2001.

PART II

Agricultural Age

Sometime in the early 1990s I had a strange encounter at an international conference in Chicago. The convention ended with a gala banquet, and I happened to be seated at a large table at which I was the only person to opt for the vegetarian meal. Following a few raised eyebrows, a broad Midwestern accent asked: "And what's wrong with the steak?" I did not intend to make anyone feel uncomfortable about their dinner choice, so I smiled and answered: "Probably nothing, I'd just like to have—for once—something in common with Einstein, da Vinci, Edison, and Schweitzer." But they did not let me off the hook that easily. One colleague stressed the importance of meat as a protein source, another claimed that we all descend from mammoth-chasing hunters, and one voice shouted at me from the other side of the table: "And how about Hitler? Wasn't he vegetarian, too? Why don't you mention *him?*"

That Hitler was vegetarian is actually a myth. His doctors put him on a vegetarian diet for health reasons (a stomach condition), but he never stayed entirely away from meat for prolonged periods of time, quite in contrast to his contemporary Mahatma Gandhi, who devoted himself to life-long vegetarianism. Nevertheless, Hitler's documented affection for animals in parallel to his disregard for human life was, to say the least, quite a paradox.

Concern about animal welfare probably ranks first among the reasons why an increasing number of people in industrialized Western countries are turning vegetarian. Consumers are now well-informed about farm animals' extremely close confinement, prophylactically applied drugs, cannibalistic species-to-species protein feeding, the use of growth hormones, conveyor belt slaughtering, and more. Many disapprove of these measures as they feel we are already living in a world of affluence, and it is unnecessary to increase animal suffering in order to boost meat growth and milk productivity ever further for the purpose of cost efficiency. Other consumers choose a vegetarian diet because they are concerned about their own health. Cardiovascular diseases are now by far the largest killer in the world, and the saturated fatty acids contained in meat products are associated with the formation of plaque that narrows or blocks arteries.[1] However, reducing your meat consumption has also been cited as the single most efficient thing you can do to reduce your negative impact (or footprint) on the environment. Large-scale animal raising consumes a lot of water and produces enormous amounts of excrement and litter. Livestock also releases significant amounts of greenhouse gases associated with global climate change. (Yes, livestock is flatulent and burps a lot. According to a 2007 study, eating one kilogram of beef creates as much climate alteration as driving an average car for 250 kilometers.[2] Though this has so far attracted little attention, the livestock sector is now responsible for a larger share of greenhouse gas emissions, 18 percent, than transportation fuels, 14 percent.[3])

However, at the core of the adverse environmental effects of meat consumption is an energy problem. Nutritional energy starts out as sunlight and is converted into chemical energy by plants, which pass it to herbivores, which pass it to carnivores. Unfortunately, every conversion and transfer of energy involves losses, and no digestive system is fully efficient. If you eat a steak today it will contain only about 10 percent of the nutritional energy the bovine has been eating to grow that meat. (Not to mention that a lot of energy goes into building body parts such as bones and horns that carnivores cannot digest.) This would not be an issue if farm animals were left to grazing or other natural forms of feeding. But cattle, pigs, and poultry are fattened with grain fit for direct human consumption. This connects to a complex global humanitarian problem, as some 800 million people are going hungry every day in the early 21st century: They are too poor to bid on world markets for grain that ends up being fed to livestock. In the United States and other industrialized countries over half, and globally over one third, of the grain harvested on the fields finds its way into the stomach of farm animals, even though this grain could feed the hungry (or else serve the production of biofuel). It is a distinct feature of the current Energy Era, the Oil Age, that many regions have become so spoiled by the huge amounts of energy available through oil and coal, that they could afford to begin feeding most their grain to animals and still have plenty of cereals for direct human consumption. The technical energy now invested into the production of agrochemicals, agricultural machinery, and fuel to keep the tractors moving by far exceeds the solar energy captured by plants and converted to consumed nutritional energy.[4] In short, we Oil Age people have turned agriculture into an energy sink. And this is why we tend to forget how it all began, with agriculture being our greatest source of energy, and field work creating the initially small energy surpluses that allowed for settled, literary civilizations to emerge.

NOTES

1. World Health Organization, *Cardiovascular diseases* (Geneva: WHO, 2007), http://www.who.int/mediacentre/factsheets/fs317/en/index.html.

2. Akifumi Ogino, Hideki Orito, Kazuhiro Shimada, and Hiroyuki Hirooka, "Evaluating Environmental Impacts of the Japanese Beef Cow-calf System by the Life Cycle Assessment Method," *Animal Science Journal* 78 (2007): 424, http://www.blackwell-synergy.com/doi/abs/10.1111/j.1740-0929.2007.00457.x. Find more on this topic in: Vegetarian Society, "Information Sheets," http://www.vegsoc.org/info; People for the Ethical Treatment of Animals (PETA), "Factsheets," http://peta.com/mc/facts.asp.

3. H. Steinfeld et al., "Livestock's Long Shadow—Environmental Issues and Options," *LEAD-Livestock Environment and Development Initiative*, coordinated by the UN FAO's Animal Production and Health Division (Rome: FAO, 2006), http://www.virtualcentre.org/en/library/key_pub/longshad/A0701E00.htm, http://www.virtualcentre.org/en/library/key_pub/longshad/A0701E00.pdf. Read more about this in this book's global climate change chapter in "Beyond the Oil Age."

4. In the United States, about 10 to 15 calories of fossil fuel energy are currently used to create 1 calorie of food. The food system uses about 17 percent of the U.S. annual energy budget, and is thus the single largest consumer of petroleum products when compared to any other industry. In 1945 one calorie of energy input into corn production yielded 4 calories of energy output, but this return diminished to 2.4 calories output for every 1 calorie input by 1979. Energy use is higher for fruits and vegetables and highest for animal products. Fruits and vegetables required two calories input to yield one calorie of output while animal proteins required 20 to 80 calories of energy input for one calorie of energy output in 1979. J. Hendrickson, "Energy Use in the U.S. Food System: A summary of existing research and analysis," Center for Integrated Agricultural Systems, College of Agricultural and Life Sciences, Wisconsin Institute for Sustainable Agriculture, University of Wisconsin-Madison (1996), http://www.cias.wisc.edu/wp-content/uploads/2008/07/energyuse.pdf.

CHAPTER 1

WHAT IS AGRICULTURE?

For early farmers agriculture[5] was the struggle to carry more nutritional energy from the fields than was expended during field work and food processing. However, there were principal limitations as to how much nutritional energy any area of land could possibly provide. First of all, the amount of solar radiation shining onto any region per day is limited. The sun's radiation output is in itself limited; areas are turning away from the sun during the night; and clouds may block sunlight during daytime. Second, plants can capture only a tiny fraction of the solar radiation that does indeed reach the Earth's surface. (This limitation is inherent to the chemical reactions that are promoted by sunlight during plant growth.) And third, lack of such growth factors as water, nitrogen, phosphor, or potassium may restrain plant growth long before the limits of photosynthesis are approached.[6]

On the other hand, even the earliest farmers made a tremendous difference in terms of capturing solar energy by directing the available sunlight towards the growth of biomass that humans can eat. They cleared natural landscapes and turned them into managed fields on which only small shares of the arriving sun radiation fueled the growth of indigestible stems and leaves or unwanted weeds, while much solar energy was captured in form of digestible plant material. This radically increased the amount of nutritional energy available to humans per acre.

Plant seeds were especially important. They contain carbohydrates, oils and often quite a large share of protein. They have a high energy density and are easy to store, which is critical for complex societies thriving on seasonal harvests. Both the high energy content and the limited spoilage are

The Food of Agricultural People Just three grass species, wheat, rice, and maize (corn) cover over half of the nutritional energy now globally consumed by humans. All plant food combined accounts for some 85 percent. (Photograph of wheat (top left) by Aviad Bublil. Photograph of rice (bottom left) by David Nance, Agricultural Research Service (ARS), U.S. Department of Agriculture, Image Number K2958-2. Photograph of maize (top right) by Manfred Weissenbacher.)

founded in the low water content of cereals, which differentiates them from fruits and tubers. Currently, the seeds of just three grasses, wheat, maize (corn) and rice, account for more than half of the nutritional energy globally consumed by people. All plant food combined accounts for about 84 percent. (Vegetables, oil crops, fruits, and sugar provide for nearly a fifth of the nutritional energy globally ingested by people; roots, tubers, and pulses provide for about 10 percent.[7])

Meat was a lot less important for sedentary civilizations, because far more people can be sustained per acre if they choose a vegetarian diet. Plants utilize sunlight directly to fuel their life, while animals (including humans) depend on the chemical energy that plants store in their tissue during growth. Herbivores are therefore sustained by second hand energy obtained from plants, and things look even worse for carnivores, which acquire their energy third hand, after it was passed from sunlight to plants to herbivores and finally to the meat-eater. As every step involves losses, carnivores can access only a tiny percentage of the solar energy initially captured by plants in any given area. (The reality of these energy flows is quite visible in nature. Most of the planet's biomass is made up by plants, while the combined mass of herbivores is much smaller, and that of carnivores is very small. The largest terrestrial animals, such as elephants, rhinos, and hippos, are plant-eaters.

There would simply not be enough energy in the form of meat to sustain herds of large carnivores.)

Nevertheless humans may gain nutritional energy by eating animals, given that these animals feed on plant material that humans cannot digest or do not have access to. Ruminants, for instance, such as cattle, antelopes, or giraffes, have a digestive system that allows them to feed on grass or leaves that humans cannot eat. But humans can very well digest the meat of these grazers. Besides, hunters may not permanently share the same habitat as their prey. Animals may migrate into a hunter's area or hunters may migrate into their prey's habitat for just part of the year. In this case, the prey will collect and store nutritional energy in the absence of humans, and deliver it to the hunter at a later time or in a different area.

Animal husbandry, when compared to hunting, provides more food energy within a given area much the same way that crop farming provides more energy than gathering. Domesticated animals are usually kept in much closer confinement than they would choose in the wild, and they ideally eat either naturally occurring food that humans cannot digest, or leftovers (stalks) from crop farming. Livestock supplied valuable protein, while the time and energy devoted to looking for game, hunting it down and carrying it home was saved. The risk of getting injured or killed during hunting was eliminated as well. Instead, animals were kept and bred in captivity, and slaughtered whenever it was convenient. Moreover, some domesticated animals continuously delivered nutritional energy in form of milk and eggs.

On the other hand, no complex literate civilization has ever emerged on the basis of meat consumption. Even now, in the energy-rich Oil Age, meat contributes only some 7 percent to the nutritional energy globally consumed. All animal products combined (meat, fish, milk, and eggs) account for about 16 percent. In terms of protein supply animal products are somewhat more important. Animal products currently account for about one third of the protein globally consumed by humans, with meat, milk, and fish contributing evenly. (In comparison, cereals provide for about 43 percent of the total global protein supply, and pulses and other plants for 23 percent.)[8]

But livestock has always delivered more than just nutritional energy. Wool, hides, and bones have been important raw materials for many civilizations. What is more, animal dung has traditionally been used as fertilizer on fields, as construction material, and as a fuel in regions where wood was scarce. In fact, dried dung once was the most important fuel for cooking in many areas of interior Asia, the Indian subcontinent, the Middle East, Africa, and both Americas. The European colonizers of North America used "buffalo wood" (also known as Nebraska oak) for heating purposes, and camel dung still remains a highly valued fuel in desert areas of Africa and western Asia. The droppings of cattle, water buffalo, yak, and llama have served as fuel as well, while sheep dung was generally avoided because it produces an acrid smoke when burned.

Even more importantly, domesticated animals have been playing a critical role in human societies as prime movers. Large, heavy mammals are much stronger than people and revolutionized human power command. (Power refers to energy available per time unit.) Cattle, donkeys, horses, and camels were all initially domesticated as a food source, but eventually served for draft and portage. Animalback riding redefined human transport and warfare. Some animals were used to drive water pumps, millstones, and other machinery. Perhaps most critical, oxen, donkeys, and horses were used to pull scratch plows, which transformed crop farming and boosted the energy harvested on fields to fuel human muscles and brains.

NOTES

5. The word agriculture derives from Latin *agricultura*, from *ager*: field, and *cultura*: cultivation.

6. C. T. de Wit, "Photosynthesis: Its Relation to Overpopulation," in *Harvesting the Sun*, ed. A. San Pietro, F. A. Greer and T. J. Army (New York: Academic Press, 1967): 315–20, cited in Joel E. Cohen, *How Many People Can the Earth Support?* (New York: W.W. Norton & Company, 1995).

7. Robert E. Taylor, *Scientific Farm Animal Production—An Introduction to Animal Science* (Upper Saddle River, NJ: Prentice Hall, 1995). Ray V. Herren, *The Science of Animal Agriculture* (Albany, NY: Delmar Publishers Inc., 1994).

8. Ibid.

CHAPTER 2

HOW DID AGRICULTURAL
TECHNOLOGY EMERGE?

If gatherer-hunters had known how much work full-blown agriculture involves, they probably would not have started it in the first place. But agriculture was not invented through a sudden stroke of genius. It emerged very
slowly in a long series of small innovative steps, each of which seemed to be a
good idea to improve the energy return in comparison to the additional work
invested. Gatherer-hunters tend to accumulate a very intimate knowledge
of the plants and animals in their environment. Current foragers frequently
intervene in the lifecycles of plant and animal species to promote the growth
of edible plant material and the prospering of preferred prey. Australian aborigines, after digging out the tubers of a certain wild yam, replace the stem-
attached top of the tubers so that more yams will grow to be dug up another
time.[9] And more generally, foragers all over the world may have discovered
that occasional weeding favored the growth of preferred food plants.

In some especially food-rich regions gatherer-hunters began to settle
down long before they started to actually produce food. When plant seeds
were collected and brought back to permanent (or seasonal) campsites, they
may have been accidentally dropped around these settlements over and over
again. As people removed other plant material around the campsites to create
space for their daily life, the dropped seeds may have found excellent light
conditions to develop into fully-grown plants. And one way or another, the
realization that seeds were not just edible but could be used to grow a whole
plant must have been the most critical leap forward.

When people started to plant stored seed stock deliberately, they also
began protecting their plants. This changed the evolutionary pressure that

Teosinte, the Ancestor of Maize   To increase the density of edible seeds, generations of farmers have slowly bred the seed spike of a wild grass called teosinte (shown on the left), with just one single row of kernels, into the familiar large maize ear, which has many rows of kernels that are not protected by fruit cases and adhere to the cob. (Image by Nicolle Rager Fuller, National Science Foundation.)

these food plants experienced, as they no longer had to survive in a natural environment. Instead, people created a new environment for them, and selected for other characteristics than nature previously had. Seeds recovered at archaeological sites clearly show that early farmers selected for larger seeds and thinner seed coats. Thick, impermeable seed coats are often essential for seeds to survive in a natural environment because the seeds of many wild plants remain dormant for months until winter is over and rain sets in (or conditions become otherwise suitable for germination). But under human management thick seed coats are unnecessary, as farmers take over responsibility for storing seeds away from moisture and predators. In fact, seeds with thinner coats were preferred as they are easier to eat or process (into flour), and they allow seedlings to sprout more quickly when sown.[10]

Another characteristic humans selected for was the extent to which ripe seeds stick to the stalk. The seed of wild grass plants does not cling too much to the stalk because seeds need to fall to the ground for the next generation to sprout. Farmers, on the other hand, like to carry seeds home on the stalk and hence prefer plants to retain their seed long and tight enough to make sure not too many grains are lost during the harvest and subsequent handling. Plants that matched this requirement had a much better chance for their seed to make it into storage and being planted in the following season. With time, the percentage of seed that falls off the stalk when ripe declined, and crops were domesticated to a point where they could no longer reproduce without human assistance. Modern maize, for instance, has kernels that are very difficult to get off the cob.

Maize also exemplifies how farmers selected for seed density on the stalk. Maize probably descends from a wild grass called teosinte, which has a seed spike consisting of but one single row of kernels, with each kernel enclosed in a hard shell-like case. (On ripening, the teosinte spike shatters, scattering the seeds.) Countless generations of maize farmers developed the seed spike into the much larger maize ear, which has many rows of kernels that are not protected by fruit cases and adhere to the cob. Since maize is now occupying vast areas of the Earth's land surface it may well be considered one of the most successful plants in the history of evolution. And yet, any maize field left unattended will soon be overgrown by other plants, as maize is incapable of dispersing its seed.

DOMESTICATION OF ANIMALS

The domestication of animals was a gradual process as well. In the beginning, people may have just slightly altered the environment in ways that benefited their game. They may have burned woodlands to create grassland for grazing animals, and they may have protected herds from other predators while following them. People then may have started to constrain the movement of their game and later isolated a herd or flock of animals. Alternatively, people may have separated single animals from wild herds. Current gatherer-hunters capture young wild animals, raise them as pets, and rear them to adulthood. This kind of human behavior would have provided auditioning opportunities to acquire knowledge helpful to manage and breed animals. One way or another, hunters gradually turned into herders while they asserted more and more control over animals. Once a herd or a large enough number of individuals had been separated from populations in the wild, humans assumed responsibility for managing the size of the area the animals occupied, their food supply, and their reproduction.[11]

The ideal animal species to be targeted for domestication was any wild animal that had already been a food source, did not run away, was a dietary generalist, tolerated breeding and feeding in close confinement, and lived in herds of highly social dominance hierarchies. Wild goats and sheep, the first game to be domesticated by humans, fit this profile pretty closely. Other animal species, though they lived in herds, lacked a follow-the-leader social structure (this is true for kangaroos and the North American big horn sheep). Or they turned out too skittish (gazelles and antelopes run extremely fast and sophisticated fences would be necessary to confine them). Some species would not reproduce under crowded conditions (as individuals are used to separate from their herd for mating or to give birth), others take too long to mature (25 years in the case of gorillas). Others again had feeding habits so specialized that it would have been impractical to raise them (koalas, for instance, eat mainly eucalyptus leaves). Relatively solitary and strongly territorial species are not suited for domestication either, as they are adapted to defend their territories against intruders. In contrast, human herdsmen could easily step into the position of the leading dominant male as far as species were concerned that in the wild form gregarious, highly social groups comprising both sexes. (These have a preexisting capacity for submissive behavior that dramatically increases the ability of human herders to communicate commands to the captives.) By the end of it all, there are actually only few animal species that are truly suited for domestication. Of the planet's 148 large terrestrial mammal species only 14 have been domesticated.[12] These are goat, sheep, cattle, pig, donkey, horse, reindeer, Arabian camel (dromedary), Asian camel (two-humped or Bactrian camel), water buffalo, Bali cattle, yak (the large, long-haired type of cattle of the elevated parts of central Asia), gaur (Bos gaurus, the large cattle native to India and southeast Asia), and llama.

Just as the increasing size and thinning coats of seeds unearthed at archaeological sites testify to the beginning of crop farming, certain changes

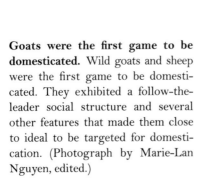

Goats were the first game to be domesticated. Wild goats and sheep were the first game to be domesticated. They exhibited a follow-the-leader social structure and several other features that made them close to ideal to be targeted for domestication. (Photograph by Marie-Lan Nguyen, edited.)

in the bones of animals indicate the shift from hunting to animal husbandry. One such indication is a decrease in the size of animals. Perhaps this was due to general malnutrition suffered by captive animals, or due to inadequate nutrition of pregnant females. The levels of disease and parasites under confinement may have been higher, and earlier weaning may have contributed to reduced growth, especially of those farm animals whose mother's milk people began to drink. Quite possibly, farmers also actively selected for smaller body size as smaller animals are easier to handle, or because large aggressive males (that would otherwise have left a lot of tall offspring) had to be removed to ease herd management.[13]

Again, humans changed the selective forces acting upon species in the wild. Depending on the primary objective, animal breeders variably optimized their domesticates for meat production (fast growth), milk productivity, egg-laying productivity, wool growth, skin quality, or body strength and endurance (packing or draft). And some domestic animals, like plants, lost their ability to survive unattended in the wild.

WHERE AND WHEN AGRICULTURE EMERGED

Agriculture emerged for the first time about 10,000 years ago. This is not especially surprising. We know from records stored in nature that planet Earth has been experiencing cyclical climate patterns for at least the past one or two million years: Long glacial periods of slowly falling temperatures and growing glaciers kept on being interrupted by short interglacial periods of rapidly rising temperatures. We are now somewhere close to the end of such interglacial period. The last cooling period, often referred to as the last Ice Age, began some 120,000 years ago and came to an end about 18,000 years ago, when global temperatures had reached a minimum. Back then, nearly a third of the planet's land area, including large parts of North America and Europe, lay under several kilometers of ice, and the sea level was some 120 meters lower than it is today (because so much water was stored away as ice). But some 13,000 years ago, that is around 11000 B.C.E., the ice cover began to shrink quite rapidly, and the global average temperature (though with fluctuations) has kept on rising since.[14]

From all we know, anatomically fully modern humans emerged nearly 200,000 years ago, but remained confined to Africa still by the onset of the last Ice Age. While temperatures were still falling some 50,000 years ago, anatomically modern people then began spreading into Eurasia. The low sea level then allowed them to move on to Australia and the Americas. People presumably reached Australia some 40,000 years ago, and crossed from Siberia into North America some 20,000 years ago. (The Bering Strait is today a stretch of ocean but was back then a dried up 1,000-mile-wide stretch of open tundra.) The oldest unquestioned human remains in the Americas, from sites in Alaska, have been dated to around 14,000 years ago, but much of North America was still covered by ice at this stage. A millennium later the ice was retreating and opened up a corridor that allowed people to migrate towards the area that is now the contiguous United States and further on, all the way to Patagonia in the south of South America.

In short, the current interglacial period is the first warm period in which anatomically modern people were present all over the world. This human species (to which all people now living on the planet belong) is the only human species of which we know for certain that it was smart enough to start producing food rather than picking it up the way it occurs in nature. Agriculture began to emerge in various regions just a few millennia after humans had completed their spread from Africa through Eurasia to Australia and all of the Americas. But there were actually only few farming centers where agriculture arose independently.[15] They all had their specific regional set of domesticates, and they all initially saw farming emerge as a sideshow to continued gathering-hunting.

The Original Farming Centers

Agriculture first emerged in a relatively small region of southwest Asia that became known as the Fertile Crescent. It curved over some 2,000 km, from present-day Jordan, Israel, and Palestine through Lebanon, Syria, southeast Turkey, northern Iraq, and into western Iran. Gatherer-hunters settled in permanent villages in this region from about 10500 B.C.E., and began to grow wheat and barley from about 8000 B.C.E..[16] Goat, sheep, pig, and cattle were domesticated in the Fertile Crescent between 7000 B.C.E. and 6000 B.C.E.[17]

The world's second and third oldest independent centers of agriculture were both located in what is now China. Agriculture emerged in the Yangtze River corridor of South China about 6500 B.C.E. based on rice, and at North China's Yellow River about 5800 B.C.E. based on millet.[18] These two cereals are quite the opposite of one another. Rice is the only major cereal that grows in water, while drought-resistant millet has one of the lowest water requirements of all cereals. Pigs and chickens were the principal farm animals of the region, domesticated perhaps around 5400 B.C.E. The Asian water buffalo was domesticated much later, when large farming societies had long been established. There are two principal types, the river buffalo (which was probably first bred on the Indian subcontinent, perhaps around 3000 B.C.E.), and the swamp buffalo (which was probably first domesticated in China, perhaps around 2000 B.C.E.).[19]

In Mesoamerica agriculture emerged on the basis of maize sometime between 4300 B.C.E. and 3500 B.C.E.[20] Teosinte, the probable ancestor of maize, does not naturally grow in such dense stands as wild wheat and barley. Per-

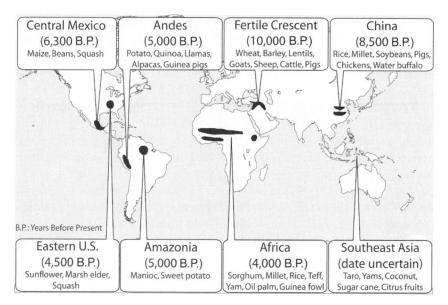

Central Mexico (6,300 B.P.) Maize, Beans, Squash	Andes (5,000 B.P.) Potato, Quinoa, Llamas, Alpacas, Guinea pigs	Fertile Crescent (10,000 B.P.) Wheat, Barley, Lentils, Goats, Sheep, Cattle, Pigs	China (8,500 B.P.) Rice, Millet, Soybeans, Pigs, Chickens, Water buffalo

B.P.: Years Before Present

Eastern U.S. (4,500 B.P.) Sunflower, Marsh elder, Squash	Amazonia (5,000 B.P.) Manioc, Sweet potato	Africa (4,000 B.P.) Sorghum, Millet, Rice, Teff, Yam, Oil palm, Guinea fowl	Southeast Asia (date uncertain) Taro, Yams, Coconut, Sugar cane, Citrus fruits

The Original Centers of Agriculture   There are relatively few regions where agriculture emerged independently, each with its own set of particular plant and animal domesticates. The onset of agriculture in these different centers was separated by millennia: Western Asia (8000 B.C.E.), China (6500 B.C.E.), Mesoamerica (4300 B.C.E.), South America (from 3000 B.C.E.), Eastern North America (2500 B.C.E.), sub-Saharan Africa (2000 B.C.E.).

haps this is why village life started in Mesoamerica only after agriculture had emerged. Squash farming may predate Mexican maize farming (and indeed Fertile Crescent farming), but squash contains a lot of water and is not nearly as energy-dense as cereals.[21] Hence, squash could have hardly provided enough nutritional energy to serve as the principal domesticate for settled farming communities. Beans were probably first domesticated in western Mexico, around 2000 B.C.E., and the turkey remained Central America's only farm animal. (The precise area of initial turkey domestication remains unknown. Currently the earliest signs in the archaeological record come from the Yucatán peninsula from after 100 B.C.E.)

In South America's Andes, agriculture emerged sometime between 3000 B.C.E. and 2500 B.C.E. based on potatoes (and three other similar types of tubers) as well as quinoa, a goosefoot plant with tiny seeds forming large clusters at the end of the stalk. Tubers did not generally become a staple food for humans as they contain more water than cereals and spoil more easily when stored. Hence, tubers mainly sustained people in the tropics where their year-round harvest obviates storage. In the Andes, potatoes served as a staple as people came up with a unique way of food preservation: potatoes were processed into chuñu, a dehydrated foodstuff produced by trampling potatoes and alternately freezing and drying them. This product was storable for years. People in the Andes also domesticated two animal species: the llama (as well as its smaller cousin, the alpaca) and the guinea pig (or cuy). Both were domesticated sometime between 3000 B.C.E. and 2000 B.C.E.

In what is now the eastern part of the United States, agriculture may have emerged independently as well. From around 2500 B.C.E. people in this region planted goosefoot and sunflower, but both these domesticates were grown in Mesoamerica too. (Their farming may therefore have spread to the U.S. region.) One type of squash, as well as the marsh elder, may also have been independently domesticated in eastern North America.[22] Meanwhile the native bison, or American buffalo, was difficult to handle. It is less stress-tolerant in confined areas than Eurasian cattle, and has never been fully domesticated. Even present-day bison breeders need to use very high fences, as a grown bull is capable of making a standing six-foot jump.

As far as Africa is concerned, it is uncertain whether or not agriculture emerged truly independently on the continent at all. Already around 6000 B.C.E. farming had spread from the Fertile Crescent into North Africa, where a moist climate prevailed at the time. By 3000 B.C.E., when the climate had become much drier, the famous ancient Egyptian civilization emerged along the Nile River. And around 2000 B.C.E., farming began in the Sahel zone, the semiarid grassland strip that stretches west to east across Africa, immediately south of the Sahara desert. People living in the Sahel zone either imported cattle of Fertile Crescent origin or independently domesticated it. Thereafter they adopted a more sedentary lifestyle and domesticated sorghum (about 2000 B.C.E.), pearl millet (1000 B.C.E.), and African rice (200 C.E.). Guinea fowl, which looks similar to turkey, was domesticated in ancient times in sub-Saharan Africa as well, while the ostrich, a large, flightless bird native to southern Africa, has been farmed there only since the later Coal Age.[23]

Two further African agricultural centers may or may not have started out with imported domesticates. One is located in tropical West Africa, where African yam, oil palm, and kola nut have been domesticated. The other is located in the highlands of present-day Ethiopia, where teff (a tiny-seeded cereal), okra (a tall-growing, annual vegetable), kenaf (a relative of okra), chat (Arabian tea), and coffee have been domesticated.

WHY DID AGRICULTURE EMERGE SOME PLACES EARLIER THAN OTHERS?

The fact that agriculture emerged first in southwestern Asia (about 8000 B.C.E.), in China (6500 B.C.E.), next in Mesoamerica (4300 B.C.E.) and South America (from 3000 B.C.E.), followed by eastern North America

(2500 B.C.E.) and sub-Saharan Africa (2000 B.C.E.), has been interpreted as highly critical for the later development of political history, because people from earlier agricultural regions (continents) came to dominate people from later agricultural regions. Based on the fact that humans have been living in Africa the longest, and thus have had by far the most time to accrue knowledge about the plants and animals in their environment, it should be assumed that agriculture should have emerged in Africa first. Instead, people grew cereals in the Fertile Crescent six and a half millennia earlier than in sub-Saharan Africa. In Australia anatomically fully modern humans have been present nearly as long as in East Asia, but agriculture did not emerge at all. And in the Americas, agriculture was delayed by some 4,500 years compared to Eurasia. If these dates had been reversed, Australian aborigines might have made Britain their colony, the Aztecs might have conquered Spain, and Africans might have enslaved Europeans for field work. It is therefore important to ask what was behind the timing of the onset of agriculture on the different continents. Why did some regions get an energetic agricultural head start over others?

As it turns out, environmental specifics seem to have determined the likelihood of agriculture to emerge more easily in some regions than others. Plants and animals that lend themselves well to domestication are relatively scarce, and were rather unevenly distributed throughout the world. Some areas exhibited extremely high levels of biodiversity while others were comparatively impoverished in this respect. Thus, not every region offered the environmental conditions that favored an onset of agriculture. The Fertile Crescent, however, was predestined to become the cradle of agriculture due to its climate, its geography, its soil fertility, and its biodiversity. One of the Fertile Crescent's most striking features is its wide range of elevations, varying from the lowest spot on Earth, the Dead Sea, to mountains over 5,000 meters (nearly 17,000 feet) high, which provides a corresponding range of diverse environments.

Since agricultural societies tend to base their nutrition on grain, natural occurrence of the right type of cereal was most critical. The Fertile Crescent was extremely well-stocked with cereals suitable for domestication. Of the world's 56 species of large-seeded grasses, 32 are found wild in the Mediterranean area. By contrast, South America boasts just seven, East Asia six, North America and sub-Saharan Africa four each, and Australia just one.[24] The chief reason why the land around the Mediterranean Sea shows such immense variety in large-seeded grasses is its climate, with mild, wet winters and hot, dry summers. To be sure, Mediterranean climate is also found in a few other regions: in coastal California and Chile, parts of southern Australia, and the Cape of South Africa. But all of these are significantly smaller, and none of them exhibits such accentuated climatic variation between seasons and years as the Fertile Crescent. Like nowhere else, the extreme climate of the Mediterranean favored the evolution of annual plants that survived the long, dry summer by putting much of their energy into big (edible) seeds, while leaving the (inedible) remainder of the plant to die back and regrow each year.

What is more, the large-seeded Fertile Crescent grasses wheat and barley occurred in the wild in large, dense stands that could be easily harvested and would support a sedentary gatherer-hunter population even before agriculture emerged. In contrast, stocks of teosinte, the ancestor of maize, do not naturally grow in concentrated stands, and true village life started in Mesoamerica only after agriculture had emerged. (Maize actually derives from the one and only large-grained annual genus found in its region.) To be sure, stand density, seed density on the stalk, and ease of processing may actually have been more important for grasses to be targeted for domestication than seed size. After all, small-seeded millet was the principal domesticate of the North China farming center, which may be nearly as old as that of the Fertile Crescent. Similarly, plant species with tiny seeds have been the principal cereal in the Andes (quinoa), in the Sahel zone (sorghum and pearl millet), and in Ethiopia (teff). Small-seeded grass species are much more evenly distributed among the continents than the large-seeded grasses that some scholars tend to associate with the early onset of agriculture. They thus speculate that North China is only a secondary agricultural center, where people began to domesticate millet after they had learned about farming from migrating South China farmers, who based their nutrition on large-seeded rice. They also emphasize the importance of Andean potato farming and Sahel cattle raising over small-seed cereal farming in these regions.

Another advantage for people living in the Fertile Crescent was that this region's annuals tend to belong in the group of self-pollinating plants that usually pollinate themselves but are occasionally cross-pollinated. This was advantageous for early farmers because occasional cross-pollination generated several strains to choose from, while the predominant self-pollination ensured that varieties selected for their desired qualities usually perpetuated themselves unchanged and were not immediately lost by hybridization with less desirable strains. Maize, in sharp contrast, outcrosses at high rates, which may explain the seemingly slower evolution of maize toward preferred traits in the face of gene flow from unselected populations.[25]

NOTES

9. Joel E. Cohen, *How Many People Can the Earth Support?* (New York: W.W. Norton & Company, 1995).

10. Bruce D. Smith, *Emergence of Agriculture* (New York: W. H. Freeman & Co, 1995). Find more on this topic in: G.C. Hillman and M.S. Davies, "Measured Domestication Rates in Wild Wheats and Barley under Primitive Cultivation, and Their Ar-

chaeological Implications," *Journal of World Prehistory* 4 (1990): 157–222; B.D. Smith, "Documenting Plant Domestication: The Consilience of Biological and Archaeological approaches," *Proceedings of the National Academy of Sciences* 98 (2001): 1324; M.A. Blumler and R. Byrne, "The Ecological Genetics of Domestication and the Origins of Agriculture," *Current Anthropology* 32 (1991): 23–53; Mark A. Blumler, "Evolution of Caryopsis Gigantism and Agricultural Origins," in *Research in Contemporary and Applied Geography: A Discussion Series XXII (1–4)*, Department of Geography, Binghamton University, State University of New York, Binghamton, NY, 1998, http://geography.binghamton.edu/pdf/Caryopsis.pdf; Mark A. Blumler, "Seed Weight and Environment in Mediterranean-type Grasslands in California and Israel" (Ph.D. diss.), University of California, Berkeley, 1992).

11. Bruce D. Smith, *Emergence of Agriculture* (New York: W. H. Freeman & Co, 1995).

12. Jared Diamond, *Guns, Germs, and Steel: The Fates of Human Societies* (New York: W.W. Norton & Company, 1997).

13. Bruce D. Smith, *Emergence of Agriculture*.

14. J. D. MacDougall, *A Short History of Planet Earth: Mountains, Mammals, Fire, and Ice* (New York: John Wiley & Sons, 1996); Robert Claiborne, *Climate, Man, and History* (New York: W.W. Norton & Company, 1970); Illinois State Museum, "Ice Ages," http://museum.state.il.us/exhibits/ice_ages/.

Find a record of the past ice age cycles in: John Houghton, *Global Warming: The Complete Briefing* (Cambridge: Cambridge University Press, 1997).

15. Paul Gepts, "Evolution of Crop Plants: The Origins of Agriculture and the Domestication of Plants," Department of Agronomy and Range Science, University of California, Davis, http://www.plantsciences.ucdavis.edu/gepts/pb143/pb143.htm. Paul Gepts, "Lecture 10: Where Did Agriculture start? Centers of Origin and Diversity," http://agronomy.ucdavis.edu/gepts/pb143/lec10/pb143l10.htm.

16. D. Zohary and M. Hopf, *Domestication of Plants in the Old World* (Oxford: Clarendon, 1988); D. Zohary and M. Hopf, *Domestication of Plants in the Old World: The origin and spread of cultivated plants in West Asia, Europe, and the Nile Valley* (Oxford: Oxford University Press, 1993). M. Nesbitt, "Clues to Agricultural Origins in the Northern Fertile Crescent," *Diversity* 11 (1995):142–143; G. Willcox, "Archeobotanists Sort Out Origins of Agriculture from Early Neolithic Sites in the Eastern Mediterranean," *Diversity* 11 (1995):141–142. The three principal crops were Einkorn wheat, Emmer wheat, and barley. Barley tended to provide lower yields, but withstood drier conditions, poorer soils, and some salinity. Emmer wheat needs mild winters and warm, dry summers and was the crop grown most often in the region, at least from ca. 7500 B.C.E. to 6000 B.C.E. Einkorn wheat provided comparatively lower yields and prefers cooler conditions, but can tolerate somewhat poorer soils.

17. Gordon Luikart et. al., "Multiple Maternal Origins and Weak Phylogeographic Structure in Domestic Goats," *PNAS (Proceedings of the National Academy of Sciences of the United States of America)* 98 (2001): 5927–5932. www.pnas.orgycgiydoiy10.1073ypnas.091591198, http://www.pnas.org/cgi/reprint/98/10/5927.pdf; B. Bower, "Domesticated Goats Show Unique Gene Mix," *Science News*, May 12, 2001, http://www.findarticles.com/cf_dls/m1200/19_159/75309403/p1/article.jhtml; Daniel G. Bradley, "Genetic Hoofprints: The DNA Trail Leading Back to the Origins of

Today's Cattle Has Taken Some Surprising Turns along the Way," *Natural History*, Feb 2003, http://findarticles.com/p/articles/mi_m1134/is_1_112/ai_97174195/; Oklahoma State University; "Breeds of Livestock," http://www.ansi.okstate.edu/breeds/cattle/.

18. Bryan C. Gordon, "Preliminary Report on the Study of the Rise of Chinese Civilization Based on Paddy Rice Agriculture," Canadian Museum of Civilization, Hull, Quebec, February 1999, http://www.carleton.ca/~bgordon/Rice/research__resources.htm; Charles Higham, Tracey L.-D. Lu, "The Origins and Dispersal of Rice Cultivation," *Antiquity* 72 (1998): 867–877.

19. C. H. Lau et al. "Genetic Diversity of Asian Water Buffalo (Bubalus bubalis): Mitochondrial DNA D-loop and Cytochrome b Sequence Variation," *Animal Genetics*, 4 (1998): 253–64, http://www3.interscience.wiley.com/journal/119119543/abstract?CRETRY=1&SRETRY=0.

20. B. Bower, "Maize Domestication Grows Older in Mexico," *Science News* 159 (2001): 103, http://www.findarticles.com/cf_dls/m1200/7_159/71191553/p1/article.jhtml; B.F. Benz, "Archaeological Evidence of Teosinte Domestication of Guilá Naquitz, Oaxaca," *Proceedings of the National Academy of Sciences* 98 (2001): 2104; D.R. Piperno and K.V. Flannery, "The Earliest Archaeological Maize (Zea mays L.) from Highland Mexico: New Accelerator Mass Spectrometry Dates and Their Implications," *Proceedings of the National Academy of Sciences* 98 (2001): 2101; "Early Agriculture Flowered in Mexico," *Science News*, June 16, 2001, http://www.findarticles.com/cf_dls/m1200/24_159/76653912/p1/article.jhtml.

21. Wade Roush, "Squash Seeds Yield New View of Early American Farming," *Science* 276 (1997): 894–895, http://www.sciencemag.org/cgi/content/abstract/276/5314/932; Bruce D. Smith, "The Initial Domestication of Cucurbita pepo in the Americas 10,000 Years Ago," *Science* 276 (1997): 932–934, http://www.sciencemag.org/cgi/content/abstract/276/5314/932.

22. To be sure, maize became the most important domesticate in the region, once it had spread there from Mesoamerica. J.C. Vogel and N.J. van der Merwe, "Isotopic Evidence for Early Maize Cultivation in New York State," *American Antiquity* 42 (1977): 238–242.

23. Rebecca S. Thompson, "Raising Emus and Ostriches," Special reference briefs 97–06, Alternative Farming Systems Information Center, Information Centers Branch, National Agricultural Library, Agricultural Research Service, U.S. Department of Agriculture, 1997, http://www.nal.usda.gov/afsic/AFSIC_pubs/srb9706.htm; Joseph Batty, *Ostrich Farming* (Midhurst: Beech Publishing, 1995); Claire Vandervoodt, *The Dasana Ostrich Guide: A Practical Handbook* (Devonport: Nova Creative Publishing, 1995)

24. M.A. Blumler and R. Byrne, "The Ecological Genetics of Domestication and the Origins of Agriculture," *Current Anthropology* 32 (1991): 23–53; Mark A. Blumler, "Evolution Of Caryopsis Gigantism and Agricultural Origins," in *Research in Contemporary and Applied Geography: A Discussion Series XXII (1–4)*, Department of Geography, Binghamton University, State University of New York, Binghamton, NY, 1998, http://geography.binghamton.edu/pdf/Caryopsis.pdf; Mark A. Blumler, "Seed Weight and Environment in Mediterranean-type Grasslands in California and Israel," (Ph.D. diss.), University of California, Berkeley, 1992); Jared Diamond, *Guns*,

Germs, and Steel: The Fates of Human Societies (New York: W.W. Norton & Company, 1997).

25. Maize joined wheat, rice, and barley as a truly major global crop only after its conversion to self-pollination combined with hybridization between favorably interacting inbred lines increased the yield of maize several-fold in the 20th century. Robert W. Allard, "History of Plant Population Genetics," *Annual Review of Genetics* 33 (1999): 1–27, http://arjournals.annualreviews.org/doi/abs/10.1146/annurev.genet.33.1.1; Jared Diamond, "Spacious Skies and Tilted Axes," *Natural History* (1994), http://www.mc.maricopa.edu/dept/d10/asb/anthro2003/lifeways/hg_ag/agspread.html.

CHAPTER 3

THE SPREAD OF AGRICULTURE

Farming spread rapidly from the original centers of agriculture to neighboring regions. Migrating farmers took their domesticates to new regions, and gatherer-hunters acquired agricultural technology from nearby farmers. However, there were environmental limitations as to how far, and how fast, agriculture could spread. Wheat, for instance, which is native to a temperate climate zone, cannot be grown in the tropics. Environments are generally similar on an west-east axis, because the intensity (angle) of incoming solar radiation, and the seasonal day-length variation, depend on the distance from the equator. Regions in similar latitudes thus tend to have similar climates. In Africa, sorghum, pearl millet, and African rice spread rapidly over a vast expanse, from the Sahel zone's very west to its very east. But Africa stretches about as far north-south as it does west-east. The north-south axis boasts such divers climate zones as the Sahara desert, the Sahel, and the tropics, and it took some 8,000 years for cattle, sheep, and goats to spread from northern Africa to South Africa. Similarly, the pronounced north-south orientation of the Americas slowed the spread of American domesticates. Some 4,000 years after llamas had been domesticated in the Andes, the Aztecs and Mayas of Mesoamerica still did not have any draft animals at all. And it also took a very long time before Mexican maize finally spread through the tropics to the temperate zones of present-day Argentina.

Eurasia, in contrast, has a very pronounced west-east orientation. It is the world's largest landmass, has hardly any physical barriers to human migration, and a climate that allowed farmers to take their domesticates over great distances without encountering any major problems of adaptation. In fact, nearly all of the Eurasian landmass is situated on the same side of the equator, in the northern hemisphere. By Roman times wheat was thriving in all of Eurasia's vast areas of temperate climate. It was cultivated along a 10,000 mile expanse, from Ireland to southern India to Japan.[26]

But first of all, the art of food production had to radiate out of the environmentally privileged Fertile Crescent region into its immediate vicinity. The Crescent actually curved right around southern Mesopotamia (in present-day Iraq), where the world's first empire was going to emerge. In this area the Euphrates and Tigris Rivers delivered fertile soil through flooding, but there was too little rainfall for original farmers to introduce their trade. Then, around 5500 B.C.E., people solved the problem by building irrigation canals that diverted water from the rivers onto the fields. This innovation may be connected to the onset of a cool and dry climate period at around 6200 B.C.E. Reduced rainfall may have prompted farmers to migrate from Fertile Crescent areas (especially from northern Mesopotamia) to the floodplains of Mesopotamia's south, where irrigated agriculture, rather than rain-fed agriculture, supplied nutritional energy for several hundred villages by 4300 B.C.E. Meanwhile agriculture also spread rapidly from the Fertile Crescent to the Indus valley (in present-day Pakistan). Wheat and barley was grown there from around 6000 B.C.E., and eventually people of this region domesticated sesame and cotton. Wheat then kept on spreading towards the east and was grown in China by 2500 B.C.E.

Wheat and rice, Eurasia's two most important domesticates, probably first met in India. The fertile Ganges river basin (of present-day India and Bangladesh) is situated roughly halfway between the world's two oldest agricultural centers, the Fertile Crescent and the Yangtze River corridor of South China. However, the region's wet monsoon climate made it better suited for rice than for wheat and barley. Hence, the Ganges river basin became part of the vast rice-dominated area, which stretched from India to Japan, and south to Indonesia.

INTO EUROPE

On the other side of the Fertile Crescent the art of food production rapidly radiated into Mediterranean Europe, where the climate (with dry summers and wet mild winters) was especially well suited for Fertile Crescent domesticates. From about 6000 B.C.E. the gatherer-hunters inhabiting the caves of southern Europe began limited herding of sheep and goats, and they started to plant wheat and barley in addition to their usual foraging activities. Food production technology then spread quickly around the Mediterranean, and between 5000 B.C.E. and 4000 B.C.E. through nearly all of Europe.

The spread north into continental Europe followed upstream the Danube River, which disembogues into the Black Sea, the large body of water situated right north of the Fertile Crescent. The Danube reaches all the way into

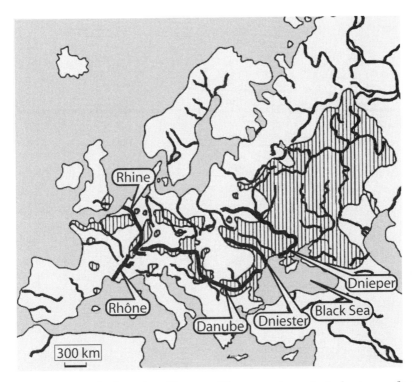

The Spread of Agriculture into Europe There were two principal gateways from the Fertile Crescent into Europe. One was along the shores of the Mediterranean, the other upstream along the rivers that rise in Europe's interior and flow into the Black Sea, primarily the Danube and the Dniester. People then settled in Europe's vast, fertile Loess Belt. (Note that this illustration is based on a loess map of the later 20th century, compiled by the INQUA Loess Commission, http://www.loessandust.org/history.htm.)

what is now southern Germany, from where agriculture spread further north along the Rhine, which disembogues into the North Sea in what is now the Netherlands. However, farmers soon discovered that Europe's best soil was found neither along the coasts nor along a major river, but in the regions covered by a wind-blown silt called loess, which is found in a belt stretching right across Europe from near Cologne, on the Rhine, all the way to the southern Ukraine. Starting from about 5000 B.C.E., almost the entire loess belt was being rapidly colonized by a surprisingly homogeneous group of people known as the LBK farmers. (LBK stands for German *Linearbandkeramik* describing the distinctive linear-banded pottery produced by this civilization.) These early European farmers lived in large timber-framed longhouses that were five to seven meters (16 to 23 feet) wide and 12 to 45 meters (39 to 148 feet) long. They planted wheat, barley, and flax, and domesticated opium poppies. However, the Fertile Crescent system of growing wheat during a

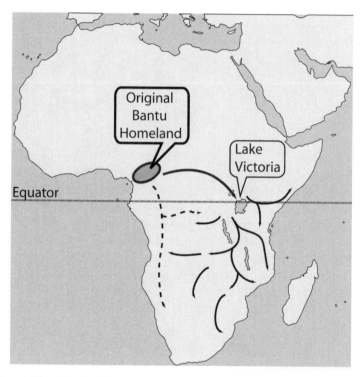

Turning Africa Black: The Bantu Migrations   Bantu-speaking, black-skinned farmers colonized most of sub-equatorial Africa, replacing other African peoples.

wet mild winter and harvesting it in spring had to be changed. Adapting to the local climate, the LBK farmers instead took advantage of the high spring rainfall and had their wheat grow during spring and summer, harvesting it in fall. Some of the harvest had to be used to feed domestic animals during the long winter months, which offered little in the way of cold-season grazing. Cattle became a significant source of meat and manure (to fertilize the fields after the fall harvest). Sheep and goats, which preferred the milder winters and open grasslands of western Asia and Mediterranean Europe, were less important.

As the agricultural system developed by the LKB farmers did not spread beyond the loess belt, it did not reach the Atlantic or North Sea coast. Northern and extreme western Europe, including the British Isles, showed soils and a climate that were less optimum. Hence, farming took hold more slowly.[27] Gatherer-hunters of these areas adopted cereals, legumes, and livestock (especially sheep and goats) only as minor additions to their traditional way of life. First signs of agriculture appeared in Britain and

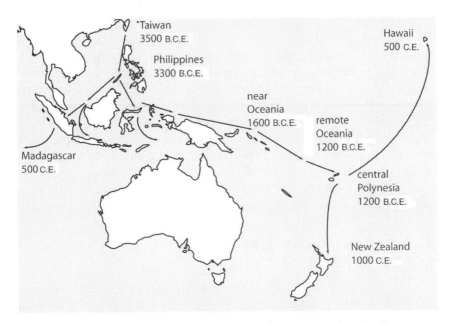

Austronesian Expansion   Farmers from a small coastal region in what is now China colonized an area that spanned half the globe. Thus Austronesian languages are now spoken over wide parts of coastal Asia and Oceania.

Scandinavia only from about 4000 B.C.E., with full-scale agricultural communities flourishing only after 2000 B.C.E.

BLACK AFRICANS AND CHINESE POLYNESIANS

Two of the early agricultural expansions are especially well documented as they left a severe genetic and lingual imprint on the planet. One was the migration of black-skinned Bantu speakers, who were initially restricted to an area of present-day Nigeria and Cameroon, but later spread through nearly all of sub-equatorial Africa. As late as 1000 B.C.E. Africa was occupied by farmers and herders exclusively north of the equator. Thereafter Bantu speakers, equipped with wet-climate crops such as African yam, replaced most of the south's original gathering-hunting inhabitants. They effectively turned much of Africa into *Black Africa*.

Similarly, the offspring of an initially small group of farmers living in a coastal area of what is now southern China spread out in an amazing expansion that eventually spanned halfway around the globe. People of this civilization left mainland China for the island of Formosa (Taiwan) sometime after 4000 B.C.E., and their descendants (over the course of several millennia) moved on to colonize much of southeast Asia, virtually all Pacific Ocean islands (reaching as far south as New Zealand and as far east as Hawaii), and even reached Madagascar, off the coast of Africa.[28] These people were not Sino-Tibetan-speaking Han Chinese (who today make up the great majority of the Chinese population), but spoke Austronesian (formerly known as Malayo-Polynesian) languages, which now comprise about 1,200 languages spoken in the colonized areas, including the Philippines, Indonesia, Laos, Thailand and Myanmar (Burma). The expansion was fueled by rice and millet, but these farmers also carried chickens, pigs, and dogs with them. And in some areas they domesticated new species. The dry-land or wet-land crops taro and yam (which may be storable for several months), as well as the tree crops breadfruit, banana, and coconut, played a central role on many islands, while fishing remained important almost everywhere.

HOW BRUTAL WERE THE FIRST ENERGY CAMPAIGNS?

To be sure, most of the early agrarian expansions are not well documented at all. We therefore know little about the extent at which agriculture was carried by farmers into new areas as compared to being adopted by local foragers (who acquired domesticates through trade or theft). Farming societies certainly had an incentive to expand. They harvested more nutritional energy and experienced accelerated population growth, which pushed some into neighboring gatherer-hunter areas. Killing or displacing native foragers was easy, because farmers enjoyed a higher population density (more warriors) and more advanced technology that provided them with better weapons and tools to exploit the environment.

On the other hand, migration and sedentary farming do not go well together. Farmers arriving in a new area had to sustain themselves by foraging, herding, or substantial grain reserves at least during the first growing period, which is normally several months. Ötzi, the oldest mummy in the world, was preserved inside a glacier of Tyrol's Ötztaler Alps. This man, dated to 3300 B.C.E., revealed that at least some people living in the Alps at this time still relied on a mix of farming and foraging. Ötzi had einkorn wheat for his last meal, but his stomach also contained plums and other plants growing in the wild. It also contained some meat, likely from wild deer rather than a domesticated animal. On the other hand, the iceman's outfit included a calf leather belt, a goatskin loincloth and goatskin coat, next to a pair of deerskin shoes and a bearskin cap. Apparently, this fellow either was part of a society that used both plant and animal domesticates of Fertile Crescent origin (wheat, cattle, goat), or he was a gatherer-hunter trading with such community.[29]

Ötzi also revealed that people, if they ever had been, had at this stage definitely stopped being friendly to one another. He had a stone arrowhead in his shoulder, which probably caused him to die, and traces of blood from four other people on his clothes, suggesting that he had earlier been involved in an infight. Generally, we have no way of knowing how brutal the displacement of original gatherer-hunters really was, while people with power in later, literate times often bragged about their brutalities in written records. Christians loved to mutilate Muslims during the Crusades, and Byzantine emperor Basil II capped a victory over the Bulgarians by blinding 14,000 of them, leaving one man per hundred with one eye so they would find their way home. In contrast, we know little about the first wave of energy campaigns, conducted by small, illiterate farming communities. We know about their migrations if they left their genetic and lingual imprints on regions, or we follow their traces by unearthing their skeletons, houses, pottery, and garbage. But we do not know how they interacted with indigenous gatherer-hunters.

Usually the population of original native foragers completely disappeared, but there are exceptions. The sub-equatorial Bantu expansion was survived by Pygmies, who still now live widely spread through the Central African rain forest, and by Khoisan speakers, who either foraged in areas too marginal for agriculture or became herders. The Austronesian invasion was survived by some of the original people of the Philippines, whom the Spanish named Negritos for their dark skin and small stature. Similar people are found on the Andaman Islands, a remote archipelago east of India. These may be somehow related to the Samang of Malaysia. But most places the aboriginal occupants were overwhelmed by the agricultural productivity and population growth of arriving farmers (unless they were already agrarian themselves). We know practically nothing about the fate of these indigenous victims, since their population density was low and archaeological sites are rare. Incidentally, in Africa and Eurasia these native gatherer-hunters were not the first human occupants either. Homo erectus migrated out of Africa more than a million years ago and Neanderthals lived in western Eurasia until a short 35,000 years ago.

THE JUNGLE AND THE STEPPES

In addition to the fertile areas that were soon occupied by settled farmers, and those relatively inhospitable environmental niches where foragers still survive, our planet also boasts marginal areas that allow for a kind of semi-nomad farming life. In tropical rain forests, for instance, most nutrients are bound in the plants, not in the soil. Farmers therefore adopted a system of slash-and-burn agriculture. They cut and burn trees and natural undergrowth, and grow crops for only a few years on the bare soil that rapidly loses fertility. Thereafter the whole village moves on and repeats the process

Pastoralism Nomadic pastoralism is a lifestyle between foraging and sedentary farming, and did not trigger many of the feedbacks associated with agriculture. The picture shows a Bedouin camp with a herd of goats. (Library of Congress image LC-DIG-matpc-06051, edited.)

on new virgin land. (Another disadvantage associated with such regions is the year-round warm climate. Nothing works better for pest and weed control than winter frost as it is experienced in central-northern Europe, North America, and other temperate zones.)

Nomadic pastoralism is another weak form of agriculture. This lifestyle, which was unknown in the Americas, is practically between gathering-hunting and sedentary farming. It is mobile and does not allow for bulky property. Population densities are low, typically around 1 or 2 people per square kilometer (2 to 5 people per square mile). Pastoralists migrate with their herds of domesticated or semi-domesticated animals (such as goats, sheep, cattle, camels, horses, or reindeer) in search of pasture, usually following the same seasonal route year after year. These herders enjoy adequate nutrition from their herds (milk, meat) and gather plant food along their migration routes. In addition, they acquire food and other supplies through trade from sedentary agrarians by selling hides, meat, and live animals. The only way to find enough feed for their herds is by constantly moving around. However, they generally have little incentive to settle down, as the labor requirements for traditional nomadic pastoralism are very low. In Central Asia, two mounted Mongolian shepherds can handle 2,000 sheep, while a single herder in East Africa may take care of some 100 camels or 200 cattle or 400 sheep or goats. (The most laborious activity of a Somali camel herder, for instance, is to draw water for a herd of 50 to 100 camels from wells up to 20 meters (66 feet) deep during the dry period of the year.)

Nomadic pastoralism generally did not trigger the feedbacks typical for sedentary civilizations. A social organization capable of managing seasonal collection, storage, and redistribution of agricultural produce was unnecessary, and professional specialization remained low as all men were herdsmen. Nomads did not develop writing systems and did not accumulate technology the way agrarians did. Nevertheless nomadic pastoralists lived safely on the distant outskirts of agricultural areas, typically in semi-arid regions, because crop farmers had no incentive to expand into these areas. On the contrary, it was often nomadic herders who troubled even well established empires, simply because they tended to be skilled animal-back riders (kinetic energy), had little property to lose, and were difficult to fight as they could easily strike and move on. Hence, wealthy sedentary societies feared these mobile raiders, who were perceived as rough and uncivilized. In several instances nomads conquered the empires of sedentary farmers, but the invaders in turn had to adopt a fully agricultural, grain-based lifestyle. Otherwise they would have been forced to move on in order to provide feed for their animals.

NOTES

26. Jared Diamond, "Spacious Skies and Tilted Axes," *Natural History* (1994), http://www.mc.maricopa.edu/dept/d10/asb/anthro2003/lifeways/hg_ag/agspread.html; The Science Show, "Jared Diamond Lecture," Radio National, Broadcast October 28, 2000, http://www.abc.net.au/rn/science/ss/stories/s199676.htm.

27. H. Tauber, "13C Evidence for Dietary Habits of Prehistoric Man in Denmark," *Nature* 292 (1981): 332–333.

28. Fraser Weir, "A Centennial History of Philippine Independence: Austronesian Expansion—Taiwan 4,000 BC," University of Alberta, http://www.ualberta.ca/~vmitchel/rev3.html.

29. South Tyrol Museum of Archaeology, "The Iceman," http://www.archaeologie museum.it/f01_ice_uk.html.

CHAPTER 4

FRUIT, MEASLES, AND BACKACHE

Early farmers were shorter than gatherer-hunters, according to skeletons dating to the time right before and after the onset of agriculture. Male foragers in regions of present-day Greece and Turkey had an average height of 5' 9", which declined to 5' 3" by 3000 B.C.E. (The average height of women likewise declined from 5' 5" to 5' 0".)[30] Similarly, Denmark's late foragers were taller than the first farmers of about 4000 B.C.E.[31] With our knowledge about agricultural energy, we could immediately conclude that shorter but more numerous and technologically advanced farmers arrived to kick out or kill off the original gatherer-hunters. Alternatively we can speculate that the onset of agriculture was generally accompanied by increased fighting that killed many tall young warriors before they had a chance to reproduce. But there is another answer, too. Early farmers may have become shorter because they suffered from decreased quality of nutrition, more severe diseases, and harder work. Native American skeletons from the Illinois and Ohio river valleys, for instance, revealed that maize farmers living in that area around 1150 C.E. had a nearly 50 percent increase in enamel defects indicative of malnutrition, and a threefold rise in bone lesions reflecting infectious disease in general, when compared to earlier gatherer-hunters of that region. And their life expectancy had decreased from about 26 years to about 19 years.[32]

IMPROVING THE AGRICULTURAL SYSTEM

Early agricultural systems were far from perfect. They may have delivered plenty of nutritional energy, but they probably did not at all match the quality of the diverse and healthy diet that foragers enjoyed. One problem was protein. Cereals tend to contain little of it, and farmers all over the world eventually began to grow protein-rich pulses. Lentils, peas, chickpeas, and bitter vetch complemented the wheat-based diet of the Fertile Crescent, while various bean species were domesticated in China, Mexico, and the Andes. (China's soybeans actually contain as much as 38 percent protein.) In Africa, cowpeas and groundnuts were domesticated to supply protein.

As nothing is as filling as fatty food (which contains lots of nutritional energy), farmers also had a strong preference for plants that develop seeds with a high oil content. Flax, sesame, mustard, and poppies were all domesticated for their edible oily seeds. Flax was grown in the Fertile Crescent as early as 7000 B.C.E. and soon took on a second role. It became the source of linen, the principal fiber used for clothing in the Fertile Crescent and all of Europe nearly up until the Coal Age. Both flax and its Chinese counterpart hemp were selected for long and straight stems that were processed into fiber. The cotton plant, which was domesticated in both South Asia and America, had useful fibers arranged in fuzzy balls that protect oily seeds at the end of the stalks. Presumably, ancestral cabbage plants were initially grown for their oil seeds as well. In turn they underwent great diversification as they were variously selected for leaves (cabbage and kale), stems (kohlrabi), buds (brussels sprouts), or flower shoots (cauliflower and broccoli). These vegetables then provided valuable vitamins.

What may have affected the health of early farmers the most was lack of fruit, the main food staple of our distant arboreal ancestors. Only around 4000 B.C.E. did farmers of the Fertile Crescent begin to domesticate fruit and nut trees. This was a project suitable only for securely settled agricultural people, because it takes several years for trees to grow and become productive. Olive trees, planted for the high oil content of their fruit, probably came first, but trees that yielded figs, dates, pomegranates, and grapes followed just a little later. All these trees were self-pollinating and could easily be cultivated by directly planting cuttings that developed into a tree that was genetically identical to the original plant. Unfortunately, this method did not work for cross-pollinating fruit trees such as apple, pear, plum, and cherry. Only a rather difficult technique (known as grafting), developed by the Chinese in classical times, made the domestication of these trees possible.[33] (This was some eight millennia after agriculture had first emerged, and four millennia after the first olive trees had been planted.)

To be sure, general malnutrition was not the only problem faced by early farmers. Many lost their life due to their incompatibility with the new food staples. People who were intolerant to gluten, a wheat component, had problems surviving in areas where agriculture was based on Fertile Crescent domesticates. (As a result, gluten intolerance is now relatively rare in the

Western world.[34]) Similarly, most people in the world (over 70%) are not able to drink milk in adulthood as they are intolerant to lactose, a sugar contained in milk. Drinking the mother's milk of another animal species is actually quite a bizarre custom. Humans, like all mammals, naturally lose the ability to digest milk when they are about five years of age, as mammals drink mother's milk only until they can digest other food. However, in some parts of the world people apparently became heavily dependent on animal milk for food, and in these populations people with the unusual genetic mutation that confers the ability to digest milk in adulthood survived better and left more offspring. In result, most people now living in India, East Africa, Europe, and the Americas (as far as those of European origin are concerned) are actually able to drink milk in adulthood.

INFECTIOUS DISEASES

A lot more disastrous was the selective pressure imposed on agrarian societies by infectious diseases. When population densities increased, people began to live in unprecedented close contact to one another, domestic animals, and all kinds of wastes. Pathogens that would have been unable to persist in small foraging bands found new hosts over and over again and caused devastating epidemics. Smallpox, tuberculosis, cholera, malaria, typhus/typhoid, measles, scarlet fever, yellow fever, leprosy, dysentery, and many other diseases all took their toll. Many pathogens were transferred to humans from farm animals (and rodents attracted by granaries), which carried everything from virus to worm. Of more than 1,400 known viruses, bacteria, and other pathogens that affect humans, over half are likely to derive from animal hosts.[35] Some may have entered the human population through the consumption of wild animals, while others came from domestic animals. The measles virus may be a form of the rinderpest virus. Smallpox may have been acquired from cattle, which carry the similar cowpox virus. And influenza is still now transferred from chickens and pigs to people.

Over many generations, at the cost of countless premature deaths, agrarian populations gradually acquired immunity against many of these newly emerging infectious diseases. Those who coincidentally had genetic features that rendered them immune against one disease or the other survived better and left more offspring, also immune. Hence, many infectious diseases with time lost their lethal effects on human populations, and some of them are now known as relatively harmless childhood diseases (think mumps, measles, and chicken pox). However, when migrating agrarian societies carried their pathogens to new foraging or farming regions, these diseases often wiped out the vast majority of the native population. The epidemics that European diseases caused in the Americas in post-Columbian times are the best-known example.

HARDER WORK

The malnourished Native American maize farmers of the Illinois and Ohio river valleys also experienced an increase in degenerative conditions of the spine. This is indicative for the hard work that farmers generally had to put up with. Harnessing additional sun energy required investment in form of human labor. Often this work was so hard as to affect the health of farmers negatively and to decrease their life expectancy. Why then, we need to ask, did these people not remain foragers or return to gathering-hunting? The answer to this question is probably simple. It was not those very early farmers living in extremely food-rich areas who were working hard. For them, a tiny bit of extra effort returned a lot of additional energy. Those farmers who had no choice than to work hard were those who had migrated to less optimal regions, in which a return to foraging would not have delivered enough energy to sustain their societies.[36] The first empires did not emerge in the Fertile Crescent, but in the areas adjacent to it. People who migrated to Mesopotamia's south, where rain-fed agriculture was impractical, eventually harvested more energy than any other region in the world—but only after they had invested a lot of work into building irrigation schemes and creating fields. Once these investments began to pay off, people probably

Hard Work: Agricultural Scenes from Ancient Egypt During the Agricultural Age, field work was hard, much harder than the kind of work gatherer-hunters had been used to. However, the workload was now unevenly distributed in society. This illustration from the fifteenth century B.C.E. shows an official by the name of Nakht overseeing farmers. The scene is depicted on his tomb in Thebes. (Facsimile prepared by Norman de Garis Davies and Nina M. Davies in 1907.)

worked no more on average than people had previously in foraging communities. But what really happened was that society had entirely changed, and "on average" now had a different meaning, as some labored very hard, while others did not work at all.

NOTES

30. Jared Diamond, "The Worst Mistake in the History of the Human Race," *Discover*, May 1987, 64–66.

31. H. Tauber, "13C Evidence for Dietary Habits of Prehistoric Man in Denmark," *Nature* 292 (1981): 332–333.

32. The study of native American skeletons from the Illinois and Ohio river valleys has been documented by George Armelagos et al. at the University of Massachusetts, quoted in Jared Diamond, "The Worst Mistake," 64–66.

33. Jared Diamond, *Guns, Germs, and Steel: The Fates of Human Societies* (New York: W.W. Norton & Company, 1997).

34. Celiac Disease and Gluten-Free Diet Support Page, http://www.celiac.com.

35. Louise H. Taylor, Sophia M. Latham, and Mark E. J. Woolhouse, "Risk Factors for Human Disease Emergence," *Phil. Trans. R. Soc. Lond.* B 356 (2001): 983–989, http://www.epa.gov/ncer/biodiversity/pubs/phil_vol356_983.pdf; Louise H. Taylor and Mark E. J. Woolhouse, "Zoonoses and the Risk of Disease Emergence," *Int. Conf. Emerg. Infect. Diseases*, Atlanta, GA, July 16–19, 2000–Program & Abstracts Book, Late Breaker Poster Session, Board 122 (2000): 14.

36. Alternatively, the climate may have changed in some regions to the same effect.

CHAPTER 5

TRANSFORMED SOCIETY

Sedentary agrarian societies were organized entirely differently from egalitarian gatherer-hunter societies. Initially, the small family-based foraging bands turned into larger tribes of a few hundred individuals. These consisted of several clans, that is, formally recognized kinship groups. Within a tribe, differences in respect to wealth and status still remained low, and virtually all people were related by either blood or marriage or both. But as sedentary agrarian populations kept growing, tribes turned into chiefdoms of a few thousand people, many of them unrelated. Hence, for the first time in human history people were regularly encountering unrelated strangers and had to learn to respect and tolerate them, and not to attempt killing them. One way to avoid violence between people was for one person, the chief, to exercise a monopoly on the right to use force. Typically, the leader of the most powerful clan became the chief, while the leaders of the other clans tended to form an aristocratic elite. The position of the chief soon became an office that was inherited, and a class emerged that did not work physically at all. It merely assumed managing roles.

Centralized governance was necessary to provide the infrastructure and leadership that allowed larger communities to function. The rulers organized the collection, storage and re-distribution of agricultural goods, and made sure that enough of the seed was preserved for sowing in the following season. The governing body also claimed labor (especially during the agricultural off-season) for construction of public works such as streets, storage buildings, irrigation schemes, temples, and city walls. And as ruling elites commonly do not redistribute all they collect, they became very rich and powerful.

However, agrarian societies also developed inequalities with respect to wealth and social esteem within the non-ruling population. The food surplus nourished professional specialists who had different incomes, and the sedentary lifestyle allowed for a lot of bulky personal property. Knowledge accumulated and was held within a family or a small guild of craftsmen without being shared with the general public. The egalitarian structure of foraging societies disappeared, and a complex societal composition emerged, with classes based on hierarchical divisions of status, power, and wealth.

The lowest segment of such stratified societies was made up by slaves, that is, people who were officially owned by other people. Perhaps a sort of slavery already existed among early gatherer-hunters, where it may have been created as pay-off for personal debt. (This type of slavery persists in some present-day African societies in which men retain a sense of ownership over their family, putting up their wives or children as bargain against debt.) However, in the Agricultural Age slavery became a very common feature of society. Often, it affected a large percentage of the population. Agrarians could easily source slaves from neighboring populations of gatherer-hunters or less developed farmers. And when empires started to border one another, people were enslaved in the course of warfare or captured in more remote regions.

Interestingly, slaves actually had rights in some early agrarian civilizations. The fully preserved legal code of King Hammurabi of Babylon (1792–1750 B.C.E.) attests that slaves, though themselves considered merchandise, were allowed to own property; to enter into business; and to marry free women. Manumission (that is, the formal release from slavery by the owner) was possible either through self-purchase or adoption. Meanwhile, "free" married women (and their children) were considered the property of their husband.

Written law eventually organized most aspects of life in complex sedentary societies. It assumed an independent character as it also applied to those who had written it down or administered and executed it. Laws defined what sort of behavior was considered unacceptable, and provided a framework for an orderly form of retaliation to avoid an escalating cycle of mutual revenge between individuals. In principle, the development of written legal systems was beneficial for societies. But there was one major problem involved: until quite recently, most people were illiterate and depended on a literate elite to interpret legal texts for them. Hence, there was ample opportunity for rulers to bend the law and manipulate their subordinates.

Priests did the same with religious texts. Organized religions were actually a convenient way to impose rules on society. It was a lot cheaper to maintain order by exploiting the superstition of people than to build up an effective police force. Most religions provided incentives (such as promising afterlife in heaven or paradise) to make people follow rules, and posed threats (such as afterlife in hell) if people disobeyed them. Religious doctrines typically included regulations that facilitated life under crowded conditions, including respect for fellow citizens and provisions of personal hygiene. Religions were also designed to keep the ruling class in power, and the ruler himself was usually declared divine. The Egyptian pharaohs claimed they were living gods, the Inca ruler was considered a divine descendant of the Sun, and the Chinese emperor ruled by the grace of the gods. Roman emperors ruled by divine right and were elevated to the status of gods after their death, with some emperors insisting on being honored as gods even during their lifetime. The monotheistic faith as formulated by the founders of Judaism was basically a power-to-the-people doctrine. The divine nature of the pharaoh and the Roman emperor was undermined by the proposed existence of an almighty invisible god. Similarly, Buddhism defied Hinduism and its caste system, opposing especially the powerful priestly Brahman caste. In virtually every complex society the priesthood eventually wielded considerable power and sometimes managed to separate from the actually governing rulers to become an independent political force in its own right. This was certainly true for the Christian church with its regional network of powerful, land-owning bishops.

To be sure, religions existed already among gatherer-hunters, at least in a less organized form. Humans are intelligent enough to anticipate their own death and that of their family members and friends. The moment we are born we are sentenced to death. Early humans knew as well as we do that nobody lives forever. The concept of some kind of continuity in the form of an afterlife helps to cope with this reality, and even Neanderthals buried their dead with gifts or tools. In many later societies wives, servants, and farm animals were killed and buried together with deceased men of high social rank to serve them wherever they were going. And the death of a king was often accompanied by an outright mass slaughter of people and animals.

CHAPTER 6

MUSCLE POWER: THE
MAMMALIAN MACHINE

It shouldn't surprise us that kings of the Agricultural Age demanded to be buried with people and draft animals. After all, these were the principal prime movers of their time. (If their belief system had prevailed, then Mad King George of England might have similarly demanded to be buried alongside a steam engine in 1820.) Part of the reason why people and animals are such wonderful prime movers is that they can work *on demand*, whenever things need to be done. Windmills, in contrast, are as moody as the weather, and waterwheels may stand still during dry periods or when rivers freeze over. Besides, the human body, the principal engine of the Agricultural Age, is really quite special. Humans show remarkable endurance. For example, they are capable of chasing even the fastest herbivores to exhaustion, a quality ascribed to the high density of sweat glands that allows humans to sweat more amply than any other mammal. On the other hand, even excess nutritional energy cannot change the metabolic imperatives that principally limit how much work humans (or draft animals) can do within a given time. There is a maximum rate of food ingestion and digestion, as well as a restricted mechanical efficiency of muscles. The amount of oxygen people can inhale is limited as well. By the end of it all, no more than 20 to 25 percent of ingested food energy is actually converted to kinetic energy during peak aerobic performance, and sustained delivery of human power can usually not exceed 100 watts.[37] (This kind of power would keep no more than a single bright light bulb lit.)

During the Agricultural Age the total amount of work executed by human societies soared for several reasons. First of all, the availability of plenty of food removed the limitations that shortage in nutritional energy would have imposed on the work of individuals. Second, excess nutritional energy allowed for the human population to grow, creating a lot of additional prime movers. And finally, new innovations provided better ways for people to turn muscle power into useful work. People can normally lift and move only loads that are substantially lighter than their own bodyweight, a problem overcome by such inventions such as the lever, the inclined plane, and the pulley. Wedges, screws, wheels, treadwheels, gearwheels, cranes, capstans, windlasses, cranks, and wheelbarrows all helped to utilize muscle power more effectively.

Human legs are principally much stronger than arms. Two people treading a traditional Chinese water ladder could lift enough irrigation water to grow additional food worth 30 percent in excess of the nutritional energy they needed to eat to accomplish the task.[38] In the Mediterranean, the Archimedean screw was used to lift water to high points of irrigation schemes (from where it could flow far afield). This device, hand-cranked or treaded, was also used to pump water out of flooded mines or the hulls of large ships from the third century B.C.E. Internal vertical treadwheels were a common sight in medieval Europe. They were used to pump water from mines and for powering cranes, with the largest ones being powered by eight men.

Chinese Muscle-powered Water-lifting   These illustrations show that it was meaningful in Chinese agriculture to invest muscle work to lift water for the irrigation of fields which in turn delivered additional nutritional energy in excess to that spent during the operation of the foot-treadle-operated chain pumps. The drawings are contained in *Tiangong Kaiwu* (The Exploration of the Works of Nature), published in 1637 by Song Yingxing.

Treadwheels In the Middle Ages human muscle power was utilized through treadwheels. Internal treadwheels, as shown in the upper picture, were used to lift water from mines and building materials at construction sites. Horizontal and, less common, inclined treadwheels were used to crush ores and to mill grain. (Upper image reproduced from Georgius Agricola, *De Re Metallica*, 6th of the 12 books, 1556. Lower image based on Agostino Ramelli, *Le diverse et artificiose machine*, Figure CXXIII, 1588.)

Massing labor in form of teams of humans (or animals) was generally the solution to multiply the power of the single-worker labor unit. On the water, synchronized human power was especially efficient. Ancient Greek, Carthaginian and Roman vessels were powered by hundreds of oarsmen (who generally were slaves). These ships were usually constructed for warfare, typically designed to ram holes into the hulls of enemy ships. Still in the 17th century C.E., the Venetians operated galleys that had 56 oars, each crewed by five men. And the Maori of New Zealand used dugout canoes oared by up to 200 warriors in the 19th century C.E. (And yet, even the strength of hundreds of synchronized humans would be small in comparison even to small earth-moving machines of the Oil Age.)

POWERFUL ANIMALS

The other option people had if they needed a source of higher power output was to utilize large domesticated (or tamed) mammals. These were much stronger than humans, even though they were initially small by modern standards. It took the efforts of generations of animal breeders to slowly increase their size, weight, and strength. Besides, draft animals were initially poorly harnessed, which severely limited the degree to which their power could be utilized.

All draft animals were initially domesticated for meat consumption. Goats and sheep, which came first, were small and hardly useful as beasts of burden. Presumably they were used as pack animals here and there, but it was cattle that gained real importance for draft. Cattle were heavy, at least three times as powerful as humans, and easy to harness by attaching yokes (in form of straight or curved wooden bars) to their horns or neck. Wheels appeared in Mesopotamia before 3500 B.C.E., and four-wheeled carts and chariots soon thereafter. From about 3000 B.C.E. ox-drawn wagons had achieved enough prestige to be deemed worthy to be included in ceremonial burials with their owners.

Donkey-Drawn Battle Cars

In subsequent centuries, donkeys with throat-and-girth harnesses were used in Mesopotamia to pull chariots. Donkeys descend from the African wild ass and were domesticated in Egypt for meat consumption, probably around 4000 B.C.E. Donkeys resemble small horses, but have longer ears, a short erect mane, and a tuft of long hair only at the tip of the tail. Donkeys are very sure-footed and therefore well suited as pack animals even on mountainous trails. Besides, they live quite long, between 25 to 50 years, and they are very tolerant to hot and arid climates. Donkeys are relatively small and may not have been much more powerful than strong adult men, but the Af-

Donkey-drawn War Chariot   Donkey-drawn cars were once the fastest means of transportation available to people. The picture comes from a Sumerian wooden box known as the Standard of Ur. Dated to circa 2500 B.C.E., this box has two main panels depicting scenes of "Peace" and "War." The "War" side shows 4-wheeled chariots in different situations, during marching and fighting.

rican wild ass can run up to 50 km/h (30 mph), while humans reach no more than 37 km/h (22 mph) even on very short distances. (This is based on Jamaican Usain Bolt's 2008 100 meter world record of 9.69 seconds.) Hence, when donkey-drawn chariots appeared in Mesopotamia, they were the fastest means of transportation that had ever been available to people. Typically these cars transported two men, one to drive and one to fight. The Standard of Ur, a mysterious Sumerian wooden box dating to circa 2500 B.C.E., depicts four-wheeled chariots, each pulled by four donkeys that are trampling enemy soldiers.

Horses: Meat, Milk, and Speed

Thereafter horses appeared on the scene. Cave paintings in southern France, apparently 30,000 years old, clearly indicate that horses were hunted and appreciated as a food source by gatherer-hunters long before these animals were domesticated.[39] Still 15,000 years ago horses were widespread, thriving in large parts of Eurasia, Africa and North America. Thereafter, the warming climate replaced much of the world's open tundra with forest, and by 5000 B.C.E. horses survived only in the still-open grassland steppes of the Ukraine and Central Asia. People of this region probably began domesticating horses well before 3000 B.C.E. for meat and milk consumption, but many centuries later, when oxen and asses were already drawing carts in the Fertile Crescent, hunting rather than herding horses was likely still commonplace among the steppes people. The domestication process was probably delayed because horses, though they were small by modern standards, were skittish and fast. But shortly after the horse domesticators of the steppes north of the Black Sea had began herding these animals for milk and meat consumption, they may also have started to use horses as pack animals.

When the domesticated horse spread south into the Fertile Crescent, its size, strength, and speed predestined it to replace much slower oxen and donkeys entirely for the purpose of drawing chariots. (Modern horse breeds run up to 70 km/h or 45 mph.) Horse-drawn cars appeared in the Fertile Crescent after 2000 B.C.E. These speedy and maneuverable vehicles had to be lighter than ox-and donkey-drawn carts, which was in part achieved by the invention of the spoked wheel at about this time. From what we know, the Hyksos introduced the horse-drawn chariot into warfare and used this advantage to establish themselves as the rulers of Egypt in circa 1648 B.C.E. Where exactly the Hyksos and their technology had come from is unknown. Apparently they immigrated from regions of present-day Palestine and Syria, but before they may have lived even further north, in areas where they might have enjoyed cultural exchange with the Caucasus region. In the Caucasus region horses were adopted from the steppes people to the north, and the region also tends to be credited with the invention of hard metal smelting. Perhaps diffusion of technology from this area allowed the close-by Hittites, a people that began consolidating their kingdom in central Anatolia in the mid-17th century B.C.E., to become the very first, or nearly first, iron workers.[40] When the Hittites in later centuries battled the Egyptians for dominance in the western part of the Fertile Crescent, their chariots were the lightest and fastest, and 2-wheeled horse-drawn chariots were dominating warfare in the region. In China, chariots were introduced by Central Asian tribes. The oldest known specimen of a Chinese horse-drawn chariot comes from the grave of Emperor Wuding, who died in 1118 B.C.E.

Horseback riding started surprisingly late. But it revolutionized human travel and warfare, which were dominated by this form of locomotion for the remainder of the Agricultural Age and most of the Coal Age. Unambiguous evidence puts the beginnings of horseback riding at around 1500 B.C.E. on the steppes north of the Black Sea and Caucasus, where horses were first domesticated. But only around 1000 B.C.E. did mounted soldiers begin to appear in historical records of the Fertile Crescent, roughly one millennium after horse-drawn chariots had come into use. This long delay may have been caused by the inadequately small size of the horses of the time, or by lack of the necessary riding equipment.

Most importantly, riders did not use stirrups even beyond Roman times, which made it difficult to keep balance on the horse, especially if a heavy weapon was to be carried. Meanwhile, the lightly-clad archers of the Central Asian steppes did extremely well without them. They began to harass the

Tutanchamun Fighting with Bow-and-Arrow in a Horse-drawn War Chariot, ca. 1355 B.C.E. This illustration showing "Tutanchamun fighting Asians" comes from a chest from the Valley of the Kings. Light, two-wheeled, horse-drawn war chariots were at this stage dominating warfare in the region. Note the spoked wheel.

Chinese Middle Kingdom in the 4th century B.C.E., and the Chinese reacted by introducing mounted soldiers for their defense. However, it took about another seven centuries, until the 4th century C.E., for the Chinese finally to invent stirrups. By the 8th century C.E. these foot-rests, which hang in a pair from each side of the saddle, had spread all across Eurasia from China to Europe, where stirrups in turn revolutionized warfare, as they allowed heavily armed knights to mount horses.[41] Stirrups enabled these warriors to merge themselves into an integrated fighting unit of rider and horse, and to use a lance to deliver a blow that fully reflected the combined kinetic energy of man and animal. This development also depended on the breeding of tall strong horses. The type of horse that Vikings introduced in Iceland in the late 9th century was so small that it is now merely considered a pony—it still had not advanced over the size of early Fertile Crescent horses. However, Vikings ate their horses and did not ride them in warfare, while breeders further south in Europe had by this stage already begun to develop horses that could carry a fully-armored knight.

Horseback riding was also very important in communications, simply because it was the fastest way for messengers to move around. Still in 490 B.C.E., a Greek messenger is said to have run about 25 miles (40 km) to Athens, announcing before he dropped dead of exhaustion that the Greek army had defeated the Persians in the Battle of Marathon. (The modern long-distance footrace of 42.2 km commemorates this event.) However, the Persians (whose cavalry was absent at this battle) ruled the largest empire at the time and still in the 5th century B.C.E. introduced relays of mounted couriers to dispatch commands from their capital. The Chinese and Japanese

The Stirrup The armed mounted knight, a picture familiar in European medieval history, was made possible by a Chinese innovation, the stirrup. (Reproduced from a miniature from a Psalter written under Louis VI, King of France from 1108 to 1137, edited.)

built up a similar communication network, and the Pony Express, operating in the United States until 1861, was not all that different either. (The Pony Express employed 157 stations on a 1,800-mile (2,900-km) stretch between Missouri and California. It took about 10 days to complete the route, with riders changing horses about every 10 to 15 miles. The service was rendered obsolete with the completion of the trans-continental telegraph system.)

Riding Camels

To be sure, people did not only ride horses. Donkey-back riding presumably predates horseback riding, and camels, initially appreciated for their meat, milk, and hair, became extremely important for this purpose too. The one-humped Arabian camel, or dromedary, is no longer found in the wild. Unambiguous evidence attests that it was domesticated not too long after 1000 B.C.E. People first used camels as pack animals, but by the mid-7th century B.C.E. had begun riding them in warfare. (A Mesopotamian relief dating to this time shows two soldiers riding one dromedary. One is an archer, the other one the chauffeur.) Later on, the armies of the Persians, Alexander the Great, and the Romans all employed dromedaries in the wider Fertile Crescent arena. The introduction of the North Arabian camel saddle sometime between 500 B.C.E. and 100 B.C.E. finally allowed for heavy loads to be transported on dromedaries. This was a major breakthrough that facilitated the expansion of trade in the area—without the construction of roads on which wheeled vehicles would have depended.

The second type of camel, the two-humped Asian camel, or Bactrian camel, is less tall but can also go several days without drinking water. However, this longer-haired camel is adapted to relatively cool rather than desert climate, and was originally distributed over dry steppes from southern Russia to Mongolia. Again, there is no solid evidence that the Bactrian camel had been domesticated before 1000 B.C.E. From the late third century B.C.E., long-distance trade was gradually established along the so-called Silk Road, which connected the Fertile Crescent and China. Camels were used to transport goods along this route and the Arabian camel consequently spread towards the east, while the Bactrian camel spread towards the west, with crossbreeds appearing along the way. Llamas, which also belong to the camel family, were important pack animals in the South American Andes, but people were never riding them.

Tamed Elephants

Elephants have been tamed and ridden since ancient times as well. However, they have never been truly domesticated, and those kept in captivity are not genetically different from those in the wild. The African savanna elephant (Loxodonta africana) is currently the world's largest land animal, weighing up to 16,500 pounds (7,500 kg) and standing up to 4 meters (13 ft) tall at the shoulder. (The African forest elephant is somewhat smaller.) African elephants can be trained, but there is no major tradition of having them work for humans. Nevertheless, it is well documented that Carthaginian general Hannibal's army of 40,000 was accompanied by a substantial number of African elephants, which were probably widespread in the Atlas Mountains of the southwest Mediterranean, north of the Saharan desert. (Hannibal took off from New Carthage (now Cartagena, Spain) and crossed the Alps to march on Rome in 218 B.C.E.)

So-called Indian elephants, which are widespread in South and Southeast Asia, may have been used for work from as early as 2000 B.C.E. (We also know that some were employed in warfare against the army of Alexander the Great in 326 B.C.E.) Still today, they are utilized in quite large numbers to harvest timber. In the wild, mating pairs often separate from the herd for several weeks, and in captivity, elephants rarely bear young at all. Hence, people acquire elephants in the wild by separating them from their families at young age. They lock them into cages, and hit and starve them in order to break their resistance and render them submissive. Adult elephants eat enormous amounts of vegetation (more than 500 lbs [225 kg] daily). They are usually taken out at night to feed and rest in nature, held by a tethering chain.

WHY HAVE MOST ANIMALS BEEN
DOMESTICATED IN EURASIA?

The above list of critical mammal domesticates is almost exclusively Eurasian. But here we can hardly argue that Eurasia has been comparatively best-stocked with large mammal species as it has been with large-seeded grass species: We know that Sub-Saharan Africa, where humans have been living longest, is teeming with mammal species of the size that people like to domesticate. But perhaps Eurasia has been better stocked with large mammals that actually lend themselves well to domestication?

There are about 4,600 mammal species in the world, which include specimens as different as humans, whales, and bats, while almost half of all mammal species are rodents. Only about 148 mammal species are large terrestrial mammals that, due to their body size, appear to be attractive candidates for use as beasts of burden or for significant meat and milk production. Africa's long list of large mammals includes zebras, rhinos, hippos, giraffes, elephants, lions, tigers, gorillas, and so forth, while Eurasia has rhinos (in Nepal and Sumatra, for instance), tigers, (Indian) elephants, bears (including pandas), orangutans, and the European moose (elk), which lately has been domesticated with limited success in Russia. North America's variety in native large mammals is much lower (deer, bear, bison), and Australia has only the kangaroo.

While this list clearly favors both Africa and Eurasia, size is but one of many criteria that would make an animal species truly suitable to be targeted for domestication. And indeed, most of the 148 large terrestrial mammal species did not lend themselves well to domestication and had to be ruled out. Some of them are not sufficiently docile, submissive or gregarious, others are not cheap enough to feed, others again are unwilling to breed in captivity, and so on. The long list of dropouts includes grizzly bears, which are not social enough; kangaroos and the North American big horn sheep, which both lack a follow-the-leader social structure; 41 out of 42 deer species, because they are too skittish; and gorillas, which take 25 years to mature. Zebras adapt quite well to life in captivity and can be trained to pull carriages, but it is generally difficult to use them for work, as they are stubborn and unpredictable, with a tendency to attack people. Besides, they are difficult to ride. Elephants, which have been utilized for work since ancient times, fall into the category of animal species that could be tamed but not truly domesticated, since it turned out difficult to breed a whole population in captivity.

In the end, just 14 out of the planet's 148 large terrestrial mammal species have been domesticated, and 13 out of these 14 are descended from wild species found only in Eurasia and North Africa (cattle, sheep, goat, horse, pig, reindeer, donkey, Arabian and Asian camel, water buffalo, Bali cattle, yak, and gaur). In fact, the only large non-Eurasian mammal that has been domesticated is the llama, native to South America's Andes. (The llama's smaller cousin, the alpaca, has not been separately listed.) Not a single one out of the 14 has been domesticated in North America, Australia, or sub-Saharan Africa, except perhaps for the cattle of the Sahel zone. We might in turn speculate that Eurasians have simply been better at domesticating animals, but no major domesticate has been added to the list since Eurasians began crossing the major oceans in the late 15th century and settled on other continents.[42] This leaves us with two questions. One, why did Africa and Eurasia end up with so much more large wildlife than Australia and the Americas? And two, why did Eurasia's large mammals lend themselves so much better to domestication than Africa's?

American and Australian Overkill?

To address the first question, a hypothesis has emerged that is based on the coevolution of increasingly skilled human hunters and their prey. Its starting point is the observation that Africa, where people have been living longest, offers the largest variety of large mammals, while Australia and the Americas, where people have been living shortest, have only few large native mammal species. It is thus theorized that large mammals have fared best wherever they coevolved with humans, but were chased to extinction wherever advanced human hunters quite suddenly appeared on the scene.[43]

Research shows that Australia and the Americas were indeed teeming with large wildlife until around the time when people first arrived. Australia had scores of exotic marsupials, the type of mammals that shelter their young in pouches. Among them were a giant kangaroo, nearly three meters (9 feet) tall, with sharp claws; a cow-sized herbivore that looked like an enormous wombat; and a fearsome predator known as the marsupial lion. There were several large reptile species, including a one-ton flesh-eating lizard that grew to a length of eight meters (26 feet), and there was a giant flightless bird, three meters (9 feet) tall and well over 100 kg (220 lbs) in weight. However, Australia experienced a continent-wide mass extinction event around 46,000 years ago, which annihilated all terrestrial mammals, reptiles, and birds of more than 100 kg (220 lbs) in weight, plus most species over 45 kg (100 lbs). (In total, at least 55 large-bodied species were affected.) It is not entirely clear what caused this event, but it took place suspiciously close to the time when humans are thought to have first migrated to Australia, around 40,000 years ago.

In the Americas, the fauna included mammoths and mastodons (both relatives of modern elephants), giant ground sloths, tapirs, a large camel, half a dozen species of horses, a large-horned bison, prong-horned antelopes, oxen, a type of mountain goat, a giant armadillo, and a type of large mammal that was covered with solid armor. Large predators such as saber-toothed cats, dire wolves, and some bears were present as well. Dozens of these species, including three quarters of America's large mammals, disappeared within perhaps a thousand years after retreating glaciers allowed people to migrate to central North America around 11,000 B.C.E.

To be sure, nobody knows for certain whether people in fact hunted the Australian and American wildlife to extinction. Climate or other environmental changes may have played a role as well, especially in America, where the mass extinction coincided with the end of the last Ice Age (or, rather, glacial period), a major environmental transformation. Besides, humans may have caused harmful changes without directly hunting species into extinction: people, or their dogs, may have carried diseases that knocked out the large animals, or they may have altered the environment through bush fires, as Australian aborigines still do now to facilitate hunting.

On the other hand, the human kill-all-you-can behavior is well documented from later extinction events on previously isolated islands. People killed off the unafraid moas of New Zealand, the lemurs of Madagascar, and the big flightless geese of Hawaii. The animals of the Galapagos Islands and Antarctica, which also evolved in the absence of humans, are still incurably tame today and have to be protected from hunters.

If the overkill theory is correct, the distant ancestors of native Americans and Australians have likely handicapped the later development of their descendants severely, because they slaughtered all potential beasts of burden into extinction. The only large mammal to persist in Australia was the kangaroo (males of the largest remaining kind can grow to 90 kg (200 lbs), while deer, bear, and bison survived in North America.

Africa: Plenty of the Unsuited

Africa, in sharp contrast, is still now offering a great variety of large mammals, but curiously, none of these species has been domesticated. Hence it is tempting to speculate that Africa's abundance in wildlife, and to some degree these animals' general unsuitability for domestication, may be connected to the fact that humans have been living in Africa longest. As humans slowly evolved into more intelligent and sophisticated hunters, Africa's animals may have learned to run away whenever people came too close. And some evolved into species too large and thick-skinned (elephants, rhinos, hippos) for humans to kill them. (Those species not fitting these categories were presumably hunted into extinction.)

However, Africa's mammals actually do not lend themselves to domestication for various different reasons. The African buffalo (caffer) and the rhino are simply too dangerous, the zebra too stubborn, and the gazelle too skittish. Of all large animals, hippos are now killing the most people in Africa. They simply run over people who get in their way, usually when they are heading for water holes or rivers. Similarly the African, or Cape, buffalo

(Syncerus caffer) is regarded one of the world's most dangerous animals, at least when wounded. It superficially resembles the massive Asian water buffalo, but has never been domesticated. Though not a mammal, one large animal species that has in fact been domesticated in Africa after Europeans arrived is worth mentioning: The ostrich, a swift-footed two-toed flightless bird referred to as *Struthio camelus*. This is the largest remaining bird species in the world, with males weighing up to 140 kilograms (300 lb).

Eurasia: The Right Balance?

Eurasia, the largest landmass on Earth, is second only to Africa in terms of its native large mammal species. As Eurasia eminently also ended up with the most domesticated large mammal species, it may be speculated that this landmass became the cradle of animal farming because it fit right in between Africa and Australia/America, not just geographically, but also in terms of the relationship between human hunters and their prey. Early human species migrated into Eurasia more than one million years ago. Perhaps the timing was critical, with humans arriving when their hunting skills had not yet sufficiently evolved for an outright overkill, but late enough to keep the time of coexistence with wild large mammals limited (until the first domestication attempts), thus keeping these animals from evolving into beasts that would not tolerate humans close to them at all (as has been suggested for Africa). But leaving these rather bold speculations aside, it was likely sheer environmental luck that favored Eurasia in terms of suitable animal species the same way it did with domesticable plant species. One way or another, these advantages helped Eurasians to become agrarian and use animal power earlier than people on other continents.

ANIMAL POWER FOR FIELD WORK

Animal power was especially important in crop farming. Farmers had to prepare the soil, distribute and cover the seed, and cut the stalks. Initially simple digging sticks, and then deer antlers, were used to loosen the ground for seeding. They were replaced by wooden hoes, which at first were nothing more than a (long) handle with a (thin) wooden plate attached to it at an angle. Hoes were also used during the growing season to keep the weeds down, aerate the soil, and separate plants. A bronze hoe dating to 2000 B.C.E. was discovered in the Caucasus, and the Romans used the swan-neck-shaped sarculum (or Roman hoe), which is still now common in the Mediterranean. Farmers also began to use harrows, an implement set with spikes, which was pulled over the field to pulverize and smooth the soil either in field preparation or to cover the seed with soil after the seed was sown broadcast over the field.

Animal Power in Crop Farming   Oxen, donkeys, yaks, water buffaloes, dromedaries, and horses have all been used for plowing, but effective harnessing made the difference. Oxen were early on harnessed at their horns, while the rigid horse collar provided the breakthrough that allowed the work horse to become dominant in temperate climate agriculture. The upper picture comes from the Egyptian burial chamber of artisan Sennedjem, dated to circa 1200 B.C.E., the lower picture, with a horse pulling a harrow, comes from the Luttrell Psalter, written in England around 1325 C.E.

A major step forward, which was not taken in the Americas and sub-Saharan Africa prior to European arrival, was the introduction of the plowshare. Plowing depended on the availability of draft animals and thus set Eurasia even more aside in terms of agricultural productivity. Plowing prepares the ground for seeding much more thoroughly than hoeing does. It breaks up the compacted soil, uproots established weed, and provides loosened, well-aerated ground in which seedlings can germinate and thrive. Wooden scratch plows, with plowshares that were nothing more than a sharpened stick of wood or a wood blade, were pulled by donkeys and oxen in Mesopotamia from perhaps 3500 B.C.E. Cattle became most important for the task, though donkeys often worked the light soils of the Mediterranean, and water buffaloes were used for tilling rice fields in South and Southeast Asia.

Simple symmetrical wooden lightweight plows, which were soon metal-tipped, were sufficient to work relatively loose soils and dominated Eurasia for millennia. In fact, they have been used in parts of the Middle East, Africa, and Asia still in the early Oil Age. However, they had a principal shortcoming as they left cut weed on the surface and created only a shallow furrow for the seeds. This situation was much improved with the introduction of moldboards. These were initially straight pieces of wood mounted on a plow behind the plowshare. This simple device made a big difference, because it lifted and turned the plowed-up soil. It therefore buried the weed, which started the decomposition of the removed vegetation and left the field clean without competition for the crop that was to be sown. Moldboards eventually appeared in asymmetrical designs that moved the plowed-up material to one side to leave a deep, clean furrow. This eliminated the requirement of cross-plowing and saved a lot of time and effort.

The Chinese introduced asymmetrical curved moldboards made of non-brittle cast iron before 100 B.C.E., while the Romans did not advance much over the ancient Mesopotamian plow design and did not know moldboards at all. Hence, Romans were still hoeing in all hilly or especially rocky regions, where the frail plows of the day would not work appropriately. Heavier plows, suited for the tough soils north of the Mediterranean, appeared only in the late (Western) Roman Empire, around 300 C.E. Moldboards emerged soon thereafter in regions of present-day Austria and Germany. However, the sophisticated Chinese curved moldboard design was first introduced to Europe by Dutch sailors in the 17th century.

Rigid Horse Collar

The horse turned into the most important working animal outside the tropics, but it assumed this role quite late because appropriate harnessing technology was lacking. Cattle were easy to harness at their horns, but the throat-and-girth harnesses used for horses worked only with relatively light loads (such as carriages). In field work, as soon as horses leaned into a heavy pull by lowering and advancing the head, the throat-strap of such harness created a choking effect, and the horse could not breathe. The breastband harness, introduced in China before 200 B.C.E., increased the efficiency of draft horses as it avoided choking them, but it was not ideal either, because its point of traction was set too far from the animal's powerful pectoral muscles.[44] Nevertheless, the design spread through Eurasia, reaching Italy in the fifth century and northern Europe in the eighth century.

The final breakthrough came with yet another Chinese innovation, the rigid shoulder collar harness. It emerged in the first century B.C.E., initially as a soft support for the hard yoke, but was developed into a single integrated device that was really nothing more than a wooden (later also metal) frame, lined to comfortably fit on the horse's shoulders. Simple as it was, this piece of equipment made a major difference: It allowed horses to deploy their powerful breast and shoulder muscles fully without any restrictions on their breathing. Hence, the useful draft of horses harnessed with shoulder collars turned out to be as much as four times (and for brief peaks up to ten times) higher than for horses in a throat-and-girth harness.

The rigid horse collar spread across Eurasia to reach Europe in the ninth century, right at the time when breeding efforts were on their way to create large heavy horses for carrying armored knights. It was also the time when horseshoes became common in Europe. The metal horseshoe allowed horses (and oxen) to plow stony fields without injury, and prevented excessive wear of the animal's insensitive but soft hooves. Horseshoes had actually appeared in the Fertile Crescent already around 400 B.C.E., but were not universally applied, as their advantage in improving traction and endurance was obvious only on hard grounds. The Greeks, for instance, did not use metal horseshoes, but encased hooves in leather sandals filled with straw.

The combination of larger breeds, the rigid shoulder collar, and the metal horseshoe turned the horse of central-western Europe into a superior prime mover that delivered about as much power as 10 men. And while animals are usually simply the stronger the heavier they are, horses outperformed oxen by this standard, because their weight is distributed differently, that is, more concentrated in the front. Most importantly, horses worked 30 to 50 percent faster than oxen on the field, except in heavy wet soil. The power output of a strong well-harnessed horse is about 700 watts, more than twice that of early draft animals. Hence, a team of 40 horses pulling a California combine (that is, a harvesting machine) in 1885 delivered 28,000 watts.[45]

ANIMAL POWER TO DRIVE MACHINERY AND PULL LOADS

In off-the-field work, the employment of horses was delayed as well. The pony-sized horses of the early Middle Ages could not compete with oxen in turning a whim (i.e. a beam attached to a central axis). This kind of horizontal treadmill was commonly used for milling grain and pressing oilseed. Later on it became important for winding ore (or rainwater and groundwater) out of mines, and to power machinery employed by various industries. Again, larger horses and better harnesses made the difference. By the 17th century very tall and strong grain-fed horses rather than oxen were preferred in Europe to turn whims. Generally the relatively large share of animal power in European agriculture must be assumed to have gone hand in hand with a larger animal power in production, because the demand for

Oxen Driving a Mill During the Agricultural Age animal power was used to drive machinery. This illustration, showing a pair of oxen powering a grinding mill, comes from the Chinese *Tiangong Kaiwu* (written by Song Yingxing in 1637).

draft in agriculture was seasonal, pushing idle prime mover capacity towards production during the off-season.

The role of oxen and horses for the transport of loads in carts was perhaps even more important. In this application animal power remained indispensable until the emergence of trucks and passenger automobiles in the Oil Age. However, wheeled vehicles required streets, and those were generally bad. (Only after 1750 were materials such as gravel, asphalt and concrete applied in road construction.) Animal-drawn carriages were thus mainly used for short-distance transport. For short spells, oxen could pull loads of up to three or four times their body weight, but they could not travel faster than 20 kilometers (12 miles) a day. Horses were a lot faster, making perhaps 60 kilometers (37 miles) a day, but with substantially lighter loads. Road specifications in the fourth-century Roman Empire limited loads to 236 kilograms (519 pounds) for horse-drawn wagons, and to 490 kilograms (1078 pounds) for the slower ox-drawn carriages.[46]

Canals

Land transport remained a problem throughout the Agricultural Age. It remained slow and expensive even on the best roads, and hence severely restricted trade, economic growth, and land development in all regions situated away from rivers and sea shores. One way to improve the situation was to construct canals. Animals walking alongside canals could quite easily pull heavy loads, because buoyancy provides an upward force on bodies floating in water, causing them to appear lighter, and because there was far less resistance to forward movement in water when compared to the primitive, hard-wheeled vehicles on bad roads. A single horse can pull a canal load of 30 to 50 tons (at least 10 times more than on the best hard road) at a speed of perhaps three kilometers per hour (just under two miles per hour).

Canals provided an important infrastructure in many prospering agrarian regions. The oldest parts of China's Grand Canal, the longest human-made waterway in the world, may date to the fourth century B.C.E. It was gradually expanded to reach its full length of 1,747 kilometers (1,085 miles) in 1327 C.E. In Europe, the first canals were constructed in northern Italy during the 16th century, modeled on the Chinese canals. The 240-kilometer (150-mile) French Canal du Midi was completed in 1681, while the longest continental and British links date to after 1750. Canals in turn reached their greatest importance in Europe during the 18th and 19th centuries, providing the key infrastructure for European industries to expand immediately before the onset of the Coal Age. In fact, many horses still worked on small European canals as late as the 1890s.

FUELING ANIMAL WORK

The horse was really the prime mover that universally filled almost every niche of the temperate zone economy in the mature Agricultural Age. It was used to drive machinery, to carry people, and to pull plows, carts, coaches, canal boats, and the first trains on tracks. Horses live long and show remarkable endurance (in comparison to humans and oxen, for instance). On the downside, horses consumed a lot of fuel. While cattle have traditionally been fed with straw and chaff, horses had to eat grain in order to perform. Horses thus competed more directly for nutritional energy that people can eat. And while the employment of horses on the fields pushed cereal output to new standards, much arable land had to be set aside for the production of hay and coarse grain to feed them. This was relatively easy to do in parts of Europe, where the maintenance of horses at times claimed up to one-third of all agricultural area, but it was unattainable in the densely populated areas of China, for instance. In China's traditional farming the cultivation of feed for draft animals claimed only about 5 percent of the annually harvested area. It was cattle and water buffalo that remained most important in East and South Asian field work.

Depending on population density, the ideal balance had to be found between sacrificing grain to fuel draft animals, and the increase in grain output achieved by employing them. Farmers initially added fodder in form of crop residues to the normal grazing, but once they harvested enough grain to use some of it for feed, animals reacted to this energy-rich diet by working a lot better. Usually quite vast areas of land had to be available before agrarian societies could make the shift towards more animal work on their fields. Such a situation came about when diseases eliminated much of the human population (as the plague did in Europe in the early decades of the 1400s), or when advanced farmers migrated to new virgin lands (as Europeans did in North America). Such societies then achieved extremely high agricultural output per farmer with the assistance of lots of animal power.

Nevertheless, even when farming was highly efficient, and there was plenty of arable land and animal power available, there were restrictions how many centralized non-farming bureaucrats, craftsmen, and soldiers, and so on could be sustained. There were the traditional restrictions, including limited sunshine and photosynthetic efficiency, but the true restraint for the growth of urban population centers was the absence of advanced transportation systems. This forced most people to live quite close to the fields during the Agricultural Age. Even based on a mainly vegetarian diet, a city of 500,000 people would have required about 150,000 hectares of efficiently cultivated cropland, which usually limited the size of cities to a much lower population number, even if they were located at a river or shoreline.

NOTES

37. Vaclav Smil, *Essays in World History—Energy in World History* (Boulder, CO: Westview Press, 1994). This book by Smil has served as a general source of quantitative information with respect to historic energy issues.

For short periods, a man's muscles can deliver a maximum of about 800 watts, or little more than one horsepower, for periods of a day or more one third horsepower is a more plausible maximum. Joel E. Cohen, *How Many People Can the Earth Support?* (New York: W.W. Norton & Company, 1995): 98.

38. Vaclav Smil, *Essays.*

39. Melinda Maidens, "Horses and History," http://users.erols.com/mmaidens/index.html; find more sources and information in: Charlie Apter, "The Horse in Art, Science and History," Truman State University, 2007, http://agriculture.truman.edu/courses/343syllabus.pdf.

40. Hans G. Güterbock, "Hittites," Microsoft Encarta Online Encyclopedia, http://encarta.msn.com/encyclopedia_761563583/Hittites.html.

41. Bernard S. Bachrach and Charles Martel, "Mounted Shock Combat, the Stirrup and Feudalism," *Studies in Medieval and Renaissance History* 7 (1970): 49–75. Kelly DeVries, *Medieval Military Technology* (Peterborough: Broadview Press, 1992).

42. It may be argued that there was no reason for Eurasians to domesticate new species—they already had a good selection of farm animals, while the domestication of a wild species is a long, strenuous effort. However, the attempts to domesticate bison in North America and kangaroos in Australia, for instance, show that these efforts have indeed been made.

43. Paul S. Martin and H. E. Wright, eds., *Pleistocene Extinctions: The Search for a Cause* (New Haven, CT: Yale University Press, 1967).

44. Vaclav Smil, *Essays.*

45. Ibid.

46. Ibid.

CHAPTER 7

WIND AND WATERPOWER

In stationary work, the Agricultural Age's alternative to people or animals turning an axle was wind and waterpower. Wind is a flow of air generated when sunshine warms the planet's atmosphere unevenly. Water flows from higher to lower altitudes after being elevated by sunshine into the atmosphere and precipitating over uneven landscapes. Wind- and waterwheels emerged in the Fertile Crescent as well as in ancient China. Both prime movers were initially used for milling grain and pressing oilseed, and both were initially horizontal wheels, driving a vertical axis to which a running millstone was attached.

In Persia (present-day Iran), horizontal windmills were used from the seventh century C.E. These were enclosed in stone structures that allowed the wind to enter only on one side of the vertical axis, moving the horizontal wheel always in the same direction. Such design is useful only in regions where seasonal winds blow reliably from the same side. Europeans learned about these windmills during the Crusades of the 12th century and began constructing them in their home regions. Before long, Europeans developed this prime mover into the vertical windmill, which could turn the rotor towards any direction in order to face the wind wherever it came from. Vertical windmills reached great importance in northwestern Europe, mainly for the purpose of drainage and milling grain. The Netherlands had at least 8,000 windmills by 1650.

Early horizontal waterwheels were used in Asia Minor (present-day Turkey) in the third century B.C.E. The more familiar, and more powerful, vertical waterwheel (with a horizontal axle) is probably nearly as old, perhaps

Tower Windmill   Powerful vertical windmills eventually became very important for drainage and milling grain in northwestern Europe. Here a tower windmill at Wijk bij Duurstede near Utrecht in the Netherlands is shown. When Jacob Isaaksz van Ruisdael painted it in 1670, nearly 10,000 windmills operated in the Netherlands.

dating to the first century B.C.E. Vertical waterwheels began to spread in Roman times, but horizontal ones were still in use in early medieval Europe. The Romans are said to have made limited use of waterpower because they had so many slaves and so much agricultural energy available. There are, however, examples of amazing Roman waterpower facilities, above all, the cascade of 16 vertical waterwheels at Barbegal (north of Arles, France), which received water via an aqueduct. This facility was constructed to mill grain into flour around 350 C.E.

Medieval Europe's horizontal waterwheels eventually disappeared in favor of two different varieties of vertical wheels: undershot, with the wheel simply dipping into the stream, and the more powerful overshot, with water being directed to the top of the wheel. Between the 10th and the 14th centuries the number of waterwheels in Europe increased dramatically, a development perhaps promoted by the fact that less true slave labor was available than in other, earlier civilizations.

Overshot Waterwheel   Waterwheels were used for various productive tasks during the Agricultural Age, including the milling of corn and the fulling of cloth. In this picture from Georgius Agricola's *De Re Metallica* of 1556, an overshot waterwheel is used to crush rocks.

UNPRECEDENTED POWER

With respect to their contribution to the total energy available to societies of the Agricultural Age waterwheels and windmills remained relatively unimportant. However, they began to make a notable difference for certain applications in parts of Europe as well as Asia between 500 C.E. and 1000 C.E. because they emerged into prime movers of unprecedented power output. Compared to the sustained power of a single human worker (100 watts) or contemporary draft animal (300 watts), a Roman vertical waterwheel turning a millstone in 50 C.E. delivered 1,800 watts. (The Barbegal cascade of 16 vertical wheels combined delivered as much as 30,000 watts.) Single European waterwheels of the 11th to 16th centuries had a power output of about 4,000 watts, a level only slowly reached by windmills towards the end of the 16th century. Nevertheless, there were large German post windmills delivering 6,500 watts in 1500 C.E. to crush oilseed. And toward the end of the Agricultural Age (1750 C.E.) large Dutch windmills, used to drain polders, delivered some 12,000 watts. However, around this time waterwheels were being constructed that delivered four to five times the power of even the largest tower windmills.[47]

Sailing: Wind Power for Mobility on the Water

People of the Agricultural Age also learned to harness the kinetic energy of wind for mobility on the oceans. Waterborne transportation was extremely important in the Agricultural Age, as it was much easier and cheaper to move goods on the water than on land. Hence, almost all complex civilizations arose along sea shores, river streams, or lakes. Earliest illustrations of sailing vessels appear on Egyptian vases from 3500 B.C.E., while the well-preserved sailboat found in the tomb of pharaoh Khufu (Cheops) dates to about 2600 B.C.E.[48] Later on, wind power was utilized for long-distance transportation of goods on open seas, but muscle-powered galleys operated in the Mediterranean still in the 17th century. The reason oaring prevailed as a mode of propulsion on open sea, in parallel to wind power, was that warships could not afford being immobilized by a dead calm. Besides, the ancient Egyptian and consecutive Mediterranean cultures rigged their ships exclusively with square sails, which provided efficient energy conversion only with the wind astern. By the year 1000 C.E. sailing technology in the Mediterranean still had not changed much, and voyages without tailwinds were still potentially uncertain and extremely lengthy.

More sophisticated sail designs emerged in China and eventually spread westward to reach Europe via Muslim countries. Notably, the Italians in the 11th century adapted the (triangular) Arab lateen rig, which made it possible to sail in a wider range of wind conditions. (This rig was set at an angle to the mast, replacing the rectangular sail set square to the mast.) Similarly, the stern-post rudder, invented in China in the first century C.E., reached Europe in the late Middle Ages to replace trailing oars as a more effective means of steering. The power of rudders was in turn enhanced by the use of cranks and pulleys, which made it much easier to maintain course in rough weather.[49]

In the 15th century, the Portuguese dominated naval technology and explored the Atlantic islands and the African west coast. They pioneered new rigging systems and built maneuverable and fast ships with more masts. Such ships were able to tack into the wind with greater ease, and were robust enough to operate successfully in the stormy winds and strong currents of the Atlantic Ocean. Commanding wind energy with such ships eventually allowed Europeans to cross the big oceans, to discover new continents, and to open the world.

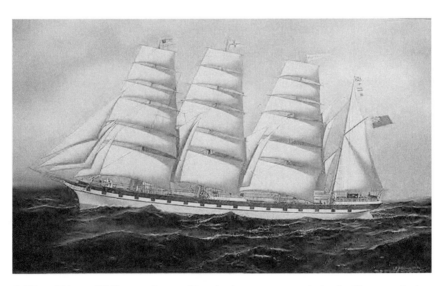

Sailing Ships Well over three millennia (over 33 centuries) of sailboat evolution are reflected in these two paintings. The upper one, of an Egyptian river-going boat, comes from the Tomb of Menna, dated to ca. 1416 B.C.E. The lower one, showing the clipper *Forteviot*, was painted by Antonio Jacobsen in 1896 C.E.

NOTES

47.   Vaclav Smil, *Essays in World History: Energy in World History* (Boulder, CO: Westview Press, 1994).

48.   Timothy J. Runyan, "Ship—Earliest Sailing Vessels," Microsoft Encarta Online Encyclopedia. http://encarta.msn.com/encyclopedia_761571524/Ship.html#s62.

49.   Find an account on "Developments in Sea Transport and Navigational Technology, 1000–1820," in Angus Maddison, "Growth Accounts, Technological Change, and the Role of Energy in Western Growth," *Economia e Energia*, sec. XIII–XVIII, Istituto Internazionale di Storia Economica "F. Datini" (Prato: Le Monnier, 2003). http://www.eco.rug.nl/~Maddison/ARTICLES/Role_of_energy.pdf.

CHAPTER 8

BIOMASS: ENERGY FOR LIGHTING, HEATING, AND METALLURGY

During the entire Agricultural Age people had no means to convert the energy derived from a turning axle (be it driven by animate, wind, or waterpower) into light or heat as was later done via electricity generators. Hence, it was biomass that provided virtually all the energy available for heating, lighting, and metallurgical needs. Like food, non-edible biomass stores solar energy captured during plant growth. This energy can in turn be released through the burning of such materials as wood, crop residues, plant and animal oils, waxes, and dried dung.

Open flames represented the only artificial indoor light available during the Agricultural Age. Both oil lamps and candles offered inefficient, weak and smoky illumination, but they were portable and safe, and allowed people to work after sunset. Candles, made of slow-burning wax or fatty material enclosing a fiber wick, were used in the Fertile Crescent in primitive versions at least from about 800 B.C.E. The first dipped candles were made by the Romans, who (like the Egyptians earlier on) relied on tallow gathered from cattle or sheep suet as the principal ingredient of candles. Such tallow candles were cheap and widely used for many centuries. Beeswax (secreted in small flakes from glands of worker honeybees to make the cell walls of the honeycomb) was used for candles in Europe from about the 16th century. They burned brighter, produced less smoke, and smelled a lot better than tallow candles, but they were expensive and reserved for churches and wealthy households.

On the downside, charcoal was a very energy-expensive fuel. Its production consumed tremendous amounts of wood. Only about 20 percent of the weight of a pile of air-dried wood is recovered as charcoal after the conversion process. About 60 percent of the original energy contained in the feedstock is consumed or lost in the process.[51] About 5 kg of air-dried wood were necessary to produce 1 kg of charcoal, and between 8 kg and 20 kg of charcoal were used to produce 1 kg of iron.[52] Hence, the production of 1 kg of iron consumed at least 40 kg of dried wood, but possibly much more. These large wood requirements severely limited the size of industries and cities that depended on charcoal as fuel. All the major iron production centers of the Agricultural Age therefore had to emerge in or near forest-rich regions.

NOTES

50. The Davistown Museum, "Steel-and Toolmaking Strategies and Techniques before 1870", http://www.davistownmuseum.org/PDFs/Vol6_SteelToolMaking.pdf.

51. To be sure, energy can neither be created nor destroyed; it merely changes its form, and it can be transferred from one body to another. If we talk about energy losses, we mean a loss of energy in a concentrated, useful form.

52. Vaclav Smil, *Essays in World History: Energy in World History* (Boulder, CO: Westview Press, 1994); Bert Hall, "Medieval Iron and Steel," Medieval Science and Technology: Original Essays, http://www.the-orb.net/encyclop/culture/scitech/iron_steel.html.

CHAPTER 9

WEAPONS TECHNOLOGY: ENERGY
USED TO KILL AND DESTROY

When people began commanding more energy during the Agricultural Age, they used it for destructive as well as productive means. Excess nutritional energy fueled professional soldiers and whole standing armies. The equipment used by the people who specialized in killing other people was initially similar to hunting gear, but gradually developed into efficient purpose-built weapons. Metals revolutionized warfare in form of knives, swords, arrow heads, chain shirts, and so on, but in principle the goal was to develop weapons that could kill from a distance. Technology that enabled one party to launch a destructive weapon from longer distance than the opposing party immediately translated into political power in all epochs. During the Agricultural Age, all such weapons were based on sending a hard object of high kinetic energy (that is, a sufficiently large mass at a high speed) towards a target that was destroyed upon impact. (Bombs that explode at the target site were first developed in later epochs, though there were a few early types of exploding bombs and hand grenades. What is more, some incendiary weapons, most famously the Greek fire of the Byzantines, were applied from a distance.)

SPEAR AND BOW-AND-ARROW

The original long-distance weapon, known already in early gatherer-hunter times, was the spear. The killing range of spears was just a few 10s of meters, but using a spear-thrower potentially increased this distance to more than 100 meters. Bow-and-arrow came next and dominated warfare for millennia. In the Mediterranean and the more northern parts of Europe bows were the primary long-distance weapon up until and throughout the Middle Ages, and in China and Japan even longer. Bows were made of a strip of wood bent and held in tension by a string. The Egyptians built composite bows from about 2800 B.C.E. Such bows typically consisted of a wooden core that had layers of bone (or horn) and sinew attached to it. The combined elastic properties of these materials in turn allowed for small bows to fire an arrow farther and with greater force than wood-only bows. The bowstring was usually catgut, a tough cord made from sheep (or other animal) intestines. Good composite bows delivered piercing arrows over distances of up to 300 meters. Huns, Turks, Mongols and other peoples of the Eurasian steppes excelled in warfare as mounted archers. The Huns developed a superior composite recurved bow, which spread to the Eastern Roman Empire through their invasions of the fourth century C.E. Later known as the Turkish bow, this powerful weapon had a great impact on Eurasian warfare in the Middle Ages. In England, noncomposite longbows were the leading missile weapon from the 14th well into the 16th century. These were made of wood (preferably yew), and owed their enormous power to their considerable size. Longbows were usually six feet (two meters) long and shot one-yard (one-meter) arrows effectively at targets over 200 yards (180 meters), thus seriously troubling mounted armored knights from the distance.[53]

CROSSBOWS

The performance of bows is principally limited by the power one extended and one flexed arm can exert to draw a bow. Crossbows (arbalests) eliminated this problem as they were fitted with a stirrup that the archer placed on the ground, setting his foot into it and pulling the bowstring back with both hands until it notched in the nut. Later crossbows also employed mechanical loading devices such as pulleys and gears. However, the main advantage of the crossbow, which at the minimum consisted of a short bow fixed transversely on a stock, was that it could be loaded ahead of time and held effortlessly while aiming. On the down side, crossbows were slow to reload and often heavy and inaccurate.

Mentioned in Chinese writings from before 1000 B.C.E., crossbows were widespread in China by 400 B.C.E. By 209 B.C.E. the Chinese army allegedly counted 50,000 crossbowmen, whose mass-produced bronze weapon had an effective range of 650 feet.[54] Crossbows were not known to the ancient Fertile Crescent civilizations, and reached western Asia only in post-Roman times. Thereafter, they spread to Christian Europe, mainly through the Crusades of the 12th century. (However, the first hand-held wooden crossbows had appeared in Italy in the 10th century.) European crossbows then evolved into a leading missile weapon of the Middle Ages. Advanced types owed their destructive power to a metal bow, which would propel a bolt to pierce chain mail from a distance of 1,000 feet (300 meters). Crossbows remained in use until the 15th century.

CATAPULTS

Catapults were developed to throw heavier objects, typically intended to destroy city walls. Catapults usually relieved the tension of a wooden beam or twisted cords made of gut, sinew, horse hair, or the like. The Greek engineers of Philip II of Macedonia, father of Alexander the Great, are credited with designing the earliest catapult, the ballista, in the third century B.C.E. This weapon practically resembled a giant crossbow, drawn by twisting a set of ropes. The Roman version of the ballista was capable of accurately hurling a projectile of 60 pounds (27 kg) up to about 500 yards (450 m). The Roman mangonel, a catapult designed for hurling larger objects, also employed twisted ropes, but these pulled down a single wooden arm. Loaded with a rock, the arm would be flung forward upon release until it hit a wooden crossbar. The medieval trebuchet, probably invented in Italy around 1400 C.E., was not powered by torsion but a counterweight. These structures, which stood some 15 meters high, were essentially giant levers, consisting of a long wooden arm resting on a pivot point. The launching end was sunk with the help of a pulley to lift the counterweight. The catapult was then loaded and the launching end released. Generally, catapults could throw stones of 20 to 150 kilograms (44 pounds to 330 pounds) some 200 to 500 meters.

CHINESE GUNPOWDER

Up until this stage all weapons, from swords to crossbows to catapults, depended on human muscle power to be employed or loaded. Hence, the energy released by weapons was really a manifestation of grain-derived energy. This situation changed when the Chinese in the ninth century invented black powder and introduced chemical energy to propel projectiles. Black powder, or gunpowder, turned into the principal explosive of the Agricultural Age. Upon ignition it produces a rapid 3,000-fold expansion of its volume in gas, which results in a violent explosion. Black powder is a mix of charcoal, saltpeter (potassium nitrate, KNO_3), and sulfur, with the weight ratio eventually being perfected to 15:75:10. Charcoal, which was already strategically important for metal production, was the fuel of the mix. It combusts rapidly within the gunpowder mix because it does not need to draw oxygen from the surrounding air. Instead, saltpeter (KNO_3) releases the necessary oxygen to turn the charcoal's carbon into carbon dioxide (CO_2) gas. Saltpeter was actually the main constituent of gunpowder. It occurs in nature (exuding from rocks and human-made walls) but was later also produced in niter beds.

These were filled with composing vegetable matter and animal waste and surrounded by an earthwork. In Sweden, for instance, every 16th century farmer had to deliver to the Crown urine-saturated soil from beneath stables and cowsheds.[55] (The nitrogen [N] salts contained in urine were transferred into KNO_3.) Sulfur was mixed into black powder to provide for easier ignition and smooth combustion. Sulfur was found in nature (most famously on Sicily), or was smelted from "fool's gold" (pyrite, FeS_2).

Black powder had many important peaceful applications, especially in mining, but became mostly associated with warfare because it increased the speed, range, and effective power of projectiles. By the 10th century the Chinese constructed firelances, which were missiles propelled by a gunpowder explosion that was appropriately confined inside a bamboo tube. Later on, the missiles were launched from metal tubes, which eventually evolved into bronze cannons. The first true cannons were cast in China before the end of the 13th century, and by the early 14th century gunpowder technology had finally spread to Europe.

EUROPEAN SIEGE CANNONS

The earliest European mention of cannons is their use in the siege of Metz (in the northeast of present-day France) in 1324. The English king Edward III apparently used cannons on the battlefield of Crecy in 1346 to create noise and panic, but the primary use of early cannons was to knock down the walls of besieged cities and castles. This seems to have been achieved for the first time when Philip II, the Bold, Duke of Burgundy, conquered the fortress of Odruik in 1377. Early cannonballs were made of stone, but they were soon replaced by cast iron balls, which were much more destructive due to their higher density, and easier to produce in a nearly perfect sphere. The cannons themselves were made of long wrought iron bars, forged together to form a barrel. They were loaded through the muzzle and mounted on wheeled carriages. From the 14th century cannons were also cast out of bronze. (Berthold Schwarz, a German monk and alchemist, is sometimes credited with being the first European to achieve this. He is also said to have discovered gunpowder around 1313.) Bronze cannons were lighter than wrought iron, more accurate and powerful, and much safer to use. But they were also more expensive than both the common forged iron cannons, and the cast iron cannons that also appeared on the scene.

Famously, Ottoman sultan Muhammad II, the Conqueror, in 1453 used cannons to breach the enormous walls of Constantinople. Apparently Muhammad hired a renegade Christian gunsmith, who cast a giant 19-ton bronze cannon on site. Around the same time cannons helped the French to drive the English out of northern France (Normandy), and the Spanish to drive the Muslims out of southern Spain. The heaviest 15th-century cannons could fire 140-kg iron balls about 1,400 meters, and lighter stone balls twice as far. However, these cannons were dangerous to handle. King James II of Scotland was actually killed when a cannon was being fired in 1460 in celebration of the successful siege of Roxburgh Castle. The cannon exploded and a wedge of wood severed his leg: the king bled to death. Fortifiers obviously had to react to the new threat imposed by advanced siege cannons. They turned to earthwork constructions that proved a far more reliable protection against cannonballs than masonry. Hence, the traditionally-walled castle disappeared and the ditch became a basic feature in fortification.

NAVAL CANNONS

Cannons also revolutionized warfare at sea, which had previously been dominated by ramming and boarding. (Grain energy and human muscle power provided for the speed of the ramming galley.) The last major battle of the closing era took place at Lepanto in 1571, nearly a century after the Spanish had discovered America. Thereafter agricultural energy (utilized through oarsmen) was less important in naval warfare than the chemical energy utilized through cannons. Close combat changed to distant combat. Sufficiently maneuverable ships engaged the enemy at a distance with broadsides from heavy naval cannons.

In 1588 a fleet of some 197 English ships attacked the Spanish Armada, which consisted of about 130 (mainly larger) ships anchoring in the English Channel. The English are said to have used relatively light projectiles to fire from long distance, while the Spaniards used heavier projectiles that could not reach as far. However, the English actually sank only two Spanish ships, and damaged seven. Contrary to the myth, it was not the English who wrecked the Spanish Armada, but bad weather. The Spanish attempted to retreat but due to heavy winds from the west were forced to sail up north along the English east coast to circumnavigate Scotland and Ireland before being able to head south. During this maneuver storms wrecked the Great Armada and only 67 ships returned to Spain.

A century later, large men-of-war were fitted with up to 100 cannons. It was actually the combination of advanced naval cannons (command of destructive chemical energy) and superior sailboats (command of wind energy) that allowed the Europeans to control the Asian waters they reached in the 16th century.

HAND-HELD FIREARMS (MATCHLOCKS)

Promoted by a high frequency of armed conflicts, European firearm designs underwent a rapid evolution not only towards larger, but also towards smaller size. Hand-held firearms first appeared between 1460 and 1480, with German gunsmiths leading the way in the development of the first trigger systems, right before Europeans reached America and the Indian ocean.

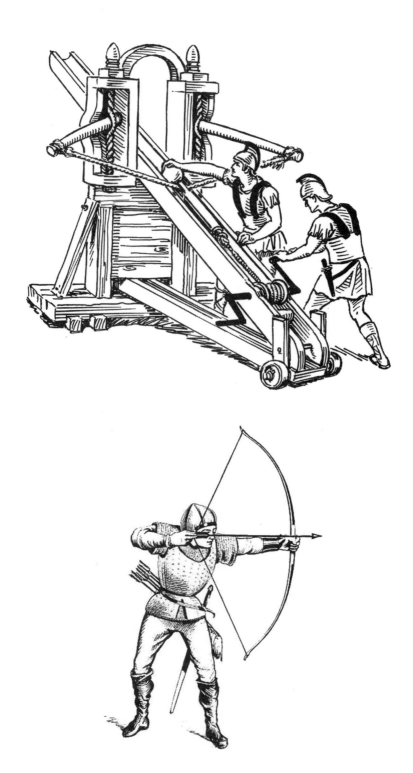

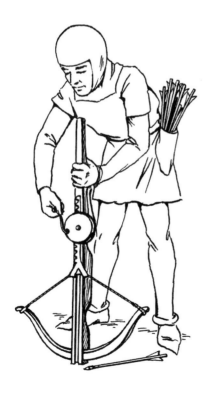

Long-distance Weapons of the Agricultural Age Spear, bow-and-arrow, crossbow, and catapult were loaded or employed by means of muscle energy, while cannons, matchlocks, wheel locks, and flintlocks utilized the chemical energy contained in black powder. The picture above shows Mons Meg, a giant siege gun made in Wallonia (present-day Belgium) around 1449. Such wall-breaking weapons in turn evolved into both more mobile, but still large field, defense, and naval cannons, as well as smaller hand-held firearms. (The illustration of Mons Meg at Edinburgh Castle is reproduced from Robert Chambers' *Domestic Annals of Scotland*, Vol. 2, of 1858. The other illustrations are contained in *The Historians' History of the World* of 1907, compiled by Henry Smith Williams.)

The matchlock consisted of an S-shaped arm that was fastened to the side of a hand cannon's stock and held a lit match that was lowered into a pan at the side of the barrel when the trigger was pulled. The (flash)pan contained a priming powder that in turn burned through a small port in the breech to ignite the main charge. Although this triggering system was slow, it was a major step forward. Previously, gunmen could not have both hands firmly gripped on the weapon at the moment of firing, because one hand had to lower a lit wick into the flash-pan. On the flip side, the matchlock's wick had to be kept lit all the time, which gave away the soldier's position in the dark and made the handling of gunpowder dangerous.

Various different weapons were equipped with matchlocks, but the musket, a heavy large-caliber muzzle-loading shoulder gun, assimilated the name of its triggering mechanism and became itself known as "matchlock." Early muskets were typically 5.5 feet (1.7 m) long and fired a projectile about 175 yards (160 m) with little accuracy. Later muskets were smaller, lighter, and accurate enough to hit a person from some 90 yards (85 m).[56] (This was considerably shorter than the 300 meters from where good composite bows or crossbows could kill.) The cartridge, a paper tube containing a charge of powder and a bullet, was introduced in 1586 to make the loading of muskets easier. (Thick paper is still now known as cartridge paper due to this application.) The base of the cartridge was ripped or bitten off, the gunpowder poured into the barrel, and the bullet rammed home.

WHEEL LOCKS

The first advance over the matchlock was the wheel lock. It was developed around 1515, probably by German gunsmith Johann Kiefuss. In wheel locks sparks are struck from a flint (or a piece of iron pyrite) by a revolving wheel, which is the same principle as used in current flint-and-wheel cigarette lighters. Muskets equipped with a wheel lock could be used in damp conditions and were the first self-igniting firearms. Each firing required the main spring to be rewound, but the wheel lock musket nevertheless took no more than a minute to reload.

On the down-side, wheel locks were difficult and expensive to produce, and the complexity of the system made it difficult to clean. The complicated mechanism also had a tendency to break and jam (especially when not cleaned properly), and the sparking material would only last for about a dozen shots before it had to be replaced. Hence, matchlocks remained in use alongside wheel locks until both were replaced by simpler, and less costly, flintlocks.

FLINTLOCKS

Flintlocks were introduced around 1635 and dominated the remainder of the Agricultural Age. Their trigger mechanism was based on a hammer striking a flint to create a spark that ignited the charge. In a typical design the hammer would hold the flint, striking a piece of steel or iron on the flash-pan lid. Later versions did not need a priming charge, because part of the gunpowder inserted through the muzzle would pass from the barrel through the vent into the pan (where it was held by the cover and hammer). Flintlock guns were the standard hand-held weapon in much of the world for more than two centuries. To be sure, all these weapons were muzzle-loaded and hence had to be smoothbore. (That is, they had a barrel with an unrifled bore.)

NOTES

53. Encyclopædia Britannica, "Longbow," http://www.britannica.com/EBchecked/topic/347452/longbow; Encyclopædia Britannica, "Military Technology," http://www.britannica.com/EBchecked/topic/382397/military-technology.
54. Michael S. Neiberg, *Warfare in World History* (London & New York: Routledge, 2001).
55. Joseph LeConte, *Manufacture of Saltpetre* (Columbia: Charles P. Pelham, State Printer, 1862), http://docsouth.unc.edu/imls/lecontesalt/leconte.html.
56. Encyclopædia Britannica, "Musket," http://www.britannica.com/EBchecked/topic/399353/musket.

CHAPTER 10

THE RISE AND FALL
OF GRAIN-FUELED EMPIRES

With the list of all the Agricultural Age's energy sources and their various manifestations complete, we are ready to start a whirlwind tour through political history. But before we begin this trip, we will have a quick look at the reoccurring patterns and the dynamics behind the rise and fall of empires during the Agricultural Age. First of all, empires had an incentive to expand throughout the Agricultural Age, because the energy commanded by people was proportional to the fertile land they controlled. Every acre of land cultivated by agricultural slaves or heavily taxed farmers would deliver additional energy to the king, emperor, or any other kind of central ruler or institution. The expansions into neighboring lands also delivered new prime mover capacity in the form of newly enslaved or subdued people. In the fifth century B.C.E., slaves probably accounted for two thirds of the population of Greece. Similarly, about half of the Italian peninsula's population was slaves under Caesar and the Roman emperor Augustus. Slaves labored not only on the fields, but in the mines, in workshops, and in temples and households. Rulers sometimes used their agricultural energy to fuel hordes of artists or harems with hundreds of women. However, they usually had to use much of their precious grain on soldiers who defended the borders or widened them. As agrarian populations grew, villages turned into city states, and city states became part of emerging empires. Empires then kept expanding until they reached oceans, infertile land, or the borders of neighboring empires. And even in the absence of external threats, armed forces became increasingly more important in large empires, as regional aspirants to power began breaking them up from the inside.

Transportation and communication systems generally set limitations to the size of empires. Central centers of governance had to be supplied with both fuel from surrounding fields and information from distant territories. Most empires emerged along rivers or coastlines because waterborne transport was far easier than transport on land. The Romans were able to create a huge empire, but it was centered around the Mediterranean Sea, which is really a quite unique body of water. It is relatively large, with a very long coastline, but it is a sea rather than an ocean, with waters calm enough to be challenged by the quite primitive galleys and sailboats of the time. The Romans were therefore able to ship grain from the fertile North African coastline to their capital, and to move armies swiftly around in the entire Mediterranean on ships propelled by enslaved oarsmen.[57]

The Chinese tackled the transport problem by constructing a vast canal system around their rivers, but otherwise inland transportation remained difficult. The low speed of wagons drawn by horses or oxen set principal limitations as to how far afield grain energy could be transported towards the urban centers of political power, and even information delivered by the fastest couriers on horseback took a long time to reach the capital's rulers from the outskirts of an empire. Thereafter it would take even longer to march an army to a distant point of trouble. But the alternative, to keep a horse for every soldier, would have been far too energy-expensive for sedentary societies. Even to keep a small mounted army absorbed enormous amounts of nutritional energy. A pair of chariot horses consumed the barley harvested from about 4 hectares (10 acres) in ancient Egypt, for instance.[58] Hence, a balance had to be found between the extent to which empires were expanded (to gain additional nutritional energy for horses and soldiers) and the energy expended to defend the additional area and to administer it centrally.

The nomadic people of the central Eurasian steppes were able to keep one horse or more per adult man without diverting nutritional energy away from people. As horses were decisive in warfare, these skilled mounted archers had the means to harass even Eurasia's most advanced sedentary regions for centuries (until the emergence of advanced firearms). Mongols, first unified in 1206 by Genghis Khan, enlisted all adult males and equipped each with a string of about five (relatively small) horses to ride into battle. This provided Mongol warriors with tremendous mobility to affect their composite bows, the best long-distance weapon of the time. The Mongols established the largest continuous continental empire the world has ever seen, stretching from China all the way to the Adriatic arm of the Mediterranean Sea. Communication via horseback couriers in combination with a decentralized organization and relatively independent regional decision-making made governance of such a huge continental empire possible. But the Mongols left no large buildings or other signs of a complex civilization save hundreds of monoliths and plenty of burial sites.

Following Genghis Khan's battlefield death in 1227 the empire became fragmented, and only those Mongol branches remained influential that had managed to establish themselves as the ruling class in such agricultural regions as China and India. In these sedentary areas they were subjected to the same energy imperatives as all agrarians. They had to let go of their horses (which had traditionally been grazing on open steppes) and relied on nutritional energy captured on fields.

SUSTAINABILITY OF ENERGY SOURCE

Maintaining the energy flow harvested on fields was actually a major challenge in the longer run. The fate of empires, their stability and persistence, often reflected nothing more and nothing less than the sustainability of their agriculture. As farmers attempted to maximize their yield, they realized that in many regions nothing worked better to increase field output than irrigation. Unfortunately, though, irrigation involves the risk of field salinization and water-logging. When water first precipitates to the surface as rain or snow, it is pure. However, when water sinks into the ground, or flows over land, it dissolves, and thus contains gradually more, minerals (salts). If such water is pumped from the ground or diverted from rivers to fields, its mineral content may become a problem. Water will to some extent evaporate from the fields under the heat of the sun, and it will leave behind all its salts, which in turn accumulate in the soil. This sort of salinization, which is similar to the clogging of a showerhead or the precipitation of salt (limestone) in a tea kettle, gradually decreases field productivity. In Mesopotamia, where the world's first empires emerged, people began as early as 5500 B.C.E. to divert water from the slow-flowing Tigris and Euphrates Rivers. The region's fields then gradually silted up over the course of the following millennia. Of the area's two principal crops, wheat and barley, barley was considered inferior as it provided lower yields. However, barley is actually much more salt-tolerant than wheat, and the share of wheat fell from 50 percent in 3500 B.C.E. to two percent in 2000 B.C.E.[59] The share of barley increased accordingly, but barley outputs declined as well, and political power shifted to neighboring regions: By Roman times Mesopotamia had largely sunk into oblivion.

The principal response to salinization was to rinse the fields by applying additional water to dissolve and remove salts from the fields. But this is actually a tricky procedure, because excess water may accumulate in the ground, raise the water table, and leave fields waterlogged. Roots of crops can literally become submerged, which may kill the plants or dramatically reduce productivity. Hence farmers eventually began to construct drainage channels, which allowed them to rinse the fields while preventing from excess irrigation water percolating downwards and accumulating in the ground.

Another challenge faced by farmers was soil loss through erosion, because routine farming activities, such as plowing or digging up the soil in preparation for planting, exposes the soil to wind and rain. The problem is aggravated if fields are left without plant cover for extended periods of time. Planting hedges around fields helped to counter wind erosion, while water erosion can be reduced by leaving forests around agricultural areas intact. (Forest lands efficiently absorb water even on steep slopes.)

All these problems affected the basic energy supply even of advanced agrarian civilizations. Often the problems were not identified as such, or it took a long time to develop solutions. On the other hand, there were those lucky enough not to face (some of) these problems at all. The Egyptian civilization remained remarkably stable for more than three millennia, as the Nile precipitated fresh fertile soil onto surrounding fields in its yearly floodings. In China, where wet rice farming did not meet the same challenges as dry wheat farming, much of the agricultural system remained robust over millennia, and southeastern Asia had a less turbulent early political history than southwestern Asia. Much of Europe had the major advantage that its climate allowed for very productive rain-fed agriculture, which avoided all the problems associated with irrigation.

FERTILIZATION

One severe problem threatening the agricultural energy base was identified, and successfully tackled, by practically all agrarian civilizations: Crop farming depletes the soil of nutrients. In natural habitats a plant would typically grow, die, and decay at the same spot, storing nutrients temporarily but eventually returning them to the soil right where they came from. Crops grown on managed fields, in contrast, extract nutrients from the soil and are in turn harvested and carried away to the villages and cities, where they are consumed.

To be sure, carbon, the main growth material of plants, is never short in supply, because it is delivered through the air. Plants capture carbon dioxide (CO_2), a gas present in low concentrations in the atmosphere, and incorporate carbon (C) into their solid tissue. (Hence plant growth is really the redistribution of matter that is invisible to the human eye—that is, CO_2 gas—into solid matter that we can see and feel.) In contrast, nutrients such as nitrogen, phosphorus, and potassium must somehow be returned to the soil to keep fields productive. The solution that ancient farmers came up with was to manure the fields with the urine and excrements of humans and farm animals. This essentially returned the very same atoms to the field that the harvested crop had extracted from the soil, though these atoms were now part of different, less energy-rich, chemical compounds. Manuring with human waste is still now common practice in many countries (including China), while the litter of chicken and other farm animals is usually used to manure fields in Western countries. However, almost any kind of organic (waste) material will do. Composts, bonemeal, blood, and fishmeal have all been applied as fertilizer by different civilizations. Potash (potassium oxide, K_2O) was obtained by burning stalks and other agricultural refuse. It is very rich in potassium and has prime fertilizing properties. Guano, which contains phosphorus (and nitrogen), has been used as fertilizer since ancient times as well. This material consists of bird (or bat) droppings, typically collected in dry caves or from rocks in arid areas.

One relatively simple way to enrich the soil with nitrogen is to include plants such as peas, beans or clover into crop rotation schemes. Without knowing why it worked, farmers observed that these plants rendered the fields more productive for the next cereal harvest. (As we now know, the clearly visible tumorous swellings or nodules on the roots of pulses are caused by invasion of beneficial bacteria that are capable of fixing nitrogen straight from the air.) Keeping a variety of different plants on the fields also helped avoiding harvest losses due to weather fluctuations (droughts, floods, frost) and infestation by diseases or pests. (Different plants tend to be resistant against different conditions.) Droughts or pest invasions often triggered famines that wiped out whole agricultural communities. In the 1840s thousands of Irish farmers and their families died of starvation during a potato famine, as the potato blight ruined the entire harvest of the one single potato species grown on Irish fields. Andean farmers, in contrast, grew a variety of potato species and were well aware of the risks associated with nondiverse potato fields.

EPIDEMICS

Famines and epidemics had a much larger impact in the Agricultural Age than in later Energy Eras. The human suffering was the same, but wiping out large parts of the population in the Agricultural Age referred to the destruction of principal prime mover capacity. As a lot of people worked in agriculture, this also removed much of the surplus energy supply to fuel nonfarming specialists. We are all aware of the epidemics that cleared the New World of its native population after the European arrival, and perhaps of those that helped bring about the Fall of Rome. But we do not hear so much of the epidemics that distressed Coal Age cities, and history books often do not even mention the influenza epidemic that killed more people at the end of World War I than died in combat in the two World Wars of the 20th century combined.[60] Epidemics simply did not have such extreme effects on the functioning of societies in the later Energy Ages as they did in the Agricultural Age.

THE HUMAN FACTOR

To be sure, the history of the Agricultural Age, and here there was no difference from the other Energy Epochs, also contained elements that were determined by human behavior and culture rather than environmental or technological factors. The Huns, the Franks, the Mongols, and other peoples divided their empires among several heirs rather than passing power from one single ruler to another single ruler. The Frankish tradition of splitting the land between the sons of the king actually created the fragmented European landscape that later on resulted in much competition and warfare. Still, during Charlemagne's rule central-western Europe was relatively homogenous. His kingdom included most of modern France, Italy, Germany and the Benelux countries, and he himself had only one surviving son, Louis the Pious. However, Louis' three surviving sons fought one another until they settled their disputes in the Treaty of Verdun (843 C.E.), which split the Frankish realm into three parts. These were the kernels of later France and Germany to the west and to the east, and a central kingdom that narrowly stretched from the Netherlands to the south to comprise what is now Italy.

Wars were sometimes started preventively to remove a threat, or for the sole purpose of retaliation and revenge. The Persians harassed Greece for centuries, and occasionally enslaved the population of whole Aegean islands. This motivated Alexander the Great to invade the Persian Empire as soon as he came to power. The interbreeding of European royal families attests one way to deal with the problem that defeated war parties tend to seek revenge or attempt to recover lost territory: close ties between intermarrying ruling families of competing countries tended to raise the barriers to start a war. But monarchy also had its traps. A recent example of personal agenda motivating warfare may have been U.S. president George W. Bush's public claim in regards to Iraq's president Saddam Hussein, "After all, this is the guy who tried to kill my dad."[61] Critics were highly irritated by this statement, as virtually all analysts had concluded that Saddam Hussein was not behind the alleged assassination attempt on the first president Bush during a Kuwait visit in April 1993. But even if this was nothing more than political maneuvering to get support for the 2003 Iraq invasion, the statement may have exposed a major weakness of democracies that has been carried over from the times of inherited monarchies: Perhaps it would be best to change the constitution to exclude a former president's close relatives from taking office.

Another issue put forward by critics of recent U.S. militarism is that standing professional armies need to be kept busy. This has lots of historic precedence, though often in a different context. Then as now the argument went that armies can only be kept in good shape if the emergence of a whole generation of soldiers without combat experience is avoided. But another problem was that professional soldiers may cause trouble at home (think military coups), if they were not employed elsewhere. The Roman legions almost constantly plundered new regions for booty. Isabella of Castile and Ferdinand of Aragon, once they had expelled the last Muslims from Spain in 1492, sent their armies to southern Italy. At all times armies were expensive to maintain, and only conflicts or real, perceived, or constructed threats justified the cost of their existence.

Alexander the Great, after defeating the Persian empire, had ambitious plans to invade India and Arabia as well. Perhaps he thought that continuing to campaign involved less risk than dissolving a huge army of professional fighters, but most likely his ambitions simply derived from his insanity and megalomania, which are but part of the repertoire of human unpredictability that introduced chaos into world history. However, the actions of all powerful persons in history have been bound by energy constraints. Only if large energy surpluses were available was there space for (continued) craziness and unreasonable behavior. Far more often, conflicts had tangible, predictable reasons. With our understanding that control over land equaled access to fuel during the Agricultural Age, we can pinpoint the main motivation for territorial expansion and claim that most wars in human history have indeed been fought over energy. Political history mostly reflects the struggle for resources and power of both kinds.

NOTES

57. To be sure, the Romans also constructed a vast network of roads. These were especially important when the Empire expanded into more northern European regions.
58. Stephen Budiansky, "Horse," Microsoft Encarta Online Encyclopedia, http:// encarta.msn.com/encyclopedia_761562654_2/Horse.html#s6.
59. T. Jacobsen and R. Adams, "Salt and Silt in Ancient Mesopotamian Agriculture," *Science* 126 (1958): 1251–1257.
60. The death toll of the 1918 influenza epidemic has been repeatedly revised upward. The following paper suggests that it was of the order of 50 million, though it may have been 100 million. The higher figure would be even well above the total death toll of both World Wars, which was perhaps 75 million. (This figure is also uncertain.) N.P. Johnson and J. Mueller, "Updating the Accounts: Global Mortality of the 1918–1920 "Spanish" influenza pandemic," *Bull Hist. Med.* 76 (2002):105–15, http://www.ncbi.nlm.nih.gov/pubmed/11875246?dopt=Abstract.
61. John King, "Bush Calls Saddam 'the Guy Who Tried to Kill my Dad'," *CNN*, September 27, 2002, http://www.cnn.com/2002/ALLPOLITICS/09/27/bush.war. talk/.

CHAPTER 11

THE EARLY WHEAT EMPIRES

The world's earliest set of shifting empires unsurprisingly emerged around the Fertile Crescent, the world's first agricultural center. These empires covered the region stretching from the Mediterranean, including southern Europe and North Africa, through the Fertile Crescent, south to Mesopotamia, and west through Persia to the Indus valley of present-day Pakistan. In this whole area the original agricultural system of the Fertile Crescent could be easily adopted without major modifications. All the empires of this region relied principally on wheat (and barley) to fuel their societies, and they all were slave cultures. Power in this arena shifted from Mesopotamia to Egypt to Persia to Greece to Rome (present-day Italy), and to Byzantium (present-day Turkey).

In the floodplains of southern Mesopotamia irrigated agriculture supplied the energy for several hundred villages by 4300 B.C.E., and by 3100 B.C.E. there were scores of cities with up to 10,000 inhabitants. A pictographic writing system emerged from around 3300 B.C.E., and the oldest known texts in the world, dated to about 3100 B.C.E., are lists of livestock and agricultural equipment from the Mesopotamian city of Uruk. (The invention of writing systems is generally associated with the keeping of accounts.[62] Simple symbols such as a stroke for units and a circle for 10s were written into clay tables to keep record of trade activities and stored goods, which were vital to agrarian societies.) From about 2900 B.C.E. neighboring city-states began fighting one another over the land around them, and city walls were erected as protection against enemies. From the following centuries sculptures are preserved that show warriors in 4-wheeled vehicles drawn by donkeys.

THE WORLD'S FIRST EMPIRE

In about 2334 B.C.E. Sargon, the king of the city of Akkad (Agade), man-aged to subjugate all Mesopotamian city-states and created the world's first empire. He organized a standing army of 5,400 soldiers and eventually con-trolled a territory that extended west to the Mediterranean and north to the Black Sea. The writing system of this empire was syllabic cuneiform writ-ing, which had emerged in southern Mesopotamia a little earlier. Previous pictographic writing employed well over 500 different symbols as it depicted every word with its own symbol. Assigning symbols to syllables rather than entire words in turn reduced the writing system to under 200 characters. The distinctive wedge-shaped characters of syllabic cuneiform writing then remained in use in this region for about two millennia.

The Akkadian Empire collapsed around 2200 B.C.E., right at the time when a temporary global cooling period set in and rainfall decreased by perhaps as much as 20 to 30 percent. The region experienced a drought that lasted for some 300 years and affected mainly the northern part of the Akkadian Empire, which relied on rain-fed agriculture.[63] Consequently, large parts of the population migrated to the cities in the south, creating severe population pressures.

Following the collapse of the Akkadian Empire power shifted between a few city-states and finally, from about 1900 B.C.E., to Babylon, the city neigh-boring Akkad. (Babylon is situated about 89 kilometers [55 miles] south of Baghdad, the capital of present-day Iraq.) The end of the 1st Dynasty of Babylon, around 1595 B.C.E., was followed by an obscure period of over seven centuries until the Assyrians entered the scene. The Assyrian Empire grew out of the cities Ashur and Niniveh in northern Mesopotamia from 1000 B.C.E. By 715 B.C.E. the Assyrians controlled the entire Fertile Crescent region, but their empire was eventually weakened by rebellions, and after 630 B.C.E. the Babylonians reassumed power in the region.

THE PERSIAN EMPIRE

Then the Persians had their go. King Cyrus the Great extended his king-dom at the Persian Gulf, southeast of Mesopotamia, by taking control of the Median Empire, which covered the northeastern part of present-day Iran, roughly around the area where goat herding had first begun. Cyrus then conquered Anatolia and thereafter nearly all of the Fertile Crescent region. He also expanded his empire towards the northeast, all the way to the Aral Lake, and towards the east, all the way to the Indus River. Hence, Cyrus ruled over the largest empire the world had seen thus far. Following his death around 529 B.C.E., his son extended the boundaries of the Persian Em-pire even further to include Egypt.

EGYPT

The Egyptians had a wonderfully robust energy base because the Nile deposited fertile sediments on their fields each year in late summer. They simply sowed the seed of wheat and barley into the moist, fertile soil as soon as the Nile's water receded (usually in early November), and began constructing irrigation canals to direct Nile flood water to more distant fields only around 3000 B.C.E. Around this time a kingdom arose at the Nile, and hieroglyphs, a pictographic writing system entirely independent from Mesopotamian writing, came into use.

The Egyptian civilization remained remarkably stable for over three millennia, but the centralized power of the Egyptian kings, the pharaohs, temporarily diminished around 2200 B.C.E. for about 150 years. Like the collapse of the Akkadian Empire in Mesopotamia, this coincided with the onset of a cooler, drier climate period. Then stability returned, initially under the rule of the foreign Hyksos, who introduced horses to Egypt, and used their horse-drawn chariots, advanced metal swords, and powerful composite bows to temporarily establish themselves as pharaohs. But eventually a local Egyptian dynasty ousted them. Generally, the Egyptians were quite well protected from outside attacks by deserts to their east and west, and they had themselves little incentive to expand out of the Nile valley. Nevertheless, they ruled over much of the Fertile Crescent as well as Nubia around 1500 B.C.E.. Half a century later Egypt was first conquered by the Nubians, then the Assyrians, and then the Persians. The Persian rule began in 525 B.C.E., and in 332 B.C.E. Alexander the Great made Egypt part of his empire.

GREECE

Alexander led the armies of Greece, where the first complex European civilizations arose. The earliest of the Greek civilizations was the Minoan culture, which emerged on the large island of Crete around 2000 B.C.E. The Minoans eventually declined commercially and militarily against the Mycenaeans of mainland Greece, who ruled out of independent city-states and traded all over the eastern Mediterranean. However, from around 1200 B.C.E. the Mycenaean cities began to disappear following wars between city-states and a long series of attacks by the Sea Peoples, who had entered the northern Aegean sea. Four dark centuries later, around 800 B.C.E., the Dorians arrived from the north to establish themselves in Greece as a class of land-holding aristocrats. Before long, their prospering city-states experienced a population growth that could not be sustained by the agricultural base of the surrounding lands. People were therefore sent out to found colonies around the Mediterranean and the Black Sea. By 500 B.C.E. Greeks had founded many colonies at the coasts of present-day Turkey, southern Italy (including Sicily), southern France, Spain and North Africa.[64]

The only competition for fertile soil in the Mediterranean area came from the Phoenicians, a people from the coast of present-day Lebanon that had founded colonies in Sicily, Sardinia, and along the Spanish and North African shoreline. It was actually a Phoenician innovation that led to the Latin alphabet used to write this book. Around 1000 B.C.E. the Phoenicians took one of the most important leaps forward in the history of communication when they reduced the number of symbols of their syllabic writing system by indicating consonant sounds. The Arabic script, which derives directly from Phoenician writing, is still now a purely consonantal system, written without vowels. But when the Greeks took over the Phoenician script, some of the Phoenician characters stood for consonantal sounds that did not exist in the Greek language and were in turn assigned to vowels. Full Greek alphabetic writing, with syllables entirely split into consonants and vowels, emerged in Greece about 800 B.C.E. and eventually provided the basis for the Latin alphabet, in which Western languages are written, as well as the Cyrillic alphabet of eastern Europe. Writing was thereafter relatively easy to learn in these regions, as alphabetic writing uses only about 30 symbols, each of which can be a simple, easily differentiated character. (In contrast, the Chinese and Japanese languages are still now written in hundreds of different, somewhat complicated word-syllable symbols, which makes it much more difficult to learn to read and write.)

Alexander the Great, who was taught by the philosopher Aristotle, was certainly literate. In 336 B.C.E. he succeeded his father Philip, who had arrived from Macedonia to conquer all of Greece. Two years later Alexander crossed into Anatolia to take on the Persian Empire, which at the time covered practically all of the wider Fertile Crescent region, including Egypt and Mesopotamia. Alexander defeated the Persians and took control of their entire territory. Thereafter he led his army further east to defeat an Indian king in the Indus valley. However, upon his return to Mesopotamia Alexander died quite suddenly (in 323 B.C.E.). His enormous empire, the largest yet to exist, broke apart immediately. Some of his generals assumed power in different regions, but their lands were soon to be absorbed by the Roman Empire.

THE ROMANS

According to legend, the village of Rome was founded at the banks of the Tiber river in 753 B.C.E. In the seventh century B.C.E. Rome grew from a village into a city, and in 509 B.C.E. the last of a series of seven mythical kings was ousted. Thereafter Rome was declared a republic, ruled by two consuls and a senate of leading citizens. The Romans gradually widened their territory, controlling all of the Italian peninsula after ousting the Greeks from

southern Italy. In turn they took on the Phoenicians (Punians), who dominated the western Mediterranean from their North African capital Carthage.
The Romans defeated the Punians in a series of three wars between 264 B.C.E.
and 146 B.C.E., and thereafter annexed Greece (146 B.C.E.), Anatolia (129 B.C.E.),
and Palestine (63 B.C.E.) to control nearly all of the Mediterranean coastline.

Once the empire had grown to this enormous size, internal political stability declined. Ambitious generals competed for power, and in 49 B.C.E. a
civil war erupted, with the army of one of the two consuls, Pompey, being
defeated by the army of the other consul, Julius Caesar, whose campaigns
had added what is now France to the empire. In 46 B.C.E. Caesar assumed
power and abolished the republic to become the dictator of the Roman Empire. However, in 44 B.C.E. he was murdered and succeeded by his ambitious
nephew Octavian, who in 27 B.C.E. assumed the title 'Augustus' ('venerable')
to become Rome's first emperor.

With the addition of Egypt (32 B.C.E.) and Britain (43 C.E.), and the Rhine
and Danube rivers as the outside borders towards the northwest, the Roman
Empire was truly huge. Yet it remained remarkably robust until about the
death of emperor Marcus Aurelius in 180 C.E. Thereafter the empire declined
and was divided into an eastern and a western part by emperor Diocletian in
286 C.E. Though the empire was temporarily reunited twice, power shifted
to Constantinople, on the site of Turkey's present-day capital Istanbul, from

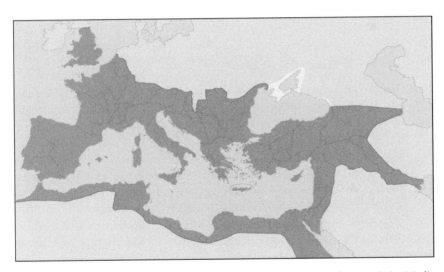

The Roman Empire   The Roman Empire was mainly centered around the Mediterranean, which allowed for water-borne transport of grain from North Africa to
Rome, for instance. However, the Romans also worked themselves through much of
continental Europe, and even into Britannia, and built up an extensive network of
roads throughout their empire.

where the eastern part of the Roman Empire, or Byzantine Empire, was ruled.

The downfall of the western Roman Empire had several reasons. From the immediate energy standpoint it was problematic that parts of the empire's bread basket, the region along the north African shore, collapsed due to massive soil erosion caused by inappropriate agricultural techniques. (The city of Rome yearly imported as much as 400,000 tons of grain transported on large ships from Egypt, other North African regions, and Sicily. The army alone consumed some 100,000 tons of grain per year.[65]) Meanwhile external pressures increased with the rise of the Huns, a mounted nomad people from central Asia that invaded southeastern Europe around 370 C.E. The Huns were skilled archers and created an empire that stretched from the Baltics to the Alps to the Caspian Sea. They repeatedly attacked eastern as well as western Roman provinces in the 440s and 450s, but their unity declined when their territory was divided among the sons of king Attila. However, for the western Roman Empire Attila's death came too late. The Huns had pushed Germanic peoples out of their eastern European homeland towards more western regions, and so it happened that Odoacer, a German chieftain, in 476 C.E. deposed the last emperor of Rome. The Byzantine Empire, on the other hand, survived the western Roman Empire by a thousand years and remained the principal cultural center of western Eurasia for many centuries. The Byzantine empire comprised Greece, Anatolia, Syria, Palestine, and Egypt, and during the reign of Justinian (527 to 565 C.E.) even included Italy and most of northern Africa.

DISEASES AND CHRISTIANITY

Probably the worst problem to ruin the original Roman Empire and its culture was an unprecedented wave of epidemics that havocked Eurasia's population centers in the first centuries C.E. Smallpox, measles, bubonic plague, and the like troubled Han China in the east as much as the Roman Empire in the west. These diseases spread along the Silk Road as trade across Eurasia had intensified and population density had increased. (All of these diseases would have been unable to persist within less dense gatherer-hunter societies.) The most severe outbreaks occurred in the second and third centuries C.E., but waves of epidemics reoccurred until the sixth century and beyond. The diseases affected everyone, from emperors to bureaucrats to soldiers, farmers and slaves, and hence diminished the amount of both human prime movers and agricultural energy to fuel them. The ruined energy base in turn had to lead to social disruption, internal conflict, and inability to react appropriately to invasions.

The first of the major Roman epidemics occurred after Lucius Verus, who was co-emperor during the first eight years of the reign of Marcus Aurelius, returned from the Persian war in 166 C.E. His troops brought a disease (probably smallpox) to Rome, which within months depopulated whole areas of Italy. The disease then spread through the entire empire. Later disease outbreaks were similarly severe, and by the time of the Hunnish and Germanic invasions the western part of the Roman Empire was too thinly settled and too impoverished to defend itself adequately.[66]

In times of hardship, when people are looking for consolation, religions tend to win new adherents. Eurasia's wave of epidemics in the first centuries C.E. is associated with the ascent of Buddhism in China, and of Christendom in the Roman Empire. Christianity started off as a Jewish sect in Palestine during the rule of first Roman emperor Augustus. Christianity is named for its founder, Jesus Christ, who commanded his followers to spread his doctrine. The Romans, like the Greeks, worshipped many deities, and they were quite religiously tolerant. They usually approved all sects and religious practices as long as they did not involve human sacrifice. However, to foster the faith in Rome the Roman emperors were elevated to the status of gods after their death, and some emperors insisted on being honored as gods even during their lifetime. Many considered this some sort of oath of allegiance rather than a religious practice, but Jews and Christians tended to refuse emperor worshipping and were thus prosecuted.

Jews were followers of Judaism, the monotheistic religion of the ancient Hebrews, who, like the Canaanites (from about 3000 B.C.E.) and Philistines (from about 1000 B.C.E.), were relatively early settlers in the region that was to become the Roman province of Palestine. (The name Palestine refers to the Philistines.) When the Romans entered the scene, the Jews were splintered into various divisions, sects, and orders, all offering different interpretations of Judaism. One branch, the Zealots, called for national independence and revolt against the Romans, but their rebellions were repeatedly put down. After every defeat more Jews were forced into exile, and following the revolt of 73 C.E., the Romans destroyed the temple of Jerusalem and enslaved thousands of Jews. (Jewish slaves then built about a quarter of the Colosseum, Rome's massive amphitheater, which was constructed between 72 C.E. and 80 C.E.) Around 300 C.E. only a few Jews were left in Palestine. Instead, they were spread out over the whole Middle East and around the Mediterranean, where they settled along the important trading routes and at busy seaports.[67]

Christians were also dispersed throughout the Roman Empire, but their fate was going to be altogether different. While other Jews kept on being prosecuted, Christians achieved gradually more acceptance and managed to differentiate Christianity enough from Judaism for it to be considered a religion in its own right. Highly critical for the different fates of the two religions was the Christian strategy to ease conversion. At the Council of Jerusalem, in 50 C.E., early followers decided that it was not necessary for non-Jewish Christians to observe all the laws of the Jews as presented in the

Torah (Old Testament). The council was actually occasioned by the controversy over whether circumcision was necessary for converts to Christianity, an obstacle that had to be removed if the new doctrine was to be successfully promoted in all of the Roman Empire.

Christians were generally very organized in promoting their religion. The apologists, writers who defended Christianity against Roman culture and gods, directly addressed Roman emperors and government secretaries, primarily in the second century. Right at this time the gradual weakening and eventual fall of the western Roman Empire clearly proved that the emperors were not divine at all, and worshipping the old pantheon of gods did nothing to keep away the devastating diseases that left the survivors in mental shock. No one knew what to do to alleviate the pain and suffering, and deprived of hope, people were looking for religious alternatives, a gap the Christian church was eager to fill. (Saint Cyprian, for instance, an important early Christian writer who was elected bishop of Carthage in 248, viewed the epidemic of 251 as a welcome event for servants of god and claimed that only non-Christians had anything to fear from the plague.[68])

A major turning point for Christianity came about when emperor Constantine, founder of the Byzantine capital Constantinople, amnestied the Christians from being outlaws in 313. Constantine was not especially pious, and had his wife and eldest son killed, but converted himself to Christianity on his death bed, when he was already delirious. (To be sure, the details are unknown. Some scholars maintain that Constantine adopted the Christian faith long before his baptizing shortly before his death.) Four decades later, in 379, Roman general Theodosius became eastern Roman emperor. He had been born to Christian parents and declared Christianity the official religion of the empire. This was nearly four hundred years after Jesus Christ was born, but followers of this religion were still shaping its most fundamental features. Arian Christians maintained that Jesus Christ was fully human and not divine by nature. Monophysite Christians claimed that Jesus Christ was fully divine, and a third version asserted that Jesus Christ was part-divine and part-human. The latter version was going to make it big, as Theodosius had it established as the universal norm of Christian orthodoxy, while being determined to exterminate all other Christian believes.

Eventually the Christian church was also being split into an eastern and western part that reflected the partitioning of the Roman Empire. The pope, the head of the Catholic church, resided in Rome, while the patriarch of Constantinople headed the Eastern Orthodox church (also known as Greek Orthodox church). The patriarch had the advantage that wealth and political power shifted to the east, while the pope (literally, father) held the claim to prestige of being the direct heir of the apostle Peter. Rivalry between the pope and the patriarch of Constantinople over church superiority actually facilitated the spread of Christianity: The two competed vigorously in their missionary activities throughout all of Europe. By the end of it, all western Europe turned out Catholic, while eastern Europe, with the notable exceptions of Catholic Poland and Hungary, became Greek Orthodox. The official break into Roman Catholic and Eastern Orthodox church came quite late, in 1054, at a time when the papacy had become very powerful. Early popes had relied on the Byzantine emperor for protection, but the later ones ruled over land they had received from Christianized Germanic kings. In fact, the pope who triggered the final break between the eastern and western church was himself a German. He was born Bruno, Graf (count) von Egisheim und Dagsburg, and reigned as pope Leo IX from 1049 to 1054.

THE ISLAMIC EXPANSION

Islam emerged on the Arabian peninsula as a third major monotheistic religion in the 7th century. Arabs traditionally believed in many different deities, and Judaism and Christianity (which was now over six centuries old) were well known in the region. Renderings of Jesus and his mother Mary have been found inside the Kaaba of Mecca, for instance, which is a small stone building around which many religious rituals were revolving at the time Mohammed (ca. 570–632 C.E.), a merchant native to the city of Mecca, founded the Islamic religion.[69] To a large extent, Islam drew upon the same beliefs and stories that appear in the Jewish scripts and Christian bible. Islam recognizes Adam, Noah, Abraham, Ishmael, Moses, Mary, Joseph, and Jesus, and promises good deeds to be rewarded in paradise and evil deeds to be punished in hell. However, Mohammed's doctrine also proposed that it was part of the holy duties to go in jihad (holy war), and that all who died for (the spread of) Islam would directly go to paradise.

In the last year of his life Mohammed managed to convert a large number of Arab tribes that had traditionally been warring with one another. United as a people under Islam, the tribal Arabs organized a rather large army and in 633 attacked the rich Fertile Crescent region, where the Byzantine and Persian empires were exhausted as they had just come out of 20 years of war between them. The region was also weakened by epidemics and unable to offer much resistance. Consequently the Arabs conquered Palestine, Egypt, Libya, Syria, and all of Persia within just 20 years. The Arab expansion then continued westward along the North African shoreline (670–698) and into to Spain (711–718). Defeats by the Byzantines right outside Constantinople in 717 and by the Franks at Poitiers in 732 put an end to the spread into Europe. Meanwhile in the East, Arab soldiers reached the Indus valley and central Asia.

In energy terms, the Arab expansion was mainly founded on camel-and-horseback riding. Horse-mounted Arab warriors moved extremely fast and enjoyed lots of kinetic energy, especially after adopting the stirrup, which merged rider and horse into a compact fighting unit. In political terms, the Arab advance was facilitated by the lack of internal coherence in the Byzantine Empire. Notably, the Orthodox church leaders of Constantinople failed to tolerate the Monophysite Christian beliefs of Syria, Palestine, and Egypt. Monophysites were frequently victims of persecution and the people of these regions had reason to welcome the Arabs as a sort of liberators. Besides, much of the Islamic expansion fell into times of horrifying epidemics that prompted people to consider new religious options. Conversion was uncomplicated, and the message of Islam was straightforward.

However, very soon after its rapid expansion the Islamic Empire became weakened by internal conflicts. Mohammed had not clearly pointed out a heir and in 680, when the empire's capital had already been relocated from Arabia to fertile Syria, a civil war was waged over the legitimacy of two lines of succession. This left the Islamic world permanently split into two major branches, the smaller Shiite or Shia branch (from *shi'at 'Ali*, party of Ali), and the Sunni, who now comprise more than 80 percent of all adherents of Islam.

In 750 the Abbasid dynasty came to power and ruled over the Islamic Empire from the Mesopotamian city of Baghdad. All members of the previous dynasty were lynched with the exception of one person, Abd al-Rahman, who escaped to Spain to found an independent Islamic state with Cordoba as its capital. By the 10th century Cordoba had become Europe's largest city, but the principal center of the Islamic world, around which the Golden Age of Muslim culture revolved, was Baghdad.

SELJUK TURKS AND CRUSADERS

By the year 1000 the power of the Abbasids had declined, and in 1055 they called in the Seljuk Turks for their defense. Like many other nomadic Turkish tribes, the Seljuk Turks had earlier inhabited the steppes of central Asia, but they eventually pushed towards the Islamic Empire and managed to establish themselves in Persia. After being called to Baghdad as mercenaries, they took control of much of the Islamic world, ruling it for about two centuries. Hence, it was critical for the fate of Islam that the Seljuk Turks decided to become Muslims.

The second sultan of the Seljuk dynasty, Alp-Arslan, inherited Persia, Mesopotamia, and Syria, and conquered Georgia, Armenia, and most of Anatolia, where the Byzantine eastern Roman Empire struggled to survive after the loss of most its territory. After Alp-Arslan had defeated a large Byzantine army, the eastern Roman emperor asked European rulers for help to regain his lost lands. Pope Urban II answered in 1096 by calling all faithful Christians to the weapons to free the Holy Land from the Muslim yoke. Western European crusaders managed to conquer Syria and Palestine, but thereafter refused to return the lands to the Byzantine Empire, founding their own states instead. What is more, the Italian cities of Genoa and Venice took full control over the Mediterranean trade, which left the Byzantines impoverished. But at least the immediate threat from the Seljuks had been averted.

OTTOMAN TURKS, MAMLUKS, AND MONGOLS

In the 14th century the remaining Eastern Roman Empire was in trouble again as the Serbs attempted to seize Greece. This time the Byzantines called upon assistance by the Ottoman Turks, who had founded their initial kingdom in northwestern Anatolia in 1299 and were very willing to help. The Ottomans beat the Serbs in Greece and proceeded towards the Serbian homeland, inflicting crushing defeats on Christian armies at Kosovo in 1389 and Nicopolis seven years later. But once they had established themselves as the new masters of the Balkans, the Ottoman Turks in 1453 also put a definite end to the Byzantine Empire: They conquered Constantinople and made it their new capital.

However, the principal power center of the Islamic world at this stage remained Egypt, where the Mamluk dynasty reigned from 1250. The Mamluks had chased the Christian crusaders out of the eastern Mediterranean and in 1260 were the first to ever defeat the Mongols on the battlefield. The huge Mongol Empire had by this stage already become fragmented, and the Mamluks actually allied themselves with the Golden Horde Mongols of southern Russia in their struggle against the Ilkhanid Mongols of Persia.[70]

Mongols had begun to diversify greatly in different regions as they adjusted to the customs of the (sedentary) areas they conquered. In the central Muslim lands, this involved conversion to Islam. One Muslim of Turkic-Mongolian descend, Timur the Lame, temporarily reconsolidated the Mongol Empire. Born in present-day Uzbekistan, he seized power in his homeland and in turn started brutal campaigns that took him all the way to India, Russia, and the Mediterranean.[71] He died on his way to invade China in 1405, but one of Timur's descendants, Babur, established himself in Kabul (in Afghanistan). Babur, who traced his descent to the original Mongol leader Genghis Khan, managed to occupy Delhi in 1525 and founded the Islamic Mughal (Mongol) Empire of India, with which the Europeans had to deal when they finally reached India by sea.

In the Mediterranean, the Ottomans in 1517 defeated the Mamluks and took over their territory.[72] By 1550 they controlled Turkey, Greece, Syria, Palestine, Mesopotamia, Arabia, Egypt, Tripoli (Libya), the Balkans, and most of Hungary. Thereafter the Ottoman Empire ceased to expand and started its slow decline following the second Turkish siege of Vienna, Austria, in 1683, which was answered by a counter offensive by Christian Europeans.

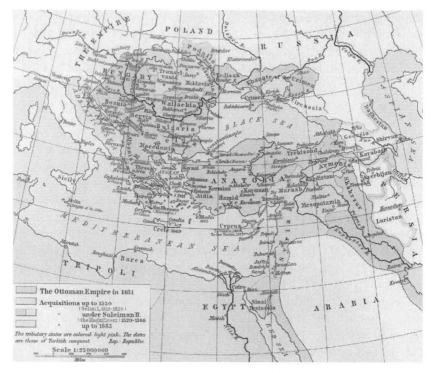

The Ottoman Empire, 1481–1683   The Ottoman Empire had substantial dimensions. It covered the southern and eastern parts of the former Roman Empire, and stretched deep into the Balkans. The map shows the Ottoman Empire's expansion up to 1683. (Reproduced from *The Historical Atlas* by William R. Shepherd, 1923. Courtesy of the University of Texas Libraries, The University of Texas at Austin.)

NOTES

62. Glyn Davies, *A History of Money: From Ancient Times to the Present Day* (Cardiff: University of Wales Press, 1994).

63. R.A. Kerr, "Sea-floor Dust Shows Drought Felled Akkadian Empire," *Science* 279 (1998): 325–326. The end of Akkadian rule is usually given with 2218 B.C.E., when the Gutians, a belligerent tribespeople from the Zagros Mountains, invaded the region. A. R. Millard, "Mesopotamia," Microsoft Encarta Online Encyclopedia, http://encarta.msn.com/encyclopedia_761559228/Mesopotamia.html#p4.

64. Most of the presented political history of the Agricultural Age is based on John Haywood, *World Atlas of the Past, Vol. 1–4* (Oxford: Andromeda Oxford/ Oxford University Press, 1999). I used the Danish edition: John Haywood, *Historisk Verdensatlas* (Köln: Könemann, 2000).

65. Ibid., 70.

66. Frederick F. Cartwright and Michael D. Biddiss, *Disease and History: The Influence of Disease in Shaping the Great Events of History* (New York: Thomas Y. Crowell Company, 1972): 12; Donald A. Henderson and Bernard Moss, "Smallpox and Vac-

cinia," *Vaccines* (Philadelphia: W.B. Saunders, 1999), http://www.ncbi.nlm.nih.gov/books/bv.fcgi?rid=vacc.chapter.3; World Health Organization, *The Global Eradication of Smallpox-Final Report of the Global Commission for the Certification of Smallpox Eradication* (Geneva: WHO, 1979), http://whqlibdoc.who.int/publications/a41438.pdf.

67. John Haywood, *World Atlas*.

68. Rodney Stark, *The Rise of Christianity: A Sociologist Reconsiders History* (Princeton: Princeton University Press, 1996).

69. Toby Lester, "What Is the Koran?," *The Atlantic Monthly* 283 (1999), http://www.theatlantic.com/issues/99jan/koran2.htm.

70. Reuven Amitai-Preiss, "Mamluks and Mongols: An Overview," in *Mongols and Mamluks: The Mamluk-Ilkhanid War, 1260–1281* (Cambridge: Cambridge University Press, 1995): 214–235, http://coursesa.matrix.msu.edu/~fisher/hst372/readings/amitai-preiss.html.

71. "Tamerlane: Lame but Never Halting," *The Economist*, August 26, 2004, http://www.economist.com/displaystory.cfm?story_id=3127398; "Letter from Samarkand: Soviet hangover," *The Economist*, July 24, 1997, http://www.economist.com/displayStory.cfm?Story_id=152626.

72. "Goodbye to the Mamluks-Millennium issue: The Turkish Empire," *The Economist*, December 23, 1999, http://www.economist.com/displaystory.cfm?story_id=347137.

CHAPTER 12

THE RISE OF EUROPE

Europe is called a continent even though it is merely a region that occupies the most western part of the Eurasian landmass. As such, it has always been open to the flow of people and innovations arriving from Asia. There were two principal gateways from the Fertile Crescent into Europe. One was along the shores of the Mediterranean, the other upstream along the rivers that rise in Europe's interior and flow into the Black Sea, such as the Danube and the Dniester. Farming had already spread into Mediterranean Europe from 6000 B.C.E., and about 2,000 years later through the entire European loess belt, which stretches from the Black Sea's north shore all the way to northern France. But outside the loess belt, and away from the Fertile-Crescent-like Mediterranean climate, efficient agriculture was more difficult to establish. This delayed European development. Bronze technology, the art of alloying copper and tin into hard metal, radiated into Europe along the usual gateways from about 2300 B.C.E., and iron smelting followed around 1000 B.C.E. The first distinctly European Iron Age civilization emerged around Hallstatt, in present-day Austria, where iron swords replaced bronze swords from about 750 B.C.E. The people of this culture were Celts, who grew wheat and raised cattle, but did not have a written language. The Hallstatt culture was sustained by the mining and trading of salt, and lasted from about 1200 B.C.E. to 500 B.C.E.

Shortly after 700 B.C.E. some Celtic tribes started to spread throughout large parts of Europe, reaching France and the British Isles in the north, and Spain, Portugal, and Greece in the south. In Greece the Celts were perceived as uncultivated barbarians, and in 390 B.C.E. they sacked Rome, perhaps because they had the advantage of superior iron swords. However, the Celts soon came under pressure almost everywhere in Europe. First the Romans expanded northward, and then Germanic tribes poured into western Europe as they were pushed out of eastern regions by the Huns.

Many of the arriving German tribes had already been Christianized while still living in eastern Europe (by missionaries sent from Constantinople). Not so the Franks. They supposedly originated somewhere along the Baltic shore, and entered the Roman provinces in 253 C.E. In turn they settled roughly around the area of present-day Belgium, and established themselves along the Rhine river. Most of the Franks converted to Christianity when their King Clovis was baptized in 503, which in light of the later expansion of the Franks was a highly critical event for the Christian church. The Franks defeated the Muslims in 732, and Charlemagne (Charles the Great, 742–814) further expanded the Frankish kingdom to comprise most of modern France, Italy, Germany, and the Benelux countries. Charlemagne believed he had re-established Rome's Golden Age and had the pope crown him Holy Roman Emperor in Rome on Christmas Day of the year 800. To be sure, the close cooperation between the papacy and the Frankish kings helped the prestige and power of both parties. Charlemagne's father, Pippin III, had been elevated to the royal throne with papal support. In exchange he provided the pope with land in central Italy, which remained under papal control until the Coal Age when it was reduced to the independent state of Vatican City (inside the modern city of Rome). Charlemagne confirmed his father's land donation to the pope and enlarged it in 774. Pope Leo III, who had been deposed by his opponents, was restored by Charlemagne on 23 December 800, and two days later returned the favor by crowning the Frankish king Holy Roman Emperor.

FEUDALISM

From the king downward, Frankish society was structured in a feudal hierarchical organization that was going to dominate Europe for many centuries. The monarch granted land (often in newly conquered territory) in exchange for war services. The soldiers became noblemen, who pledged allegiance and turned into loyal vassals. With time, the vassals stopped personally fulfilling their commitment to fight for the king whenever he demanded it, and instead provided knights whom they often paid by granting them a bit of their land. (In the later period of the feudal system, grants of land were replaced by money as rewards for war service. Those rewarded with land had turned into the European nobility, while those rewarded with money were called mercenaries.) The permission to sublet the right to hold land in the estate granted by the king created a social hierarchy from royalty to nobility down to lesser gentry, free tenants, villeins and, at the bottom of the system, serfs.

Serfs were the unfree peasants who provided the energy for medieval Europe. They cultivated their lord's manor in return for being allowed to grow some grain for themselves, and for being protected in the unsafe times of the Middle Ages. Serfdom was certainly different from true slavery, as serfs were not traded, but these farmers were subject to the will of their landlord and were not allowed to leave their estate without permission. Wheat and barley were the principal cereals in much of Europe, but rye and oats (which also originated in the Fertile Crescent) gained more importance when crop farming spread further north. (These crops thrive better than wheat in colder climates and poorer soil.) Rye, which is best known for its use to make dark breads, is actually a very close relative to wheat. Oat was often grown to feed horses, which could work longer on an oat diet (spending less time grazing) and had to be fed through the winter.

HORSES

Draft horses became common only in the later Middle Ages. The heavy wheeled plow, common along the Rhine in Charlemagne's empire, was critically important to make agriculture in northern Europe's clayey and rooty soils efficient. However, this kind of plow was pulled by oxen, not horses. Similarly, the Franks still fought on foot during the reign of Charlemagne. Stirrups had only recently reached Europe from China and the transition towards heavily armed mounted knights took a while. Taller, stronger horses had to be bred, which were also preferred over oxen for draft as soon as the rigid horse collar had reached Europe from China around 1000 C.E. Nevertheless, horses did not account for the majority of draft animals until the end of the 16th century in England, for instance. Earlier on, the build-up of a large, grain-expensive horse population was probably impeded by Europe's rapid human population growth. Between 1000 C.E. and 1300 C.E. the human population grew from roughly 7 million to 16 million in France, 4 million to 12 million in Germany and Scandinavia, 5 million to 10 million in Italy (including Sicily), 7 million to 9 million on the Iberian peninsula (Spain and Portugal), and from 2 million to 5 million on the British Isles. But between 1347 and 1400 recurring outbreaks of plague (the Black Death) wiped out perhaps 25 million Europeans, thus reducing the population by about a third, which may have paved the road for the introduction of more horse-power in European agriculture (and industry).

EUROPEAN MEDIEVAL INNOVATIONS

The Black Death, like other epidemics before, had arrived in western Eurasia from the East. However, Europe's connectedness to the rest of Eurasia also had many positive effects. It allowed for a host of innovations to arrive all the way from China, especially during the time when the space between Europe and China was bridged by the Islamic and Mongol Empires. Among the numerous innovations that reached Europe from the East were gunpowder, cannons, paper, the magnetic compass, the Arab lateen rig, the stern-post rudder, Hindu-Arabic numerals, and algebra.

But the Europeans also began to innovate themselves. The Germans invented printing independently from the Chinese, and following the publication of Europe's first major printed work, the Gutenberg Bible of 1455, the spread of information was radically accelerated in the Western world. (Already by the year 1500 more than 40,000 different books or editions had been published in western Europe in various quantities to account for millions of copies.)[73] Knowledge about optical lenses arrived from the East, and Europeans developed their own spectacles from the 13th century. Corrective lenses for the far-sighted began to extend the working years of skilled craftsmen, while mechanical clocks, which appeared in Europe around 1250, created a public sense of time.[74]

In the energy arena, medieval Europe initially relied almost exclusively on human muscle power, often pooled by means of large treadmills. Oxen were eventually used to turn a whim for grain milling, while the use of horses to drive machinery was delayed for centuries even after the introduction of better harnesses and larger horses for field work. A major achievement in medieval Europe was to shift some work to water and wind power. The limited availability (or absence) of slaves in medieval Europe supposedly helped the number of waterwheels to surge in the 10th and 11th centuries. Lots of accessory devices, such as cranks, toothed gears, and the cam (which allows the conversion of rotary motion into reciprocating motion), were either newly invented or improved in Europe, and waterpower was in turn applied for anything from grinding grain, fulling (pounding) wool cloth, beating hemp, mashing hops (for beer), hammering wrought iron, rolling and drawing sheet metal and wire, to driving bellows, pumps, and rotary saws.[75]

Windmills were used in Western Europe for milling grain from the 12th century on, after European crusaders had seen them in Muslim regions. This prime mover was rapidly adopted and improved in those windy European regions that were flat and offered little prospects in terms of waterpower. The windmills used in the Fertile Crescent area since ancient times were horizontal, with sails revolving in a horizontal plane around a vertical axis. But advanced vertical windmills appeared in the Low Countries, Germany, and northern France sometime around 1150 C.E. Dutch drainage mills reclaimed polders for crop fields throughout the 15th century and beyond. Meanwhile, waterwheels became especially important in iron production, as no other prime mover was powerful enough to drive the bellows on which the operation of large blast furnaces depended. These furnaces delivered liquid cast iron to Europeans, which was first used to produce church bells and soon thereafter to make cannons and cannon balls.

EUROPEAN NATIONAL FRAGMENTATION

European arms technology rapidly advanced due to the high frequency of wars between the numerous national entities of Europe. As the climate permitted efficient non-irrigated (rain-fed) agriculture, many different European power centers arose, even where rivers were absent. However, much of western Europe's national fragmentation initially had its roots in the Frankish custom of dividing the kingdom among the sons of the ruler. After Charlemagne's death in 814 the Frankish Empire was riven by civil war and was eventually divided into three kingdoms. The eastern, German, kingdom turned out the most powerful of the three and expanded its territory towards the east and south. The German kings held the claim to be crowned Holy Roman emperor by the pope, but German influence in Italy eventually declined as the papacy expanded its own land area and the Christian church turned into a single great European political power.

In the north, Danish and Norwegian Vikings raided the British and Frankish coasts from Carolingian times. These pirates often targeted prosperous, unprotected monasteries, but it was difficult to confront them as they avoided fighting against soldiers and instead moved on quickly in their longships after they attacked. Viking longships brilliantly combined wind-and-muscle-power. Their shallow draught allowed the pirates to smoothly row up rivers, but they were also sturdy enough to ride out the fiercest storms of the Atlantic Ocean. Thus, Vikings eventually colonized Iceland and Greenland, aided by the warm climate at the time, and in about 1000 C.E. even sailed to North America. They founded a temporary settlement in Newfoundland, but apparently did not find a place worth staying permanently. In turn their trip to America remained without significance to the Western or any other civilization.

Toward the end of the ninth century Vikings began settling in England and Ireland. In England, Alfred the Great soon restricted them to living in the northeastern part of the country, and Alfred's grandson Athelstan in 927 united the Germanic tribes of the Angles and Saxons, thus laying the foundation of the English kingdom. Vikings from Sweden pioneered the trading routes along the rivers that stretch north-to-south across eastern Europe. They penetrated the Baltic countries and Russia, and eventually proceeded through the area of the present-day Ukraine to reach the Black Sea and Constantinople. In 862 they established a dynasty in Novgorod, which in 882 was transferred to Kiev (now the capital of the Ukraine), where the Swedish Vikings laid the foundation of the medieval Russian kingdom known as the

"Kievan Rus." However, they became fully absorbed into the regional Slavic culture.

In the German-speaking areas the Habsburg family, whose early holdings were roughly the area of present-day Austria, emerged as the most powerful dynasty. Rudolph I of Habsburg was elected German King (King of the Romans) in 1273, and the Habsburgs in turn held the position of Holy Roman Emperor (with few interruptions) until 1806, when this rather loose empire of independent German states officially ceased to exist. Maximilian I of Habsburg, born in Vienna in 1459, acquired the wealthy duchy of Burgundy (eastern France) together with the Lowlands through marriage. (The Lowlands comprised 17 provinces covering the current Netherlands, Belgium, Luxembourg, and parts of northern France.) Maximilian's son Philip married Joanna, the eldest daughter of Isabella of Castile and Ferdinand of Aragon, who had united the Christian Spanish kingdoms and expelled the last Muslims from the Iberian peninsula in 1492. (This was nearly eight centuries after the Islamic expansion had first reached Spain.) Isabella and Ferdinand lacked a male heir, and Spain and Spanish-ruled southern Italy were thus added to the vast Habsburg lands.

Philip's son, Holy Roman Emperor Charles (Karl) V, Archduke of Austria, had his hands full with this enormous empire. He fought many wars against the French, who were quite unhappy about being encircled by Habsburg lands, and against the Ottoman Turks, who allied themselves with France. Eventually Charles also sacked Rome, which enabled him to keep the pope from annulling the marriage between his aunt, Catherine of Aragon, and Henry VIII of England. (This ultimately resulted in the establishment of the independent Church of England.) Charles also called German priest Martin Luther to the Diet of Worms in 1521, promising him safe conduct if he would appear. He decided to outlaw Luther's ideas, but was tied up with other concerns and hence did not stop the spread of Protestantism. Since Charles' grandparents, Ferdinand and Isabella, had sponsored Cristoforo Colombo's trip to the West in search of a sea passage to India or China, Charles also became King of the Spanish possessions in the Americas. He controlled an empire in which the sun never set, and it was during his rule that, in 1521, Hernándo Cortés, who was greeted at first as a god, conquered and destroyed the Aztec Empire in central Mexico.

NOTES

73. Jacques Barzun, *From Dawn to Decadence: 500 Years of Western Cultural Life 1500 to the Present* (New York: HarperCollins, 2001).

74. The availability of corrective lenses and mechanical clocks is emphasized as a main factor in European development by Landes in David S. Landes, *The Wealth and Poverty of Nations: Why Some Are So Rich and Some So Poor* (New York:

W. W. Norton & Company, 1998). Earlier on, large clocks on clock towers or churches had told people the time, typically also acoustically. The word clock actually derives from the same root as the German word *Glocke*, which means bell. In China, Su Song constructed a famous water-driven astronomical clock tower in 1094.

75. Find more information on medieval innovations in Paul J. Gans, "The Medieval Technology Pages," http://scholar.chem.nyu.edu/tekpages/Subjects.html. Much of the information provided at this site is based on Frances Gies and Joseph Gies, *Cathedral, Forge, and Waterwheel—Technology and Invention in the Middle Ages* (New York: HarperPerennial, 1995).

CHAPTER 13

THE OPENING OF THE WORLD

The opening of the world, which connected Europe with the Americas and South/East Asia by sea, had its starting point in Portugal. The Iberian peninsula was located at the interface of both the Christian and the Islamic world as well as the Atlantic Ocean and the Mediterranean Sea. Long before the last Muslims had been expelled from Spain, the Portuguese became pioneers of exploring the unknown seas, discovering several islands in the Atlantic ocean, including the Madeiras (1417) and Azores (1427). They used Muslim wisdom in astronomy and mathematics to develop navigation techniques, and they constructed superior wind-powered vessels while the Venetians, who dominated Mediterranean trade, still operated muscle-powered, oared galleys.

Challenging the rough waters of the Atlantic was costly and wasted valuable ships. But the endeavor was motivated by very distinct economic incentives. Most importantly, Europeans wanted to bypass the Muslim intermediaries controlling the trade routes that supplied expensive spices from the Orient to Europe. More immediately, the Portuguese used the newly discovered Atlantic islands to cultivate sugar, the crop whose production and trade in the Mediterranean was also controlled by Muslims. To man the new sugar plantations, the Portuguese began capturing or purchasing people at the West African coast line, and sold some on Mediterranean slave markets. The Spanish, who took the Canary Islands in 1402, joined the Portuguese in these activities.

The three major religions of the Mediterranean, that is Judaism, Christianity, and Islam, all originated in societies that practiced slavery and consequently accepted it.[76] Slavery persisted in Europe after the fall of the (western)

Roman Empire even though it started to give way gradually to serfdom in many regions. Vikings captured and enslaved people during raids anywhere between Ireland and Russia, and traded them widely. The Viking founders of the Kievan Rus captured Slavs and traded them either to Constantinople (via the Black Sea), or directly to the Muslim world (via the Volga River and Caspian Sea). (The word slave actually derives from the word Slav.) Jewish traders purchased Slavonic slaves in eastern Europe and walked them to France, Spain, and Italy, where they were either put to work or sold on to Muslim regions of the western Mediterranean. Venice and other Italian cities played a major role in the medieval slave trade, which made them very wealthy.[77]

Black-skinned African slaves were trafficked by Muslims from Africa to the Mediterranean, but Muslims initially used them as personal house slaves rather than for agricultural or industrial purposes. Christian crusaders brought Nubian captives back to Europe as their personal servants, and many European kings displayed enslaved blacks in court. Crusaders also observed Arabs employing East African slaves in sugar production on Cyprus and other places.[78] (The art of sugar refining had reached the Mediterranean from India in the 6th century, but Muslims greatly improved it.) In Christian Europe sugar was a real luxury item still in the 15th century, and the Portuguese and Spanish earned good money once they had established slave-worked sugar plantations on the newly-discovered Atlantic islands.

DISCOVERING AMERICA

Meanwhile the search for an alternative source of Oriental spices was still on. In 1492 then 41-year-old Cristoforo Colombo took off from the Canary Islands with three ships and 90 people to search for a sea-passage to Asia. Probably born in Italy, Columbus lived in Portugal from his early 20s and by the age of 30 was a captain with much experience in the Atlantic, perhaps sailing all the way north to Iceland and quite certainly all the way south to Guinea in West Africa. His brother was a cartographer, and Columbus, like many educated people of the time, knew that the world was spherical. In what was to become one of the greatest regrets in human history, John II of Portugal turned down Columbus' proposal for the king to finance a trip across the Atlantic. Columbus in turn relocated to Spain and persuaded King Ferdinand and Queen Isabella to support his enterprise.

Within a few weeks he reached the Caribbean Islands, encountering naked people and botanical novelties rather than spice traders and rich cities. Hence, Columbus thought he had landed on an archipelago off the coast of China or Japan, a belief he still held in 1506, on his death bed. Upon return from the first of his four trips to America, Columbus eagerly promoted the spread of the news that he had found the seaway to the Orient by utilizing a German innovation that was younger than Columbus himself: the printing press. But apparently this news was not the only thing Columbus helped spreading. Shortly after the return of the Spanish ships, syphilis, a sexually transmitted New World disease, wrought havoc in Europe, plaguing people until a cure was found in the end of the Coal Age.

For a quarter of a century the American enterprise seemed to be a commercial failure. The Spanish terrorized, tortured, raped, and killed the native Caribbeans, and shipped those to the European slave markets who survived the epidemics of smallpox and other Old World diseases to which native Americans had no immunity.[79] But the Spaniards did not find spices or gold. Then, in 1520, they encountered people of a different sort on the Mexican mainland. Erroneously named Indians, because the Spanish believed they were in Asian waters, these people lived in towns built of stone, were dressed in fine garments, and were not nearly as easy to kill or intimidate as the islanders. They had no hard metal, but used a variety of dangerous weapons, including clubs set with razor-sharp pieces of obsidian. (Obsidian is a dark natural glass formed by the cooling of molten lava.) So the Spaniards spoke softer and eventually heard about a rich empire located somewhere further west.[80]

MESOAMERICAN CIVILIZATIONS: MAYAS AND AZTECS

The Mesoamerican civilizations were based on the agricultural triad maize, bean, and squash. First efforts of maize cultivation date to about 4300 B.C.E., while village life started on the fertile plains of southeastern Mexico only around 1500 B.C.E. (In contrast to the Fertile Crescent and China, the people of Mesoamerica seem to have become entirely sedentary only after the onset of agriculture. Hence, settled village life, though of sedentary foragers rather than farmers, had started in the Fertile Crescent as much as nine millennia earlier, and agriculture had emerged nearly four millennia earlier.)

The continent's north-south orientation gets us far in explaining why American history was a lot less turbulent than Eurasian history. In Eurasia, power shifted between different empires that neighbored one another on a west-east axis, where climate and growing conditions were similar. But in Mesoamerica there was hardly enough space even for just two empires neighboring one another on a west-east expanse. (And in South America much of the widest parts of the continent are covered by the Amazon rain forest, where it would have been impossible for grain-fueled empires to emerge.) Hence, Mesoamerica saw mainly consecutive, and very similar, civilizations replacing one another in the same fertile areas.

Around 1250 B.C.E. the coastal Olmec civilization had produced a number of small states featuring ceremonial centers with monumental stone sculptures and clay pyramids for burials. The Olmecs pioneered a 260-day astronomical calendar that would become widespread in Mesoamerican cultures. By 400 B.C.E. the distinctive features of Olmec culture disappeared, while a similarly complex urban civilization, the Zapotecs, had already emerged in the Oaxaca valley. According to findings at San Andrés, Tabasco, the Olmecs (and later the Zapotecs) used a pictorial writing system from about 650 B.C.E. (This suggests that American writing postdates Fertile Crescent writing by at least 2,500 years. However, in 2006 a stone block with what is presumably a hitherto unknown system of writing was found in the Olmec heartland of Veracruz. Stylistic and other dating of the block placed it in the early first millennium before the common era.[81])

Meanwhile the Maya culture had emerged in the Guatemalan highlands and from about 1000 B.C.E. started to spread over Mexico's Yucatan peninsula, where Mayas built canals to drain and dry swamps to turn them into fields. A number of city-states arose, which were dominated by monumental temple pyramids. The Maya acquired a pictorial writing system and a calendar from regional trading partners and celebrated a complex religion with many deities and human sacrifice. The lowland Maya civilization flourished for well over a millennium, but began to decline from about 800 C.E. In the highlands, Maya cities existed still when the Spanish arrived to conquer them.

In the Valley of Mexico a civilization emerged in the first century C.E. that centered around a truly huge city, Teotihuacán, located about 32 kilometers (20 miles) north of modern Mexico City. Spreading out over more than 20 square kilometers, this city was larger than the contemporary city of Rome, but had fewer inhabitants. Teotihuacán was eventually sacked by invaders, and in the early 10th century the Toltecs took control of the Valley of Mexico. Eventually the Toltecs declined as well, and soon the Aztecs appeared on the scene. The Aztecs were a small, rough, nomadic people that arrived from primitive desert lands in the north (what is now the southwestern United States). When they first arrived, they were not exactly welcome and started to work for a more advanced people on the shores of Lake Texoco. In 1325 the Aztecs settled on an island in this lake where they founded their capital Tenochtitlán (on the site of present-day Mexico City), and gradually reclaimed more of the surrounding marshland. For a long time the Aztecs served as mercenaries for others, but they founded their own military empire around 1430, just 62 years before Columbus discovered America. By this stage at least one third of the population living in the basin of Mexico depended on irrigation for their food. On the other hand, people had no animal power at their disposal at all, and inanimate prime movers (such as windmills and waterwheels) as well as hard metals were absent in all pre-Columbian American civilizations.

The Aztecs kept expanding their empire, but had probably just passed their zenith when the Spaniards arrived. Like the Mayas, the Aztecs had to engage in constant warfare to source people for the region's tradition of human sacrifice, which the Aztecs developed into an almost industrialized mass-slaughtering. The priests cut hearts out of people every day to ensure that the god of war and the sun, the principal deity, would not lose his strength in the war against the spirits of the dark. What is more, the Aztec aristocracy apparently engaged in ritual cannibalism, eating human flesh after the priests had rolled the bodies of sacrificed humans down the stairs from the top of the pyramids, where the killing took place. (To be sure, many religions of the Fertile Crescent demanded human sacrifice as well, and people were slaughtered in the Roman Colosseum for the pleasure of the crowd. Quoting the bible, Bartolomé de Las Casas, a great defender of native American rights, wrote: "One could argue convincingly, on the basis that God ordered Abraham to sacrifice his only son Isaac, that God does not entirely hate human sacrifice".) These customs contributed to the lack of sympathetic cohesion in the Aztec Empire, and the Spanish had no trouble at all finding allies in the neighborhood. When Hernándo Cortés finally located Tenochtitlán, a city larger than any in Europe, he attempted to take control of it right away. However, the Aztec warriors drove the Spaniards out of the city and killed more than half of the small Spanish force. Months went by, and new ships with new forces arrived from Spain. Meanwhile, the first epidemic of Old World diseases was already raging in Mexico, probably depopulating much of the region. Then Cortés had his revenge. In 1521 his renewed troops destroyed Tenochtitlán and rebuilt it as a Spanish colonial city. The natives disappeared nearly entirely, but it is not clear how much of their decline was due to the introduced diseases and how much due to European brutality and falling agricultural output. The Spanish guns of the time were primitive and inaccurate, but their iron swords took their toll. Besides, the Spanish were riding horses, animals larger than any the Aztecs had ever seen.

THE INCAS: SOPHISTICATED CIVILIZATION IN THE ANDES

Motivated by rumors about the fabled wealth of a people living further south, the Spaniards immediately began to explore South America as well. Agriculture had emerged in the Andes from about 3000 B.C.E., and a fishing and farming society arose along the Peruvian coastline around 1800 B.C.E. Thereafter several civilizations appeared and disappeared. From around 600 C.E. quite a large empire existed, extending from the Pacific coast into the Andean mountains, but it disintegrated into several smaller states around the year 1000. At about the same time, a city of some 30,000 inhabitants flourished north of Lake Titicaca.

By the time the Spaniards arrived, the Incas had established a huge empire in the Andes. According to legends, the Incas came out of caves in the mountains and settled around Cuzco, a fertile valley high up in the Andes. From

Mesoamerican and South American Empires   When Europeans arrived, the two most powerful empires in the Americas were those of the Aztecs and the Incas. The former was located in what is now Mexico and based on nutritional energy provided by maize (corn). The latter covered a wide area in the Andes and relied on potatoes and quinoa.

about 1400 C.E. they began conquering neighboring lands until they ruled over 12 million subjects in an area that stretched almost 5,000 kilometers along the Andes, from present-day Ecuador to Chile, and from the Pacific coast to the Bolivian plateau. The empire was fueled by quinoa, potatoes, and a few other tuber species. As the Inca culture was the only complex agricultural civilization that did not develop or adopt a writing system, numerical records of stores were kept by means of knotted cords. Sophisticated irrigation schemes funneled water to the fields, and guano was used as fertilizer. (The guano derived from droppings of seabirds preserved on rocks in the dry climate of coastal Peru.) Llamas served for the transport of goods and as a source of wool, and llama dung was a principal fuel in the Andes. But the Incas never invented the wheel, did not ride on llamas, and abstained from coastal shipping. Hence, the emperor, called the Inca, relied on a courier system of fast runners to deliver information through the vast empire.

Potato and Quinoa   Potato, quinoa, llama, and guinea pig (cuy) were all domesticated in the Andes, but nutritional energy was mainly provided by potatoes (and three other similar tuber species) and quinoa, a goosefoot plant with tiny seeds forming large clusters at the end of the stalk. Due to their high water content, tubers are difficult to store and not especially energy-dense. Thus, Andeans freeze-dried their potatoes into chuñu, a foodstuff storable for years. (Photograph of potatoes by Scott Bauer, U.S. Department of Agriculture, Agricultural Research Service [ARS] Image Number K9152-1. Quinoa photographed by Maurice Chédel in Peru at an altitude of 3,800 meters [12,500 ft.], edited.)

Human sacrifice, including that of children, was common but not as pronounced as in Mesoamerica. Nevertheless Francisco Pizarro found a scenario of internal friction and hatred in the tributary Inca Empire that was reminiscent of the situation that Cortés had encountered in Mexico some ten years earlier. Pizarro and his small force of 180 men deceived the Incas, who agreed to meet them as friends with a party of tens of thousands of unarmed people. Apparently the Spaniards killed 20,000 of them on the spot, which amounts to more than 100 killings per Spaniard. They captured the emperor Atahualpa, demanded and obtained a ransom of gold and silver greater than any European monarch could have paid, and in turn decapitated Atahualpa. Pizarro himself founded Lima as the capital of Peru in 1535, but following a feud among other Spanish leaders (whom he had betrayed), he was assassinated. The people of the Inca Empire were enslaved in (silver) mining and food-production ventures; however, the Peruvian population was apparently less susceptible to European diseases than the Mexican: it shrank by about one fifth. Nevertheless disease may have played a larger role than this figure suggests: smallpox reached the Andes from Mexico ahead of the Spaniards, and perhaps the epidemic was a critical contribution to create the civil war situation that weakened the Inca Empire before the Spanish arrival.

THE PORTUGUESE IN ASIAN WATERS

The Portuguese were in shock when they learned in 1492 that the Spanish had apparently found a sea passage to the Orient by sailing westward. It was the Portuguese who had developed much of the naval technology that allowed Europeans to explore the Atlantic. In fact, they had tried from the 1420s to sail southward from the Canary Islands to follow the coastline around Africa, but they encountered difficulties as soon as they reached the equator: The winds blow counterclockwise around the South Atlantic and did not allow them to sail straight south. What is more, the Portuguese had to solve a serious navigation problem. In the northern hemisphere sailors used the North Star (polestar) for navigation, which happens to be (nearly) aligned with the Earth's axis of rotation and therefore appears to be fixed in the night sky. But in the south there is no such star in the direction at which the axis points. Hence the Portuguese had to use the sun's noonday altitude to determine their latitude as soon as they crossed the equator. As the sun's noon altitude changes seasonally every day of the year even at the same location, this required extensive solar declination tables, which were put together with the help of Muslim, Jewish, and German astronomical and mathematical expertise.[82]

Finally, in 1488, Bartolomeu Diaz reached the Cape of Good Hope, and in turn delivered the critically important latitude coordinates of Africa's most southern tip. Thereafter the Portuguese were able to reach the Cape much faster by sailing south to the equator, then following the counterclockwise winds toward the west and the south, and then crossing back east toward Africa in the South Atlantic as soon as they reached the latitude of Cape Hope. On one of these trips the Portuguese touched upon Brazil and founded a colony there. (That's why Portuguese rather than Spanish is now spoken in Brazil.)

In 1497, just five years after Columbus had discovered America, four Portuguese vessels circumnavigated Africa under the command of Vasco da Gama and reached the Indian Ocean. But what the Portuguese found in South Asia was quite different from what the Spanish had encountered in America. The locals were not at all interested in the junk offered as presents, they knew hard metal, and they were resistant to European diseases. And worst of all, from the Portuguese point of view, many of them seemed to be Muslims. Vasco da Gama first set foot on Indian ground in Calicut in 1498, and returned with a fleet of 20 ships on his second voyage of 1502. He capped a victory over a Muslim flotilla off the coast of Calicut by cutting off the ears, noses and hands of some 800 men, sending the parts ashore to the local ruler with the suggestion that he make curry of them.[83] In short, the Portuguese did not make friends in Asian waters. They had little to offer for trade and immediately started the sort of piracy that was going to define European activities in the Indian Ocean for centuries. It was based on superior European naval technology and cannons, manifestations of advanced command of wind power and destructive chemical energy. (Behind it, of course, was also agricultural energy to fuel the people who built and used ships and weapons, as well as charcoal and waterpower [to drive the bellows of European blast furnaces] used in large-scale iron production.)

The Portuguese business model was actually quite simple. First, they took a share of the prospering, well-established intra-Asian sea trade by forcing all merchant vessels to buy a Portuguese trading license. Then they used that income to buy spices for little money in the East and shipped them around Africa to sell them with huge margins in Europe. Hence the Portuguese did not attempt to create a traditional, or colonial, empire. They only set up but a few, strategically positioned trading posts. At the peak, 40 percent of the pepper imported into Europe went around the Cape of Good Hope, while the Venetians, who controlled the traditional Mediterranean spice trade via Muslim intermediaries, suffered substantial losses and declined. Cloves, which sold for 10,000 times the cost in Asia, were the most expensive of all spices on a weight basis.

The enormous riches Portugal gained from the spice trade invited European competition. Ferdinand Magellan, a Portuguese navigator who had been ignored for promotion, turned to the Habsburg King of Spain, Holy Roman Emperor Charles (Karl) V of Austria, and offered to sail west to the Spice Islands (the Moluccas) to underpin Spain's claim on this group of Indonesian islands that were the world's only source of nutmeg and cloves. (Run, a tiny island of the Banda Islands group, was the world's principal source of nutmeg, sold in Europe at a margin of 3,200 percent.[84]) Magellan's five ships first sailed towards Brazil and around South America. Magellan himself was killed by natives of the Philippines, but one ship proceeded through the Indian Ocean and around Africa to complete the first circumnavigation of the world in 1522. However, competition from Spain was eventually less dramatic for Portugal than competition from the Dutch and the English, who entered the Indian ocean to rob just about every ship they encountered. In turn the Portuguese from about 1600 were quite rapidly driven out of their Asian trading posts by their European competitors.

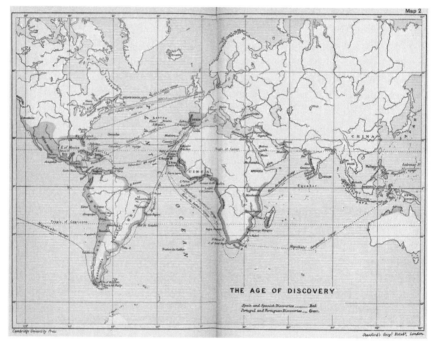

Wind-powered Explorations and Empire Building   The Portuguese attempts to sail south all the way along the African West coast dragged out over decades, because winds blow counter-clockwise around the South Atlantic. Reaching the Caribbean from North Africa was comparatively easy for the Spanish due to the easterly winds that are part of the clockwise wind rotation around the North Atlantic. The Portuguese limited their overseas activities mainly to trade and piracy, while Habsburg Holy Roman Emperor Karl (Charles) V of Austria, as he was also the King of Spain, came to rule over a true and enormous empire in which the sun never set. Karl then sponsored the trip of Portuguese navigator Ferdinand Magellan, one of whose ships completed the first circumnavigation of the world in 1522. (Reproduced from "The Age of Discovery" contained in *The Cambridge Modern History Atlas*, Cambridge University Press, 1912. Courtesy of the University of Texas Libraries, The University of Texas at Austin.)

RISE OF THE DUTCH: WIND AND PEAT ENERGY

The 17 provinces of the Netherlands, approximately covering the current Netherlands, Belgium, Luxemburg, and a region of northern France (Artois), were under the rule of the Austrian Habsburg family from 1477. This area was extremely wealthy. It produced woolen and linen fabrics and organized the trade of Scandinavian and eastern European goods (such as grain, timber, fish, tar, tallow, and hides) to western Europe. A lot of well-educated Jews, many with financial expertise, settled in these Habsburg Low Countries after being expelled from Spain in 1492, and they were allowed to stay there even though Spain was united with all other Habsburg lands under emperor Charles (Karl) V in 1519. (Charles was actually born in the Low Countries.) Later on, when Charles V in 1556 divided his huge empire into a Spanish part (handed to his son Philip II) and an Austrian part (handed to his brother Ferdinand I), the Low Countries ended up under Spanish administration.

However, the seven northern provinces of the Habsburg Low Countries, comprising roughly the area of the present-day Netherlands, increasingly rejected Spanish rule. For one thing, Protestantism had spread to the region from Germany, while the Habsburg rulers were Catholic. And more importantly, the Dutch grew gradually more rich, with their wealth promoting their quest for independence. By the mid-1560s the northern provinces of the Low Countries comprised the wealthiest region of all of Europe, and hence the entire world.[85] The area had become Europe's principal center of commerce, with the Dutch being considered masters of commercial credit and international finance. And then the Dutch managed to broaden their energy base gradually by employing more windmills and utilizing peat to develop their highly productive manufacturing industries.[86]

Peat is an accumulation of partially decayed and carbonized vegetable matter, forming in wetlands that abounded in the Netherlands.[87] Peat was used as a fuel in the region already in Roman times, but Holland Peat by the Late Middle Ages had become more salty and covered by a thin layer of marine sediments due to the frequent floodings and inundations of the peat landscape. The peat layers were thus exploited for their salt content.[88] More generally, the Dutch cut peat out of bogs, dried it (in the wind), and burned it to heat their homes and to fuel energy-intensive industries such as beer-brewing, brick-manufacturing, cloth-dyeing, sugar-refining, and glass-making. Peat was plentiful and much easier to produce than wood or charcoal, which gave the Dutch a unique advantage. On the flip side, peat actually has a very low energy content, less than half that of dry hardwood. It could not be used in metallurgy, as the peat-flame is not hot enough. Hence, peat did not make iron production cheaper and conferred no direct military advantage. Besides, the exploitation of peat in the Dutch provinces had severe environmental impacts. Peat mining created huge lakes that raised the groundwater level, which put an end to most agriculture in the peat lands. This trend prompted a rapid urbanization at the end of the Middle Ages, and perhaps contributed to the Dutch turn toward long-distance trade, which gave birth to a sea-borne empire.[89]

But there were also more immediate reasons for the Dutch to become active in overseas arenas. Following the unsuccessful Dutch Revolt of 1568, the region's prosperity was threatened as the Spanish severed the Dutch role as intermediaries in the trade of colonial goods to western Europe by closing all Spanish and Portuguese ports to ships from the Netherlands. In turn the Dutch decided to take the matter into their own hands and to import Asian spices themselves. They learned through espionage, by placing Dutch captains on Portuguese ships, and once they had acquired information about the routes, winds, and lands, they sent the first Dutch fleet to the East in 1595. The following year England and France recognized the independence of the seven northern United Provinces of the Netherlands, while the Catholic provinces covering present-day Belgium and Luxemburg remained a Habsburg colony. (Belgium belonged to Spanish Habsburg from 1467, and to Austrian Habsburg from 1713. Later on, from 1795, Belgium came under French rule before achieving political autonomy in 1831.)

The Dutch established themselves at Jakarta, the capital of modern Indonesia, and from 1605 forced the Portuguese out of several bases in the Spice Islands. In a brutal campaign they took the Banda Islands in 1621 and killed or displaced their entire native population (of perhaps 15,000 people) to replace it with imported slaves that were going to grow nutmeg. This expensive spice was then sold to Europe as well as to India, where the Dutch set up warehouses and acquired pepper, cardamom, and cinnamon. A chain of posts along the way to Europe included a settlement at Cape Town (South Africa), and a Chinese permission was received to trade at Canton. Besides, the Dutch were the only Europeans allowed to stay in Japan, though under humiliating conditions and confined to a tiny island.

Peat Cutting (Out of Bogs)   The Dutch used their abundant peat resources to fuel various industries. This painting by Cornelis Pronk dates to circa 1540. It shows peat mining operations and is now located in the Stadhuismuseum Zierikzee, Zeeland, The Netherlands.

The Dutch were also searching for a Northwest Passage from the Atlantic into the Pacific Ocean. In 1609 they hired English navigator Henry Hudson for that purpose. He happened to come off course, and after sailing too far south ended up anchoring off Manhattan and founding New Netherland. Years later, after the English had jealously watched the Dutch building up Europe's largest commercial fleet, these two nations waged a series of three wars over their conflicting commercial interests. In the second of these wars, the Dutch sunk most of the English fleet at the mouth of the river Thames. In turn, the Treaty of Breda (1667) left the nutmeg island of Run with the Dutch, who also secured Surinam (Dutch Guyana, neighboring northern Brazil). However, the British received New Amsterdam, the largest settlement of New Netherland, as a consolation prize, and renamed it New York.

THE ENGLISH IN INDIA

Failure to chase the Dutch out of the Spice Islands prompted the English to turn to India in search for alternative trade opportunities. This was eventually going to prove a lucky strike. Both Fertile Crescent and Chinese domesticates had been rapidly radiating towards the Indian subcontinent to fuel very early civilizations. Much of Indian culture is associated with the Aryans, who migrated into the Indus Valley from the north, and perhaps around 1000 B.C.E. moved on to settle in the Ganges river plains as rice farmers. They spoke Sanskrit, from which many of India's current languages derive, and sometime after 700 B.C.E. adopted a writing system. Hinduism, which is based on ancient Aryan beliefs, is considered the oldest of the world's main organized religions. It has had a profound impact on social organization in India until this day, as Hinduism promotes a caste system, a hereditary arrangement of social differences. Belief in reincarnation is also central to Hinduism. It holds that all living things are destined to go through an eternal series of reincarnations (rebirths), either into different human castes or in form of a plant or animal. The doctrine encourages adequate behavior, which will be rewarded by rebirth into a higher class. (This also allowed for the powerful higher classes to claim they had earned their position.) By the year 500 B.C.E. Hinduism had spread throughout the whole Indian subcontinent, and is now practiced by more than 80 percent of the Indian population. However, Buddhism gained ground when Emperor Asoka, who ruled over an empire that covered roughly two thirds of India, converted to Buddhism around 260 B.C.E. and made it the state religion. Buddhism, a doctrine of nonviolence, teaches how everyone can reach nirvana, a condition of total peace and happiness, by living a correct life. Asoka sent out missionaries whose work established (different variants of) Buddhism in much of Asia.

Following Asoka's death his empire disintegrated and the Indian subcontinent was repeatedly invaded by peoples coming from the north. Some invaders adopted the language and religion of the locals, but that was certainly not true for the Huns, who arrived around the year 500 C.E., nor for the Arab Muslims, who arrived in the Indus valley in 712 C.E. The first Islamic state in India was founded by Afghan Muslims in the north in the 11th century, and during the following centuries Muslims spread well into central India. In 1526, the last Muslim invasion of India culminated in the Battle of Panipat, where Babur, a Turkic invader from Afghanistan (and descendant of Timur the Lame), defeated the sultan of Delhi and established the Mogul (or Mughal, that is, Mongolian) Empire of India. However, when the English arrived in India the Mogul Empire was already showing signs of internal weakness. The English slowly bribed their way into the country and managed to acquire trading privileges for the major port of Surat in 1612. Nevertheless they remained quite inconspicuous for the time being.

CHINA

The situation was similar in China, where agriculture had emerged merely one and a half millennia later than in the Fertile Crescent. From about 3000 B.C.E. large settlements began to appear along the Yellow River, where millet was the principal grain staple. The first complex Chinese civilization arose there as well (from about 1766 B.C.E.), developing a pictorial writing system around 1300 B.C.E. This was a delay of some 2,000 years compared with the Fertile Crescent, yet northern China (based on small-seeded millet) was well ahead of southern China (where farmers grew large-seeded rice). During consecutive dynasties the initially northern Chinese Empire kept on expanding to include the western and southern parts of what was going to be known as the country of China, and developed into the largest and most advanced civilization on Earth. The Zhou (Chou) dynasty (1046 to 256 B.C.E.) is considered the dynasty that gave China its identifying political and cultural characteristics. This period witnessed technological advances such as the introduction of horseback riding, ox-drawn plows, crossbows, and iron smelting, and spawned the work of great Chinese philosophers Confucius, Mencius, and Zhuangzi. A little later, the Han dynasty (207 B.C.E. to 220 C.E.) established so thoroughly what was to be considered Chinese that Han was henceforth the term describing all Chinese people, especially as distinguished from non-Chinese (such as Mongolian) elements in the population.[90] The Han dynasty was contemporary with, but even more technologically advanced than, the Roman civilization. Epidemics contributed to the downfall of Han China just as they did to the demise of Rome. Buddhism enjoyed rapid proliferation in the chaotic times following the downfall of the Han dynasty, while Confucianism, the Han state cult, lost much of its credibility as a world view. More immediately, Daoism (or Taoism) became popular as a religion of salvation, driving out Confucianism's strong societal focus by emphasizing the natural order.

Apparently the Chinese population dropped by a quarter between the first and the seventh century C.E., but China began recovering from the wave of epidemics already from the third century and remained a large, stable, centrally governed body for nearly all of the Agricultural Age. Behind the millennia-long stability of the Chinese Empire was a well functioning, labor-intensive agricultural system that rarely collapsed. The frequent flooding of the Yellow River was usually associated with disaster, but it also replenished the soil of the flat northern Chinese alluvial plain with silt from the fertile upstream loess plateau, where early Chinese agricultural communities had thrived. (The staple crops of the north were millet, wheat, soybeans, and later maize.) The warmer and more humid south of China, where rice was the staple, was drained with water from the Yangtze River. (Rice fields need to be flooded at different stages of the crop cycle.) Paddy rice agriculture turned out very resilient, not at least because it does not exploit the soil as much as dry-land farming does. What is more, it yields more food energy per acre. Large-scale irrigation, double-cropping, and crop rotation were introduced as late as the third century B.C.E., but by Han times Chinese agriculture was much more sophisticated than Roman. It employed mass-produced iron moldboard plows, collar harnesses for horses, seed drills, and rotary winnowing fans. Sometime before 1000 C.E. rice varieties from Vietnam were introduced that allowed for two harvests per year in the flooded rice fields. Domesticated pigs were important, but densely populated China never had a strong component of animal husbandry. While wool was produced in the north, the Chinese principally dressed in hemp, linen, ramie (a plant-fiber fabric resembling linen), cotton, and most prominently silk, which is a fine but strong fiber produced from about 2400 B.C.E. from the cocoon of the silkworm (that is, the caterpillar of a moth native to northern China).

Generally, China was technologically well ahead of Europe in various fields at least until about 1350 C.E.[91] Already in Han times the Chinese constructed giant blast furnaces with waterpowered bellows to mass-produce cast iron. They also figured out methods to remove carbon from cast iron to make it steel-like. This provided the Chinese with a substantial military advantage over their neighbors and facilitated the expansion of the Chinese Empire. During the Song (Sung) dynasty (960 C.E. to 1280 C.E.) the final advances in blast furnace design were being achieved, and in the early 11th century the Chinese government had more than 16 million identical iron arrowheads mass-produced per year. Metallurgical technology also supported population growth, chiefly through the wide availability of cast iron moldboard plows. Hence, Song China was able to maintain several cities of over one million inhabitants.

Interestingly, the Chinese started major naval expeditions at about the same time as the Europeans.[92] While the Portuguese were trying to sail down the African west coast from the 1420s, the Chinese undertook at least seven major wind-powered naval expeditions between 1405 and 1431, repeatedly reaching the African East coast, probably as far south as the Mozambique Channel. However, the Chinese, in sharp contrast to the Europeans,[93] lacked economic incentives to continue these expensive voyages, whose purpose seems to have been merely the spread of Chinese prestige. The first fleet of 1405, for instance, consisted of 317 vessels with as many as 28,000 people on board, and the treasure ship of expedition leader Zheng He was 400 feet long, compared to Columbus's *Santa Maria* of 85 feet. (It has also been speculated that the Chinese did in fact reach the Cape of Good Hope to enter the Atlantic and sail to America 70 years before Columbus, but there is no hard evidence.[94] And even if this was true, the event was of no more significance than the visit by the Vikings several centuries earlier.) Astoundingly, the Chinese just a little later abandoned their long naval tradition altogether and prohibited the construction of large ocean-going ships. They could afford to do so because the Grand Canal, the largest human-made waterway in the world, was final-

Chinese Junk   This photograph, titled "A Glimpse into Past Ages—Ancient Chinese Junk, China Sea," was taken in 1906. However, the ships reaching Africa during the great period of Chinese wind-powered naval expeditions were much larger. Zheng He's largest vessels, employed from 1405 in seven major voyages in fleets of over 100 ships of all sizes, had nine masts and were four hundred feet long, compared to Columbus' *Santa Maria* of eighty-five feet. Soon thereafter the Chinese entered a period of introversion during which the huge contiguous Chinese Empire prospered internally with the help of a sophisticated canal infrastructure. (Library of Congress, LC-USZ62-91255, edited.)

ized in 1327 to connect Beijing in the north with Hangzhou in the south. The canal proceeded essentially in parallel to the coast, but it was all-weather capable and inherently secure from pirates.[95] Hence, the Chinese rivers provided for waterborne west-east mobility, while canals provided for south-north transport.[96] The Chinese were thus able to neglect their coastline and focused all resources on the (Mongol) inland threat from the northeast.

This inland focus and indifference to coastal affairs made it hard for the Europeans to leverage their naval cannons and to apply their usual black-mailing tactics. The Europeans were begging for trading permissions, as they longed for porcelain, tea, and silk, but they were viewed as barbarians and had nothing to offer that the Chinese would desire. Hence, their presence was tolerated only in a few isolated places. The Portuguese reached Macau in 1516 and started to lease land there from 1557. The Dutch failed in their attempt to conquer Macao in 1622 and in turn established themselves on Formosa (Taiwan). The Spanish, arriving from the Philippines, established a fort on Formosa as well (1624). However, the Chinese restricted all European trade to the sea port of Canton (Guangzhou), and even this was possible only because the Chinese were interested in acquiring silver that was now being shipped from the New World to pay for Chinese goods.[97]

JAPAN

The situation was practically the same in Japan. Situated at the very eastern edge of the Eurasian landmass, Japan consists of more than 3,000 islands. Four large ones, separated by narrow straits, account for some 95 percent of Japan's territory. Japan experienced much inflow of technology from the nearby mainland. The first Japanese emperor, Jimmu, ruled around 660 B.C.E., and from 300 B.C.E. rice-based agriculture was introduced, along with bronze, iron, and textile technology. During the 4th century C.E. a feudal society arose and Japan remained under strong influence of Chinese and Korean culture. A writing system was acquired from the Chinese from the fifth century C.E., and Confucian philosophy and Buddhism were introduced from China and Korea in the 6th century. In the 7th century attempts were made to diminish the power of the nobles and to set up a strong centralized monarchy based on the Chinese model. However, the power remained in the hands of the great feudal families, with feudal lords (daimyo) organizing local affairs. The 12th century saw the creation of a military government (shogunate), a form of governance that persisted until 1868. During this whole time power lay with the shogun, which was the title of a series of military strongmen who relegated the emperor's role to that of figurehead.[98]

The first Europeans to arrive in Japan were the Portuguese (1542), followed by the Spanish and the Dutch (1609). In 1549 the Jesuits established a mission that lasted for about a century, but the Japanese soon began to fear that Christian propaganda was intended to prepare for a Spanish conquest. They were well aware that the Philippine islands had been added to the Spanish colonial empire, and decided to expel the Spanish (1624) as well as the Portuguese (1639). During the following two and a half centuries of seclusion Japan's internal economy flourished, while Westerners were not allowed to trade with Japan. The Dutch, accepting humiliating conditions and sharp restrictions, were the one exception.

NOTES

76. Benjamin Braude, "Ham and Noah: Race, Slavery and Exegesis in Islam, Judaism, and Christianity," *Annales: Histoire, Sciences Sociales* (2002), Department of History, Boston College, http://www.bc.edu/bc_org/research/rapl/word/braude01.doc; Charshee McIntyre, "The Continuity of the International Slave Trade and Slave System," 1990, http://www.nbufront.org/html/FRONTalView/ArticlesPapers/CMcIntyre_SlaveSystem1.html.

77. Jewish Encyclopedia, "Slave Trade," http://www.jewishencyclopedia.com/view.jsp?artid=849&letter=S.

78. David S. Landes, *The Wealth and Poverty of Nations: Why Some Are So Rich and Some So Poor* (New York: W.W. Norton & Company, 1998).

79. Alfred W. Crosby, *The Columbian Exchange : Biological and Cultural Consequences of 1492* (Westport: Greenwood Press, 1972); William H. McNeill, *Plagues and People* (Garden City: Anchor Press/Doubleday, 1976); Sheldon Watts, *Epidemics and History: Disease, Power, and Imperialism* (New Haven: Yale University Press, 1999).

80. David S. Landes, *The Wealth and Poverty of Nations.*

81. Ma. del Carmen Rodríguez Martínez et al., "Oldest Writing in the New World," *Science* 313 (2006): 1610–1614.

82. John Law, "On the Methods of Long Distance Control: Vessels, Navigation, and the Portuguese Route to India," in "Power, Action and Belief: A New Sociology of Knowledge?," ed. John Law, *Sociological Review Monograph* 32 (Henley: Routledge, 1986), 234–263. Available online at the Centre for Science Studies at Lancaster University, http://www.comp.lancs.ac.uk/sociology/papers/law-methods-of-long-distance-control.pdf.

83. David S. Landes, *The Wealth and Poverty of Nations.*

84. "Survey: Food—Make It Cheaper, and Cheaper," *The Economist*, December 11, 2003, http://www.economist.com/displayStory.cfm?Story_id=2261831.

85. J. de Vries and A. van der Woude, *The First Modern Economy: Success, Failure, and Perseverance of the Dutch Economy, 1500–1815* (Cambridge: Cambridge University Press, 1997).

86. J.W. DeZeeuw, "Peat and the Dutch Golden Age: The Historical Meaning of Energy Attainability," *AAG [Afdeling Agrarische Geschiedenis] Bijdragen* 21 (Wageningen: Landbouwuniversiteit Wageningen, 1978), 3–31; R. W. Unger, "Energy Sources for the Dutch Golden Age: Peat, Wind, and Coal," in *Research in Economic History*, ed. Paul Uselding (Greenwich, Connecticut & London: JAI Press, 1984), 221–53.

87. Peat is usually considered a fossil fuel, though there are some countries, notably Finland, that like to classify it as a renewable biofuel. Finland is rich in peat, and currently fires it in electricity plants.

88. Such salt production started with the removing of the cover sediment layer before the peat was dug out. By burning the peat and mixing the ashes with sea water, and subsequent refining of the salt in lead, copper, and iron pans, the valuable salt was obtained. Job Spijker, " Geochemical Patterns in the Soils of Zeeland," 2005, http://spkr.hackvalue.nl/research/thesisonline/node13.html. A more detailed description of the use of peat in salt production can be found at David Bloch, "Salt Production—Boiling with PEAT fuel," http://www.salt.org.il/frame_prod1.html.

89. Petra J. E. M. van Dam, "Sinking Peat Bogs: Environmental change in Holland, 1350–1550," *Environmental History*, January 2001, http://findarticles.com/p/arti cles/mi_qa3854/is_200101/ai_n8932821/pg_1?tag=artBody,col1. Van Dam quotes: *The Tijdschrift voor Waterstaatsgeschiedenis* [Journal for Water History] 5 (1996) that devoted a special issue to "Peat Mining in the Low Netherlands before 1530." J. Renes, "Urban Influences on Rural Areas: Peat-digging in the Western Part of the Dutch Province of North Brabant from the Thirteenth to the Eighteenth Century," in *The Medieval and Early-Modern Landscape of Europe under the Impact of the Commercial Economy*, ed. H. J. Nitz (Gottingen: University of Gottingen Press, 1987), 49–60.

90. Encyclopædia Britannica, "China: History," http://www.britannica.com/EB checked/topic/111803/China/214398/History#toc214398.

91. Joseph Needham, Wang Ling, Lu Gwei-djen, "Civil Engineering and Nautics," Part III of Volume 4 ("Physics and Physical Technology"), in *Science and Civilisation in China*, ed. Joseph Needham (Cambridge: Cambridge University Press, 1971).

92. Louise E. Levathes, *When China Ruled the Seas—The Treasure Fleet of the Dragon Throne, 1405–1433* (New York: Simon & Schuster, 1994); Samuel M. Wilson, "The Emperor's Giraffe," *Natural History* 101 (1992), http://muweb.millersville.edu/~co lumbus/data/art/WILSON09.ART; Evan Hadingham, "Ancient Chinese Explorers," in *NOVA Online—Sultan's Lost Treasure*, http://www.pbs.org/wgbh/nova/sultan/explorers.html; Kwan-wai So, *Japanese Piracy in Ming China during the 16th Century* (East Lansing: Michigan State University Press, 1975), quoted by Michael L. Bosworth, "The Rise and Fall of 15th Century Chinese Seapower," http://www.cronab. demon.co.uk/china.htm; David S. Landes, *The Wealth and Poverty of Nations*.

93. The main incentive for the Portuguese to sail around Africa, and for the Spanish to cross the Atlantic, was to by-pass Muslim intermediaries to get direct access to oriental goods, most notably spices.

94. Gavin Menzies, *1421: The Year China Discovered the World* (London: Bantam Press, 2001); Adam Dunn, "Did the Chinese discover America?," *CNN*, January 13, 2003, http://www.cnn.com/2003/SHOWBIZ/books/01/13/1421/index.html.

95. It is not entirely clear why the Chinese introduced a ban on all ocean-going ships at this stage. The usual explanations include the intent to curb coastal piracy, and the decline of the eunuchs as an internal political force. (One famous eunuch, mariner Zheng He (1371–1433), had organized the great Chinese overseas explorations.) Another explanation would be that the central court wanted to avoid the emergence of coastal trading centers that would have weakened its power.

96. Find maps showing China's Grand Canal at these sites: Columbia University, "Asia for Educators," http://afe.easia.columbia.edu/chinawh/web/images/sect5/0781_5_grand_canal_600x580.jpg; NOVA online, "Sultan's Lost Treasure," http://www.pbs.org/wgbh/nova/sultan/expl_01.html.

97. Glyn Davies, *A History of Money: From Ancient Times to the Present Day* (Cardiff: University of Wales Press, 1994); Eddy H. G. Van Cauwenberghe, ed., *Precious*

Metals, Coinage and the Changes of Monetary Structures in Latin-America, Europe and Asia (Late Middle Ages-Early Modern Times) (Leuven: Leuven University Press, 1989).
98. The Columbia Electronic Encyclopedia, Sixth Edition, "Japan," http://www. bartleby.com/65/ja/Japan.html.

CHAPTER 14

THE SUPER-AGRICULTURAL ERA

The opening of the world had a profound impact on the global energy supply. No new prime mover or new energy source was introduced, but the introduction of Old World domesticates in the Americas, and New World domesticates in Eurasia and Africa, radically increased the flow of agricultural energy everywhere. The larger set of globally available crops simply allowed for a better capture of sun energy by matching them with regionally varying soil and climatic conditions. This was especially relevant, because the world went through a period of severe climate change: The long Medieval Warm Period was being followed by the Little Ice Age, with colder-than-average temperatures from about 1500 to 1850, and the absolute minimum roughly between 1650 and 1715.

Maize and potato had an enormous impact in Eurasia. Maize grew on land too wet for wheat and too dry for rice, and yielded twice as much weight in grain when directly competing on wheat fields. Potatoes, on the other hand, gave excellent yields in a cool, moist climate, under conditions inhospitable for most Old World cultivars. Potatoes were grown in Spain from 1573 and were soon found all over Europe. In England, potatoes arrived from continental Europe as well as straight from the New World. Both Francis Drake and Thomas Harriot are claimed to have introduced the potato to England in 1586. The same year Walter Raleigh planted potatoes in Ireland, where they became the principal staple. In continental Europe, Germany, Russia, and Poland were soon the largest potato producers. And from about 1600, European settlers introduced the potato in North America. The story of how maize reached Europe is more complex and not fully known. Maize spread very rapidly and appeared in Spain, Italy and the Balkans during the days of Columbus. However, the Portuguese little later also introduced in their home country a maize they had obtained in Guinea, Africa. Maize, which is typically consumed as cornbread, tortilla, or polenta (that is, a mush made of cornmeal), became especially important in south and southeast Europe.

New World staples also quickly reached East Asia. The Spanish established a profitable trade route between Mexico, the Philippines, and China. Peruvian silver was shipped to Asia, and oriental goods to Mexico, from where the Spanish transported it to Europe. Adding to the Portuguese, English, and Dutch ships that sailed to Asia around Africa, this trade route brought New World domesticates directly to the East. Hence, the Chinese began growing American crops, including potato, sweet potato, maize, peanut, chili pepper, and tomato already in the 16th century. The sweet potato, native to New World tropics, became especially popular, flourishing on hillsides unsuited for rice cultivation.

UNPRECEDENTED POPULATION GROWTH

Perhaps the best way to judge the impact of this *global agricultural revolution* is the accelerated multiplication of the human prime mover that was fueled by the increased energy flow from about 1650. Global human population growth had been very close to zero during gatherer-hunter times, but once people became farmers, the increased food supply and sedentary lifestyle accelerated population growth to rates 10 to several hundred times faster than that observed during the Foraging Age. On long-term global average, the new agrarian growth rate was perhaps between 0.02 percent and 0.05 percent per year, which means that births only slightly outnumbered deaths, with between 2 and 5 people being added annually on average per 10,000 people alive. This growth rate may seem low and indeed left the number of humans living on the planet relatively small for several millennia, but it was high enough for the global population to rise from some 5 million individuals around 8,000 B.C.E. to somewhere between perhaps 200 million and 300 million around the year 1 C.E. To be sure, there is no accurate data, and these figures are merely crude estimates. Besides, growth rates alone tell us nothing about actual birth and death numbers. Both birth and death rates were actually high in the Agricultural Age compared to the Foraging Age. Parents generally had a lot of children, but many of them did not survive until adulthood, chiefly because children are especially vulnerable to diseases. Global long-term average growth rates also disguise population fluctuations in the shorter run and on a smaller geographical scale. As it turns out, extreme fluctuations, that is, growth phases followed by population crashes, characterized much of the Agricultural Age.[99]

Population crashes were usually the result of epidemics, famines, warfare, or combinations of these. The horrifying epidemics that shattered Eurasia in the first millennium C.E. actually caused the long-term global population growth to stagnate. Bubonic plague, smallpox, measles and the like repeatedly depopulated the planet's most populous landmass, and estimates typically put the global population at somewhere below 300 million people for both the year 1 C.E. and the year 1000 C.E. In the 14th century, the bubonic plague killed about a third of the population between India and Ireland, but by 1650 C.E. the global human population had nevertheless reached some 500 million individuals. (By this stage life expectancy in Western Europe was somewhere in the 30s, maternal mortality was high, and infant mortality was perhaps 25 percent.)

In the Super-Agricultural Era the global human population began to grow about 10 times faster than before. The average annual growth rate was about 0.4 percent between 1650 and 1750 as well as between 1750 and 1850. In consequence, some 750 million people populated the planet in 1750, and the one billion mark was passed in the early decades of the 1800s. The faster population growth resulted mainly from a fall in death rates (rather than a general increase in birth rates), and the number of births thus began to significantly exceed the number of deaths. People were better nourished, which made especially children less susceptible to disease and helped to decrease the severity of epidemics. But it was also the frequency of epidemics in itself that wound down. This was probably also influenced by improved personal hygiene, the consequent removal of human waste away from households, acquired immunity against diseases, and so on. Insofar as regions with markedly increased birth rates are concerned, better nutrition may have gradually decreased the age of first ovulation and menstruation, and provided more women with the reserves of body fat required to bring pregnancy to term.[100]

Presumably, energy flows were also involved in less direct ways in the accelerated population growth than is suggested by the global agricultural revolution, because the Super-Agricultural Era (and later on the Coal Age) witnessed an increasingly larger shift away from human muscle work. Societies that managed to introduce much animal, water, or wind power (and later on coal energy) into their economies were increasingly based on more prime mover capacity that was indestructible by epidemics that affected humans.[101] Large shares of animal power in agriculture during the Super-Agricultural Era (and later on the Coal Age) would have been especially important to make certain (Western) societies a lot less vulnerable to diseases than they had previously been, as fewer people were needed to produce agricultural surplus energy. This contributed to the decrease of the observed severity of diseases, as secondary deaths due to resulting famines (as well as other secondary effects of social disruption) were reduced. This would then have helped to bring about the population growth that many historians view as highly critical for humanity to progress into more modern times.

EUROPEAN AGRICULTURAL EXPANSION

A significant contribution to the accelerated population growth in the "Super-Agricultural Era," the time period between 1650 and the beginning of the Coal Age about 1800, was the replacement of indigenous New World societies by a rapidly growing population of agriculturally highly efficient Europeans. In 1755, when just about two million people (or 0.3 percent of the global population) lived in the United States and Canada, Benjamin Franklin estimated that the population of the American colonies was doubling about every 25 years and that "before long, the greatest number of Englishmen will be on this side of the water."

The European spread to new continents was clearly an agricultural expansion, not all that different from the Bantu expansion in Africa or the Austronesian expansion in Asia-Pacific. Europeans simply took their Fertile Crescent domesticates and expanded into new lands at the cost of native gatherer-hunters or comparatively inefficient farmers. The only thing that differentiated this process from the traditional intra-continental agrarian expansions was that the Europeans crossed the large oceans to spread to new continents. And in turn they prospered wherever Fertile Crescent domesticates did.

A quick look at the map immediately reveals where this was bound to happen. The Fertile Crescent lies roughly 4,000 kilometers (2,500 miles) north of the equator, between the 30 and the 40 degree latitude lines. All areas situated approximately at the same distance from the equator, no matter if to the north or to the south, were likely to exhibit environmental conditions that would suit Fertile Crescent domesticates. In the northern hemisphere, about half of the United States (48 contiguous states) and China fall into the belt between the 30 and the 40 degree latitude lines. (These two nations are now the world's largest wheat producers.) In addition, the other half of the United States lies between 40 and 50 degrees of latitude, that is, about the same distance from the equator as Europe. (To be sure, the more northern parts of Europe in relative proximity to the Atlantic enjoy an unusual warm climate compared to their relatively high latitude, because the Gulf Stream transports energy from the Gulf of Mexico to the region.) In the southern hemisphere, a trip around the globe at a distance of 4,000 kilometers south of the equator will take a traveler to the tip of South Africa, to southern Australia, northern New Zealand, and Argentina—in short, to exactly those regions that have become major wheat producers.

The Dutch, for instance, used wheat to establish themselves in the Mediterranean climate of southern Africa's Cape of Good Hope. Founding their

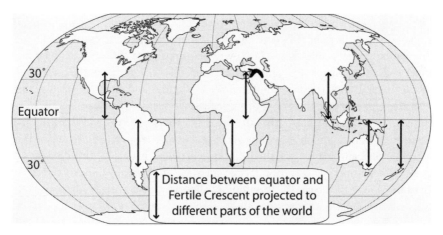

Distance between equator and
Fertile Crescent projected to
different parts of the world

European Agricultural Expansion Regions at similar latitudes, that is, at similar distances from the equator, tend to have similar climates. Europeans took their Fertile Crescent domesticates around the world and established prospering colonies between the 30 and the 40 degree latitude lines, in the northern as well as southern hemisphere. The United States, the Southern Cone (Argentina, Chile, Uruguay), southern Australia, northern New Zealand, and the tip of South Africa turned into major wheat producing regions.

first settlement (at Table Bay) in 1652, they only had to deal with the Cape's Khoisan herders. Fortunately for them, the expansion of Bantu farmers, who knew the art of iron smelting, had come to a hold at the Fish River, 500 miles east of Cape Town. The Cape's climate was simply not suited for Sahel (and tropical) crops that the Bantus had available. (Hence, the Europeans would later [righteously] claim that they had been living at the Cape long before black Bantus did.) Compared to their quick establishment at the Cape, it took the Dutch 175 years (and some coal energy) to subdue their Bantu neighbors on the other side of the Fish River.

CARIBBEAN SUGAR AND AFRICAN SLAVES

While the Europeans quickly displaced (that is, killed or expelled) the ab-originals of the world's non-Eurasian temperate zones, the native population of tropical and subtropical areas often survived. However, one part of the European agricultural expansion actually led into a non-temperate region. This was the truly insane, and incredibly cruel, story of sugar cultivation in the Caribbean and northern South America. To be sure, it had little to do with harnessing fuel for the human prime mover, but was motivated by commercial interests and had a massive impact on the global economy. Shortly after Columbus's trips to the New World, Europeans had access to plenty of gold from America, while oriental spices were shipped around Africa. However, sugar was still largely delivered by Muslim producers. The Portuguese and Spanish supplied some from the newly acquired East Atlantic islands, but this production was soon going to be dwarfed by the sugar output achieved in the West Indies (that is, the Caribbean Islands) and Brazil.[102] Sadly, this was also the story of the boom of African slave trade. The Spanish and Portuguese had initially used African slaves for their agricultural work in the eastern Atlantic, and indigenous Americans in the western Atlantic. But many American natives perished due to European brutality and disease, or they were shipped to the Mediterranean slave markets to finance the early trips to America. In turn the direction of slave trade was reversed, and physically more robust Africans were shipped to the New World.

The Portuguese had the full infrastructure in place. They operated forts on the African West coast to capture Africans (or rather purchase them from local intermediaries in exchange for alcohol and weapons) and sell them to the Mediterranean or use them for sugar production on the Madeiras, Azores, and on Capo Verde. Hence, it was only a small step to also ship African slaves to Brazil, the region first sighted by the Portuguese in 1500. Reputedly the Portuguese planted sugar in Brazil as early as 1516, but during the next four decades trade with Brazil centered around the harvest of dyewood, as sugar prices had slumped in Europe. When sugar prices picked up again in the mid-16th century, Portuguese Brazil pioneered the plantation system, growing sugar on a very large scale and processing it locally in sugar mills that delivered raw sugar and molasses for further refinement in Europe.[103]

Eventually, the focus of sugar and slave trade shifted to the Caribbean Islands. Columbus himself had shipped as many Americans as possible to the European slave markets, while his son initiated the African slave trade of Haiti in 1505, the very same year Columbus senior died in Spain. Before long the Spanish faced competition in the Caribbean. The English were especially eager to make money in the trans-Atlantic slave trade, but the Spanish viewed this as illegal smuggling. In 1568 the Spaniards thus sank several ships of a slaving expedition led by Englishmen John Hawkins and Francis Drake, who in turn started privateering activities that cost the Spanish valuable ships. The relations between England and Spain soured quickly. As the English also began supporting the rebellion (of Protestants) in the Spanish Habsburg Netherlands, and (Protestant) Elizabeth I of England was antagonizing English Catholics, Philip II of Spain in 1587 sent the Spanish Armada on its disastrous trip to England. Spain nevertheless managed to maintain its naval supremacy well into the 17th century, but from the 1620s lost gradually more Caribbean Islands to the Dutch, French, and English. Most demand for raw sugar came from the refineries in the Netherlands, where by 1662 well over half of Europe's perhaps 100 sugar refineries were located. The Dutch used peat to fuel these refineries and grew lots of sugar in northeastern Brazil and on Formosa

(Taiwan), but were ousted from the former in 1655 and from the latter in 1662. In turn they started to promote sugar cultivation in Brazil's neighboring region of Surinam (Dutch Guyana) and on such Caribbean Islands as St. Eustatius and St. Martin. The English and French copied the Portuguese and Dutch slave-based production methods on their own Caribbean sugar islands, and overall global production soared to unprecedented levels. (Perhaps the best indication of how much money was made with Caribbean sugar was that the French, after their defeat by the English in 1763, decided to cede Canada to England, and to keep Guadeloupe and Martinique instead.)

African slave trade expanded in concert with sugar production. Even though the Coal Age was getting near, humans were still enslaved on a massive scale and viewed as nothing more than prime movers. In 1658, right after the English had taken the island from the Spanish, Jamaica had less than 1,400 African slaves. A century later, in 1754, it was 86,500. Natural population growth did not contribute to the swelling African population of the Caribbean. On the contrary, work on the plantations was so hard that slaves did not live very long. It was far cheaper to get adults from Africa than to breed them as was later done in the United States. In Africa, slaves were delivered to European traders by native intermediaries who used European guns to establish their power and prestige. Wars were often deliberately started for the sole purpose of capturing prisoners who were then sold to the New World. The Dutch, French, Portuguese, Spanish, Danish, and Swedish were all trading African slaves, but no nation matched England. The English slave trade alone shipped two million people from Africa to the West Indies between 1680 and 1786. (England in 1713 obtained the right to trade an unlimited number of slaves to the Spanish colonies in the Americas for 30 years, after a broad European alliance had defeated France and Spain.) All nations combined traded a total of perhaps 15 million African slaves to America over the course of four centuries, and more than 70,000 in the single year of 1790. In addition, some 40 million Africans lost their life on board of slave ships in the notorious middle passage: wind power is not a perfectly reliable energy source and the deadly trip from Africa to the Caribbean sometimes was as short as 23 days and other times as long as 95 days.

From about 1680 slave trade became part of a lucrative triangular trade in which textiles, liquor, firearms, salt and other goods were shipped from Europe to the African West coast; slaves were shipped from Africa to America; and raw sugar, molasses and rum (and in later years cotton and coffee) were shipped from America to Europe. All of this, plus demand for European products from the Caribbean sugar islands (brass, beer, textiles, copper, candles, chairs, cider, cordage, earthenware, glass, iron, lead, looking glasses, pewter, pipes, paper, stockings, silver, salt, kettles, and so on), promoted unprecedented global economic growth. But behind it was ultimately the high demand for sugar in Europe.

Sugar is not a vital part of human nutrition, but it is addictive. In the century from 1690 to 1790 Europe imported 12 million tons of sugar, which cost, in all, about the same number of African lives. Sugar consumption in England increased tenfold between 1600 and 1700, and 15-fold between 1700 and 1800. In short, sugar turned from a delicacy into a staple food and was the most important commodity traded in the world. Sugar consumption was also stimulated by the developing tastes of Europeans for tea, coffee, and chocolate. Rum, the first mass-produced distilled spirit, became a popular drink in the 16th and 17th centuries as well. It was made from molasses, the by-product of the sugar industry that cost virtually nothing, and was therefore a lot cheaper than brandy (made from wine) and whiskey (made from grain).

The addictiveness of sugar and alcohol may explain much of this sad episode of "drug" and people trafficking that made traders and producers fabulously rich. One way or another, it was not alone a waste of human lives but also a waste of energy. Sugar hardly served as a fuel. In the late 1780s England consumed about 70,000 tons of refined sugar (astoundingly equivalent to more than half of all raw sugar shipped from America to Europe), which was replacing the food energy contained in about 80,000 tons of wheat, but cost about 7 times as much. Meanwhile, England produced as much as three million tons of wheat annually, with another 15,000 tons being imported. An additional 80,000 tons would have easily been available for import from the Baltic, the Americas, and Ireland, but this was not the point. The Super-Agricultural Era offered so much agricultural energy that those who commanded it could afford to use it for whatever they wanted. And what they wanted was sugar. So they decided to use their surplus nutritional energy to fuel the African prime mover that was condemned to produce the sweet commodity in the New World. The tragedy only ended after a German chemist in 1793 invented and published the process of how to extract sugar from beets that could be cultivated in Europe's temperate climate.

LESS RELIGION, MORE INNOVATION

At the beginning of the Super-Agricultural Era, Europe had just come out of the Thirty Years' War (1618 to 1648), which left about 40 percent of Germany's peasants dead and allowed France to establish itself as the leading Western power for about a century. Meanwhile in China the Ming dynasty had just come to an end, with the outgoing rulers calling in the northern Manchus (or Tartars) of Manchuria for help against rebels in the civil war. In turn the Manchus, though they counted just two million, took over the Chinese Empire and established their Qing dynasty in 1644. China as a whole had around 100 million inhabitants in 1650, about as much as Europe. But by the end of the Super-Agricultural Era, in 1800, China's population had soared to somewhere between 300 million and 350 million people, and Europe's population had doubled to 200 million people. (The global population was then about 920 million.)

This sort of population growth accelerated all those trends that had initially been set in motion by the actual emergence of agriculture: more agricultural energy sustained more people, who could potentially do a lot more work and develop more ideas. Hence, the Super-Agricultural Era was a time of rapid progress that in a positive feedback changed the attitude of people in ways that prompted them to innovate even more. In China (and Japan), this change in attitude is associated with the rationalistic revival of Neo-Confucianism to rebuff asceticist Buddhism (which held that salvation would come from a progressive disengagement from the real world). In Europe, leading thinkers started to question Christian religion and superstition, and instead emphasized the power of human reason in a movement that culminated in the 18th-century Enlightenment.

The development was actually quite similar at both ends of Eurasia. With the onset of devastating diseases, and the downfall of the Roman Empire and Han China, Christianity proliferated in the West, and Buddhism in the East, as religions of consolation and salvation. Around 1000 C.E., when the frequency of epidemics had wound down, and Eurasia's population began to grow, rationalistic thinking increased and both European and Chinese intellectuals turned towards the philosophies of the pre-epidemic period. Neo-Confucianism first emerged in the mid-10th century, emphasizing that humans are by nature fundamentally good, and encouraging engagement in human affairs. However, Neo-Confucianism was considered heretical until 1313, when China's Mongol rulers legalized it. In Europe, the rediscovery of classical Greek and Roman scriptures began in the 12th century (when these were translated from Arabic into Latin), and grew into the full-blown humanist movement in later centuries. Classical logic reasoning was "born again" (in French, *renaistre*), and the Renaissance guided Europe from medieval into modern times between the 14th and the early 17th century, before the real Enlightenment movement gained momentum. And in the process, modern scientific thinking emerged.

The scientific revolution that took place between the mid-16th and late-17th century fundamentally changed the view Europeans had of their planet's position in the universe. German-Polish[104] physician and astronomer Nicolaus Copernicus published *On the Revolution of Heavenly Bodies* in 1543, challenging the well-established (Ptolemaic) theory of an Earth-centered universe. For half a century Copernicus' theory did not receive much attention, but German astronomer Johannes Kepler immediately accepted it and published "Cosmographic Mystery" in Graz, Austria, in 1596. This was the first comprehensive and persuasive account of the geometrical advantages of Copernican theory. Foreshadowing the theory of gravity, Kepler also proposed that the sun emits a force that diminishes inversely with distance and pushes the planets around in their orbits. Kepler developed a full new model of the structure of the solar system and the universe, formulating his famous Three Laws of Planetary Motion, which detailed how planets move along elliptic (rather than circular) orbits with the sun at one focus, moving more rapidly when they come closer to the sun.

In Italy, Galileo Galilei in 1609 built his first telescope, a device that had been invented in the Netherlands. He used it to observe celestial bodies and consequently concluded that at least some of them do not circle Earth, which confirmed his preference for the Copernican system. However, Christian church authorities then compelled him to repudiate his beliefs and writings, and sentenced him to life-long house arrest. Galileo's final book, published in the Netherlands in 1638, reviewed and refined his earlier studies of motion, including the laws of falling bodies. These dated back to 1589, when he was professor of mathematics at Pisa and reportedly demonstrated to his students that two objects of different weight being simultaneously dropped from the Leaning Tower hit the ground at the same time, showing that the speed of fall is not proportional to weight.

In England, Isaac Newton became professor of mathematics at Cambridge in 1669 and picked up on both Galileo's writings and Kepler's theory of the sun emitting a force and how it acts upon the planets. Discussing his thoughts with astronomer Edmond Halley, Newton developed his laws of motion and the theory of universal gravitation. He argued that the whole universe was subject to the same physical laws and that an attractive force (gravitation) exists between the sun and each of the planets. This force, he proposed, depends on the mass of the sun and the planets as well as the distances between them, and hence provides the basis for the Kepler Laws. Newton is also credited for independently developing calculus, although German polymath Gottfried Wilhelm Leibniz (1646–1716) published the fundamentals of integral and differential calculus earlier and saw his superior notation and methods adopted everywhere (eventually also in England). Leibniz, who is credited with scores of diverse innovations, was still clearly of the old school, struggling to reconcile his scientific view of the world with his religious beliefs. But such disposition was overcome by the Enlightenment movement, which sought truth through the observation of nature rather than through the study of authoritative sources such as the bible. This kind of attitude increased the willingness of Europeans to experiment and stimulated the search for new technical solutions. As soon as some industries began to become more advanced and complex, they generated know-how that spread to other industries and created a lot of positive feedback—cycles that paved the way towards industrialization.

WATERPOWER IN TEXTILE PRODUCTION

The time period between 1650 and 1800 deserves being called the Super-Agricultural Era because it enjoyed unprecedented amounts of agricultural energy and human prime movers, while new energy sources such as coal and oil were still absent. (A few applications of coal energy in the later 18th century were the exception.) However, technological advances in this time period included the improvement of sailing ships, windmills, and waterwheels, which resulted in a radical increase of inanimate energy consumption as well. Perhaps even more importantly, this era also witnessed a surge in available power output: large western European windmills of 1750 C.E. delivered about 12,000 watts, while waterwheels were four to five times as powerful. Waterwheels had long been indispensable for cast iron mass production, as there simply was no other prime mover powerful enough to drive the bellows of blast furnaces, but now waterwheels became increasingly important for a variety of industries. Most notably, waterpower revolutionized textile manufacturing: mechanized, waterpowered spinning almost immediately had global consequences.

Clothing and housing are really the next most basic needs once people have access to adequate nutrition. As all clothing of the Agricultural Age was made from natural fibers, it was farmers who provided the raw materials for textiles. The availability of clothing was therefore closely connected to agricultural efficiency. The people of the ancient Fertile Crescent region and Mediterranean wore mainly flax (linen) fabrics. The LBK farmers took flax towards the more northern regions of Europe, where linen became the principal European textile of the Middle Ages. Flanders, the region that comprises the north of present-day Belgium, was renowned for its creamy flax and fine thread from the 11th century. Meanwhile many of Europe's cooler regions preferred sheep wool for clothing.

Both linen and wool fiber production was quite labor intensive. Flax stems were stripped of leaves and immersed in water or spread out on grass and exposed to the dew and sun for several weeks. The stems were then washed, dried, scutched (beaten), and combed (hackled), before shorter fibers (for coarser yarn) were separated from longer fibers (for fine linen). Wool production involved shearing the sheep, pulling the fleece apart, sorting the fibers by length, scouring (cleaning) them in a soap bath, and carding (combing) them. Thereafter, textile production (no matter which fiber is concerned) involved two principal steps: fibers had to be spun (that is, twisted into yarn), and the yarn to be woven into fabric. Woolen cloth, because its fibers are initially loose, airy and unmeshed, was further processed by fulling, which shrank and thickened the cloth. Fulling involved moistening, heating, pressing, and pounding (beating). The trampling or beating (with a fuller's bat) of the cloth was fully mechanized from the 13th century, when waterpowered wooden hammers replaced human feet in fulling mills.

Progress in the other production steps was slow. The spinning of yarn was originally done by hand with a distaff (that is, a cleft stick holding a bundle of fibers) and a weighted spindle that was spun to twist the thread. In the fourteenth century the spinning wheel, a device to turn the spindle, reached Europe from the East. A medieval improvement of the spinning wheel was the addition of a foot treadle that powered the wheel. By the 15th century the wheel was both spinning and winding the yarn onto a bobbin, but further mechanical improvements did not occur for centuries. Meanwhile weaving was very difficult as long as looms were vertical. By the 12th century horizontal looms were common in Europe, which allowed the weaver to sit down during work. (These horizontal looms were most likely adaptations of earlier Chinese silk looms.)

England had a climate that was ideal for sheep breeding and eventually supplied much of the wool for the continental European textile centers. However, England soon initiated policies to transfer textile production know-how to the British Isles. Already in 1331 King Edward III prohibited the export of unwashed wool, and asked expert weavers and dyers from Flanders to relocate to England. In subsequent centuries England began to produce textiles at lower cost than continental Europe as the entire production chain was kept on the landowner's manor, far away from guild-dominated, wage-regulated cities. By the 16th century England had become a leading cloth manufacturer, and this position was strengthened after 1685, when hundreds of thousands of Huguenots, many of whom were experts in working flax and silk, fled France and Flanders (where they were persecuted for their Protestant religion) to settle in regions of England, Ireland, and the Netherlands.

Another pool of textile expertise was to be found in Italy. Here, silk throwing (spinning) technology had arrived from the East by the 13th century, and waterpower was used to run quite complex silk manufacturing machinery. Hence, Englishman John Lombe targeted Italy with a remarkable scheme of industrial espionage. He made a trip to Italy and returned with a number of Italian textile workers and detailed drawings of the Piedmontese throwing machines. He then set up England's first waterpowered silk throwing mill in Derby in 1721. (However, silk clothing remained expensive, as raw silk is produced from the cocoon of the silkworm that thrives exclusively on mulberry trees.) Spinning of the more common, cheaper fibers remained slow. In fact, preparing yarn consumed far more labor than weaving did. To keep a single wool weaver busy required about four people to prepare yarn, a ratio that worsened to 10 spinners per one weaver once the *flying shuttle* accelerated weaving. In traditional looms the shuttle was thrown, or passed, through the threads by hand, but John Kay's flying shuttle of 1733 was mounted on wheels in a track to be shot with paddles from side to side, which enabled a single person to weave even very wide fabrics more quickly than two workers could before.

The first machine to improve on the spinning wheel was finally forwarded in 1770 by James Hargreaves, whose *spinning jenny* allowed for a number of threads to be spun simultaneously by one person. The year after, in 1771, Richard Arkwright introduced the *water frame* and in turn organized several waterpowered spinning factories. Celebrated and knighted, Arkwright was apparently a crook who stole all his ideas from others. Like the original spinning jenny, the water frame was probably invented by reed-maker Thomas Highs, or Heyes, but Arkwright became fabulously rich, even though his patents were eventually overturned.[105] In 1779, Samuel Crompton combined the jenny and the water frame in a machine known as the mule. This device ended the supply shortage in yarn as it allowed a single mule operator to work over 1,000 spindles simultaneously. Henceforth, high-quality thread and yarn was produced in Britain[106] on a large-scale in waterpowered factories.

The situation was now reversed. In the textile boom that immediately set in, scores of hand-loom weavers, rather than spinners, were needed. And since reliable machine-weaving was not developed until well into the Coal Age, this situation from the 1780s accelerated Britain's urbanization. Investors often set up cottages with extra large windows (to provide light for weaving) close to the spinning factories, or handloom weavers labored in manufactories where they could easily be overseen, while all raw material was delivered to, and finished product picked up at, a single site of production.

COTTON BOOM

When inventors had started to experiment with devices for mechanized spinning, it soon became clear that some fibers were a lot better suited for machine processing than others. Cotton, exhibiting the ideal combination of strength and elasticity, by far outperformed the other cloth fibers on the early, rough machines.[107] Flax exceeds cotton in strength and length of fiber, but its inelasticity causes the fiber to break easily under tension. Wool, on the other hand, is overly flexible, showing a tension rate of about one third of its actual length. Silk, though it was known to be machinable, remained too expensive for mass production. Hence, cotton was the big winner, combining low fiber price with good machinability.

Cotton was probably domesticated four times: on the Indian subcontinent, in Ethiopia, in Mesoamerica, and in South America. From the Indian subcontinent, where cotton was grown from very early agricultural times, it rapidly spread towards the east and west to clothe people in ancient Egypt as well as China. From the first century C.E. fine cottons (muslin and calico) were traded to Roman Italy, and after the seventh century Muslims began growing cotton in Sicily and Spain. In the 13th century cotton-weaving technology spread from Sicily to northern Italy, but in more northern Europe cotton fabrics became available in substantial quantities only when the Dutch and English began importing them straight from India in the 17th century. Since cotton was light and pleasant to wear, and easy to wash, dye, and print, imported cotton fabrics caused an outright fashion revolution in Europe. Demand was especially high for coarse cotton fustians, cloth with a short pile that was worn by working men in England.[108]

The cleaning, carding, and spinning of cotton was a lot more labor-intensive than the production of wool, but as soon as waterpowered spinning was introduced in the 1770s, many investors realized that lots of money was to be made with cotton: a fine muslin cotton from India made from a pound of raw cotton was worth nearly 700 times the price of a pound of raw cotton. The industry took off. Merely 500,000 pounds of cotton were spun in Britain in 1765, all by the hand of artisans working at home, while 16 million pounds of cotton were spun in Britain in 1784, all by waterpowered machines in factories. Lancashire, the center of England's new textile industry, had 40 waterpowered cotton spinning mills in 1787, and by 1800 the British total had risen to 900.[109]

This boom demanded a lot of raw material. British consumption in raw cotton increased from 4.8 million pounds in 1771 to 61 million pounds in 1802 to 1.2 billion pounds in 1875. Somebody had to produce this fiber. Wool could have been supplied from within Britain and flax from within Europe, but cotton had to be imported from overseas, as it grows in warm climates only. The first solution to the problem was to divert raw cotton away from the traditional Indian cotton industry. Britain had already built up the necessary trading structures to India and, most importantly, had taken control over Bengal in 1765. But this was not enough. Most of the raw cotton to supply the expansion of the English textile industry was going to be grown in North America. To be sure, it was a serious threat for the English cotton industry that the United States of America declared their independence in 1776, just five years after Richard Arkwright had set up the first waterpowered cotton spinning factory in Britain. The British hence no longer controlled the coastal areas of what was to become the southern United States, a region well-suited for cotton cultivation. So Britain straightened things out with the former American colonies and made sure trade relations were going to be smooth. Even better, the *cotton gin* (cotton-engine) was invented in the United States in 1793, just six years after the formation of an American government as laid out by the brand-new constitution. This simple device was going to have a dramatic impact on the young nation. Designed to mechanize the tedious, time-consuming task of separating the seeds from the fluffy white cotton balls, the cotton gin initially consisted of nothing more than a wooden box with a hand crank, capable of spinning cotton around a drum while wire hooks combed the seeds out of the cotton boll. The device was patented by Eli Whitney in 1794, but was so easy to copy that it spread like a bushfire. What is more, it was easy to upscale into horse- and waterpowered versions.

The cotton gin reduced the labor requirement for removing seeds from fiber 50-fold. Even more significantly, it allowed cotton agriculture to expand further inland, where only short-staple cotton could be grown. (The climate was less humid and the growing season not as long as along the coast.) Short-staple cotton was especially rich in sticky seeds and in contrast to its long-stapled coastal cousin delivered fiber that was virtually impossible to clean before the invention of the cotton-gin. Cotton plantations therefore spread from the coast into the southwestern backcountry, and westward through South Carolina, Georgia, Alabama, Mississippi, and Texas. So it happened that America was quite suddenly able to meet the huge demand for raw cotton from the waterpowered cotton industry of Britain.[110]

RESURGENCE OF SLAVERY
IN THE UNITED STATES

Much like the spinning mule in England, the cotton gin was a labor-saving device that caused an expansion that actually increased labor requirements. In Britain, the demand for additional handloom wavers created paid jobs, but in America the demand for additional cotton field workers was met with increased imports of African slaves. English settlers had been using African slaves in North America from the time they first arrived in Virginia in 1619. Slavery then became an integral part of the plantation system of the more southern regions. On the eve of U.S. independence, in 1775, as much as one-fifth of the 2.5 million people living in Britain's 13 North American colonies were Africans.[111] However, as of 1790 slavery in the United States was actually on the decline. Slave-based sugar production was not well suited for the North American mainland climate, and the price of slave-produced tobacco had collapsed due to oversupply. Meanwhile cotton production was still restricted to coastal areas and remained so labor-intensive that it was hardly profitable, despite the use of slave labor. In short, the price of the human prime mover and the fuel to run it, that is, the cost of grain to nourish slaves, was more expensive than the value of the raw cotton produced by the slave worker. But the cotton gin, by increasing output per worker and by widening the workable resource base, changed these imperatives and led to a resurgence in African slave trade after 1793.

Between 1619 and 1775 about 250,000 Africans had been shipped to the 13 North American colonies. But contrary to the Caribbean sugar slave population, which had to be sustained by imports, the slaves of these colonies were not as rapidly worked to death. The imported slave population actually expanded to about 567,000 on the eve of independence. Then, within just 18 years, an additional 80,000 Africans were imported into the United States. After the 1790–1808 period Congress banned the importation of African slaves, and focus changed to slave breeding. By 1860, approximately one in

Industrialization at the End of the Agricultural Age Cotton-picking slaves (muscle power) and waterpowered mechanized textile production were behind the initial industrialization phase that gave rise to true factories. The illustration on the top of the page, called "The First Cotton-gin," romanticizes the hard labor of cotton slaves in the United States. (It was drawn in 1869, three-quarters of a century after the cotton gin was introduced in 1793.) The bottom photograph (taken in ca. 1927) shows the Slater Mill, the first significant waterpowered cotton-spinning mill in the United States, built in 1793 at the Blackstone River in Rhode Island. As the weaving was then still done in small shops and private homes on hand-operated looms, the true "Birthplace of the Industrial Revolution in America" is often considered the textile mill of the Boston Manufacturing Company, set-up in Waltham, MA, from 1813. The technology of this firm's waterpowered looms, just as waterpowered spinning earlier on, was acquired through industrial espionage in England. ("The First Cotton-gin" by William L. Sheppard: Library of Congress image LC-USZ62-103801. Slater Mill: Library of Congress image LC-USZ62-116492, edited.)

three Southerners was a slave. And since American output in raw cotton had doubled each decade after 1800, cotton was by far America's most important export product by mid-century, accounting for nearly three-fifths of all exports.

BRITISH SUPER-AGRICULTURAL EFFICIENCY

In Britain, where most of this cotton was imported, the substantial rise in economic strength during the Super-Agricultural Era was not the effect of advances in waterpowered production alone. This was still the Agricultural Age after all, and grain remained the most important fuel. It was thus agricultural efficiency that initially set England aside. Despite experiencing rapid population growth Britain remained self-sufficient in grain supply, producing some three million tons of wheat annually in the 1790s and importing less than one percent of its grain needs. What Britain achieved was much more than the adoption of New World domesticates. New spheres of agricultural productivity were entered through a mix of farming innovations and radical land reforms, which made England the most agriculturally efficient country of all.

A four-course rotation system with nitrogen-fixing leguminous crops obviated the necessity of leaving a third or half the land fallow. Turnips served as fodder crops to feed livestock that was transformed through radical new breeding methods known as breeding in-and-in (that is, close and repeated incest). Additional agricultural innovations included the horse-drawn hoe (which replaced weeding by hand from the 1730s) and the horse-drawn seed drill, which from 1701 economically sowed the seeds in neat rows and allowed a much greater proportion of the seed to germinate by planting it below the surface of the ground, out of reach of birds and wind. (These two innovations were promoted by Oxford-educated agronomist Jethro Tull.)

Between 1750 and 1810 Britain underwent yet another period of land enclosures, which were part of an ongoing process that had began in the 14th century and was accelerated in the 15th and 16th centuries. Open land that was formerly subject to common rights, and used for common cropping or grazing, was fenced off from public access and used for more efficient agricultural production. On the downside, small farmers were deprived of their common land. Many could not afford to put fences or hedges around their fields (as the new enclosure laws required) and had to sell out to larger landholders. Hence, the field enclosures caused poverty, homelessness, and rural depopulation. Farmers looking for new jobs moved to the cities, where they contributed to the rapidly growing urban population that provided manpower for early factories and depended on the higher outputs of the increasingly productive agricultural system. To a large extent, this efficient agricultural production became possible because owners of large pieces of land reorganized it and subjected it to intensive cropping or fenced pasturage.

PASSING THE DUTCH

The restructured energy base soon also allowed Britain to achieve substantial international political power. At the start of the Super-Agricultural Era, the Netherlands were still the wealthiest region in Europe (and on Earth): They were amidst their Golden Age period of great wealth and power that lasted nearly until the end of the 17th century. It was based on wind-powered overseas trade, a powerful navy protecting Dutch commerce, peat-fueled Dutch industry, and plenty of windmills operating on the flat open land. (By 1650 the Netherlands counted as many as 8,000 windmills.) However, from about the start of the 18th century, Dutch prosperity slowly began to decline, mainly at the expense of England, where energy consumption, productivity, and prosperity began to pass Dutch standards.

Dutch Protestant ruler (stadtholder) Willem Hendrik, better known as William of Orange, actually became King of England after opponents of England's Catholic King James II had invited a Dutch army into their country. (This was King James II, Duke of York, for whom New York is named. He was William of Orange's uncle as well as father-in-law. William of Orange and his wife Mary then had equal claims to the throne and are known in the United States for chartering America's second oldest college in Virginia in 1693.) But for the Netherlands the union with England actually had negative effects. Much of the Dutch merchant elite relocated to London, which slowed Dutch economic growth, and England economically surpassed the Dutch more easily as an ally than it would have as an enemy. By the 1780s, Britain had become richer than the Netherlands and defeated them in the Fourth Anglo-Dutch War (1780 to 1784), when the Dutch supported New England insurgents seeking independence.

TAKING CONTROL IN INDIA

The mix of super-efficient agriculture, population growth that increased the number of prime mover units, improved waterpowered production facilities, advanced sailboats to harness wind energy for mobility on the oceans, and superior naval cannons to turn chemical energy into deadly kinetic energy, also allowed Britain to become more dominating in global overseas arenas during the Super-Agricultural Era. In India, the English received permission to set up a base at Bombay, valued for its deep harbor, in the 1660s. (Bom Baia, the Good Bay, had been in Portuguese hands from 1534, but was part of the dowry when Charles II of England married the daughter of King John IV of Portugal.) Under English administration Bombay's population in turn increased from 10,000 in 1661 to 60,000 in 1675. Similarly, the English on the Indian east coast from 1690 built their own commercial city on the territory of a then tiny village called Calcutta. The English operated in Asia through the East India Company, a joint-stock firm chartered by Queen Elizabeth I in 1600. The company's main objective was to make a profit for shareholders rather than to conquer Indian regions or build an empire. However, the firm bought land from Indian rulers to set up trading posts and in turn established its own army and navy to protect these posts against Dutch, French, and local competition. The Mogul emperors based in Delhi kept European merchants on good behavior and out of local politics, but after 1707 Mogul power went into steep decline.

India's decentralization posed a threat to foreign merchants, because it increased instability, but it also created opportunities. The East India Company successfully used the divide-and-conquer strategy to extend its control over increasingly more regions of the Indian subcontinent. This tactic entailed fanning the flames of religious division between Muslim and Hindu groups, and taking advantage of the political rivalries that existed among local native rulers. And once the English had gained strength, they openly resorted to force to pursue their interests. Threats of naval blockade that would hurt Indian trade and pilgrimage to Mecca, and seizure and ransom of Indian vessels, were but part of their repertoire.

Some Indian aspirants to power sought alliance with the Europeans, others wanted them to leave. In 1756, the East India Company's position in Bengal (a province now divided between India and Bangladesh) was being threatened by a new ruler, who was irritated by (among other things) the British fortifying their post at Calcutta. He decided to attack, which gave the English a reason to retaliate. After bribing the ruler's subordinates by offering them a true fortune and considerable power, they defeated and ousted him. Following political turmoil during the subsequent years, the East India Company in 1765 took full control of the rich province and collected land taxes, which in the last instance were pressed out of local farmers. The early English rule of Bengal is famous for the incredible fortunes that British governors made through taxation, trading on favorable terms, and sheer loot. However, it was also the period when a third of the Bengali population, roughly 10 million people, died in a famine. (Until this day it is debated to which extent British policies were responsible for this disaster.) Control over Bengal was a major turning point in the East India Company's history, as most of the firm's enormous income now came from taxes, not from trade. It was also the first major step towards making India a British colony, which would be achieved during the Coal Age. However, for the time being the Marathas, a native militant Hindu people that had led the resistance against the now powerless Mogul, ruled over an empire in central India that stretched from the West to the East coast.

CHINA—STILL TOO WEAK FOR THE DRAGON

The situation was quite different in China. Highly organized central power prevailed, and European thinkers of the time admired the Chinese emperors as enlightened despots, recommending their own kings to copy the Qing dynasty's methods of government. China prospered during the entire Super-Agricultural Era, producing agricultural energy that fueled a population that grew from 100 million in 1650 to 300 million in 1800.[112] Despite this expansion, the standard of living remained remarkably high.

For the Europeans the stability of the Chinese Empire made it a lot more difficult to achieve their economic goals in the region than it was in India. They were longing for trading privileges, while the Chinese kept them on a short leash. The Jesuits were able to charm their way into the country through their knowledge of mathematics and astronomy, plus their skills in manufacturing cannons. However, their religious ideas were soon perceived as an attack on Chinese claims to moral superiority.[113] (Eventually, in 1723, Christianity was officially banned.) From the early 17th century the Dutch dominated European-Chinese trade, but they were ousted from the island of Formosa (Taiwan) in 1662 by a Chinese war party seeking a secure refuge after opposing the new Qing dynasty. The English East India Company established a trading post at Canton (Guangzhou) in 1699, and the French built up commercial relations with China around the same time. In the 18th century other coastal European nations, including Spain, the Netherlands, Sweden, and Denmark followed suit: they all had factories (posts) in the foreigners' quarters outside the city walls of Canton, and they all were not allowed to trade anywhere else in China, especially not in the north. Nevertheless Chinese-European trade increased significantly between 1760 and 1770, and in 1793 George Macartney, the first envoy of Britain to China, arrived in Beijing with a large British delegation on board of a 64-gun man-of-war to ask for a permanent British embassy in Beijing, a relaxation of the British-Chinese trade restrictions, and permission for the acquisition of "a small unfortified island near Chusan for the residence of English traders, storage of goods, and outfitting of ships."[114]

Emperor Qianlong flatly denied these requests, and the Europeans could do nothing about it.[115] They had better cannons, but their mobility was restricted to windy open seas. Soon, improved land transportation (trains) and mobility on rivers (steam boats) were to give Europeans a decisive advantage over the Chinese. But for the time being they were no threat to China, which prospered away from the shoreline based on a vast interior infrastructure of rivers and canals. (China did not depend on any commodities arriving by sea.) Hence, the Europeans had to accept the Chinese terms of trade in order to be allowed to trade at all. The situation was quite similar in Japan, which also prospered in seclusion during the Super-Agricultural Era. The Spanish were expelled in 1624, the Portuguese in 1639, and the Dutch were allowed to stay under humiliating conditions. And just as in China, technology to produce firearms largely fell into oblivion in Japan during this period.

THE VERY FIRST WORLD WAR (1754–1763)

As the activities and commercial interests of Europeans became more global, their previously regional wars did as well. In the Super-Agricultural Era a war was fought that set the stage for Britain to become a superpower and for English to become a globally dominant language. Being a conflict of the Agricultural Age, this war was still dominated by grain fueling people and horses; wind power for mobility at sea; and gunpowder for destructive chemical energy. (Flintlocks were the principal hand-held firearm of the Super-Agricultural Era.) But this war took place at the dawn of the Coal Age, and Britain had some strategic advantages that foreshadowed the upcoming era: Plenty of coal-fired Newcomen steam engines were pumping water in the mining districts; cast iron was produced by use of coal rather than notoriously scarce charcoal (to make cannon balls and thin-walled Newcomen cylinders, for instance, or to process it into wrought iron for gun barrel production); Hunstman had introduced a process for making crucible steel in 1740; and the first iron rolling mill was established in Hampshire in 1754, the year the war started.

Sometimes referred to as the "Very First World War," this conflict was truly global.[116] But at its core was a European war, the Seven Years' War (1756–63), which was an extension of the War of the Austrian Succession (1740–48). The latter had started after Holy Roman Emperor Charles VI died without leaving a male heir. The succession of his daughter, Empress Maria Theresa of Austria, was disputed by a number of European powers. Prussia, supported by France and Spain, attacked Austria, which was supported by England and the Netherlands. After an Anglo-Austrian army was defeated in 1745, Maria Theresa had to formally accept a peace treaty that handed Silesia (a region of the Austrian Habsburg Empire that is now divided between Poland and the Czech Republic) to Prussia, and the Italian duchies of Parma, Piacenza, and Guastalla to Spain. However, Maria Theresa immediately began to forge new alliances, reshaping the strategic structures of Europe within a few years by bringing the century-old enemies France, Austria, and Russia into a single alliance against Prussia and Britain. In addition, France allied itself with Spain, where the last Habsburg ruler, childless Charles II, had named Philip Bourbon of Anjou, grandson of Louis XIV of France (who had married one of Charles' sisters), to inherit the Spanish throne upon Charles' death. Notably, France and Britain remained in opposing blocks before and after Maria Theresa's maneuvering, while the English and Prussians were now fighting on the same side. The British rulers were actually German themselves: In 1714 the Duke of Hannover, who did not speak English at all, became King George I of England, because Queen Anne did not leave any closer Protestant relatives. (George's descendants still now reside in Buckingham Palace, though they changed their name to House of Windsor in 1917, when Britain was fighting Germany in Word War I.)

Maria Theresa's main agenda was to recover Silesia, a region rich in hard coal and thus critically important for the upcoming Energy Era. However, she was unable to pull Britain into her broad alliance against Prussia, because the English and French were by this time clashing over their colonial interests in North America. In fact, the North American segment of the Seven Years' War started two years earlier than the European. Known as the French and Indian War (1754–63), this British-French conflict involved Native Americans, so-called Indians, who were primarily fighting on the French side. British settlements had spread up and down what was to become the U.S. East Coast, while the French had settled in Canada, as well as in the Mississippi Valley and along the Mexican Gulf coast, in what are now the states of Alabama, Mississippi, and Louisiana. From the south, the French began exploring the Mississippi and Ohio rivers all the way north to the Great Lakes, where they touched upon their Canadian possessions that stretched from the Atlantic coast along the St. Lawrence River valley. Hence, they systematically encircled New England and contained the westward expansion of the English colonies. Tensions between Britain and France in North America finally escalated over the lucrative fur trade and over fishing rights off the coast of Newfoundland. But the immediate cause of the war was the struggle for control of the Ohio Valley, where British traders established fur trading posts on land that the French had claimed.

India was another overseas arena for the British-French conflict. The French had followed the British into India in 1675, establishing their first trading post some 60 years after the English had first shown up. The French East India Company set up its principal base at Pondicherry, on the Coromandel Coast, and during the first half of the 18th century emerged as a serious rival to the English East India Company. Both the English and the French offered their services to regional rulers in exchange for trading privileges and political influence, and both built up large armies in alliance with local troops. It was merely a question of time until their rivalry would lead to open conflict. A series of three English-French wars, named Carnatic Wars (for the southeastern Indian region where they took place), began during the War of the Austrian Succession (in 1746), and ended with the Third Carnatic War (1758–63), which is considered the Indian segment of the Seven Years' War.

In Europe the Seven Years' War eventually reached a stalemate, but globally Britain gained the upper hand by attacking French possessions in every corner of the world. British forces seized French Senegal in West Africa, the French sugar islands of Martinique and Guadeloupe, and the Spanish colonies of Cuba and the Philippine Islands. In North America, where the English colonial population was ten times larger than the French, the British conquered all of French Canada by 1760. And in 1761 English forces captured Pondicherry, the capital of French India. When global warfare ended in 1763, France decided to recover the profitable sugar islands Guadeloupe and Martinique, and to forgo Canada, which brought a quite large Francophone population under English rule. The English also kept the land east of the Mississippi River, and with the French ousted, Britain firmly established itself as the principal European power in India. Prussia emerged from the war as a major European power and kept (most of) Silesia. Spain retained Cuba and the Philippines, and recovered Florida within twenty years. France eventually recovered the huge western part of Louisiana including New Orleans.

To be sure, Britain's victory came at a price. Seven years of global warfare had been tremendously expensive, and the British government was strained for money, which prompted it to turn to the colonies for revenue. The heavy taxes collected from the colonies were supposed to help pay off British debt and to maintain the troops that occupied Canada and Florida. However, with the French decisively eliminated from the region, the English colonies in North America no longer needed military assistance from overseas. Thus, they strongly resented the taxes imposed on them, and the "Very First World War" helped to bring about the American Revolutionary War (1775–1781), which gave rise to the United States of America, a future superpower.

NOTES

99. Joel E. Cohen, *How Many People Can the Earth Support?* (New York: W.W. Norton & Company, 1995). (This is a principal source used for historical population issues and figures.)

100. Ibid.

101. Animals had their own, sometimes epidemic, diseases, and various diseases, including anthrax (which is caused by *Bacillus anthracis*), were shared by humans and their (grazing) animals.

102. Mark Johnston, "The Sugar Trade in the West Indies and Brazil Between 1492 and 1700," James Ford Bell Library, University of Minnesota, http://www.bell.lib. umn.edu/Products/sugar.html#n6.

103. The information on sugar cultivation and trade presented in this section is based on: Jules Janick, "History of Horticulture, Lecture 34: Horticulture, Politics, and World Affairs: Sugarcane and Plantation Agriculture," Department of Horticulture and Landscape Architecture, Purdue University, 2002, http://www.hort. purdue.edu/newcrop/history/lecture34/lec34.html; In "Sugar & the Slave Trade," http://www.hort.purdue.edu/newcrop/history/lecture34/r_34–1.html, Janick quotes H. Hobhouse, *Seeds of Change: Five Plants that Transformed Mankind* (New York: Harper & Row, 1986).

104. Both Germans and Poles have claimed Copernicus as *their* scientist. Copernicus' mother tongue was certainly German, as his mother was German, but he was born on territory of present-day Poland. His father, sometimes described as a Germanized Pole, died early, and the family consequently came to live with his mother's brother (Lukas Watzenrode). Copernicus joined the German Nation of Bologna University when he studied in Italy, which was only possible for German native speakers. However, the writings of Copernicus are all in Latin, as was the custom at the time, and it is really quite unimportant to which extent Copernicus was German or Polish. Edward Rosen, Erna Hilfstein (Eds.), *Copernicus and His Successors* (London: Hambledon Press, 1995).

105. "Making History—Thomas Highs," BBC Radio 4, http://www.bbc.co.uk/edu cation/beyond/factsheets/makhist/makhist7_prog10a.shtml; Cotton Times— Understanding the Industrial Revolution, "The Unsung Thomas Highs," http://www.cottontimes.co.uk/highs.htm.

106. The United Kingdom of Great Britain was formed when Scotland in 1707 formally joined England and Wales. In turn the term "British" came into use to refer to all the kingdom's peoples and affairs. Ireland formally united with Great Britain in 1800.

107. David S. Landes, *The Wealth and Poverty of Nations: Why Some Are So Rich and Some So Poor* (New York: W.W. Norton & Company, 1998).

108. "A Fashion Revolution," *Making the Modern World*, The Science Museum, http://www.makingthemodernworld.org.uk/stories/manufacture_by_machine/01. ST.01/?scene=6.

109. William E. A. Axon, ed., *The Annals of Manchester—A chronological record from the earliest times to the end of 1885* (Manchester: Manchester Central Library, Salford Local History Library, 1886); "Cotton—A great yarn," *The Economist*, December 18, 2003, http://www.economist.com/displaystory.cfm?story_id=2281685; "Textiles: From domestic to factory production—The Industrial Revolution and the textiles industries," *Making the Modern World*, The Science Museum, http://www.making themodernworld.org.uk/learning_modules/history/01.TU.01/.

110. William H. Phillips, "Cotton Gin," EH.Net Encyclopedia, edited by Robert Whaples, February 11, 2004, http://eh.net/encyclopedia/article/phillips.cottongin.

111. James A. Henretta, "American Revolution," Microsoft Encarta Online Encyclopedia, http://encarta.msn.com/encyclopedia_761569964/American_Revolution. html.

112. "China's Population Growth," International Institute for Applied Systems Analysis, http://www.iiasa.ac.at/Research/LUC/ChinaFood/data/pop/pop_21_m. htm.

113. David S. Landes, *The Wealth and Poverty of Nations*.

114. Joseph Banks, "Papers of Sir Joseph Banks, Section 12: Lord Macartney's embassy to China," Series 62: *Papers concerning publication of the account of Lord Macartney's Embassy to China*, ca. 1797, State Library of New South Wales, http://www2. sl.nsw.gov.au/banks/series_62/62_view.cfm; Immanuel C. Y. Hsü, *The Rise of Modern China* (New York: Oxford University Press, 2000); James Louis Hevia, *Cherishing Men from Afar: Qing Guest Ritual and the Macartney Embassy of 1793* (Durham, NC: Duke University Press, 1995).

115. Read the letter by the Chinese emperor to the King of England at this site: "Ch'ien lung's Letter to George III," posted by Joseph V. O'Brien, Department of History, John Jay College of Criminal Justice, The City University of New York, http://web.jjay.cuny.edu/~jobrien/reference/ob41.html.

116. H.V. Bowen, *War and British Society 1688–1815* (Cambridge: Cambridge University Press, 1998). On page 7 Bowen writes, "Indeed, historians have argued that the unprecedented geographical range of the Seven Years' War was such that, as Winston Churchill once remarked, it should properly be regarded as the 'first world war' (Kennedy, 1976: 98–107)." P. M. Kennedy, *The Rise and Fall of British Naval Mastery* (n.p.: Scribner, 1976). Tom Pocock, *Battle for Empire: The Very First World War, 1756–63* (London: Michael O'Mara Books, 1998).

BIBLIOGRAPHY TO PART II

Allard, Robert W. "History of Plant Population Genetics." *Annual Review of Genetics* 33 (1999): 1–27. http://arjournals.annualreviews.org/doi/abs/10.1146/annurev. genet.33.1.1.

Amitai-Preiss, Reuven. "Mamluks and Mongols: An Overview." In *Mongols and Mamluks: The Mamluk-Ilkhanid War, 1260–1281*. Cambridge: Cambridge University Press, 1995. http://coursesa.matrix.msu.edu/~fisher/hst372/readings/ amitai-preiss.html.

Apter, Charlie. "The Horse in Art, Science and History." Truman State University, 2007. http://agriculture.truman.edu/courses/343syllabus.pdf.

Astill, Grenville G., and John Langdon. *Medieval Farming and Technology: The Impact of Agricultural Change in Northwest Europe.* Leiden: Brill, 1997.

Axon, William E. A. ed. *The Annals of Manchester—A chronological record from the earliest times to the end of 1885.* Manchester: Manchester Central Library, Salford Local History Library, 1886.

Bachrach, Bernard S. and Charles Martel. "Mounted Shock Combat, the Stirrup and Feudalism." *Studies in Medieval and Renaissance History* 7 (1970): 49–75.

Banks, Joseph. "Papers of Sir Joseph Banks, Section 12: Lord Macartney's Embassy to China." Series 62: *Papers concerning publication of the account of Lord Macartney's Embassy to China,* ca 1797, State Library of New South Wales. http://www2. sl.nsw.gov.au/banks/series_62/62_view.cfm.

Barzun, Jacques. *From Dawn to Decadence: 500 Years of Western Cultural Life 1500 to the Present.* New York: HarperCollins, 2001.

Batty, Joseph. *Ostrich Farming.* Midhurst: Beech Publishing, 1995.

BBC Radio 4. "Making History-Thomas Highs." http://www.bbc.co.uk/education/ beyond/factsheets/makhist/makhist7_prog10a.shtml.

Benecke, Norbert. *Der Mensch und seine Haustiere: Die Geschichte einer jahrtausendealten Beziehung.* Stuttgart: Theiss, 1994.

Benz, B.F. "Archaeological Evidence of Teosinte Domestication of Guilá Naquitz, Oaxaca." *Proceedings of the National Academy of Sciences* 98 (2001): 2104.

Bloch, David. "Salt Production—Boiling with PEAT Fuel." http://www.salt.org.il/frame_prod1.html.

Blumler, Mark A. "Evolution of Caryopsis Gigantism and Agricultural Origins." In *Research in Contemporary and Applied Geography: A Discussion Series XXII (1–4)*. Department of Geography, Binghamton University, State University of New York, Binghamton, NY, 1998. http://geography.binghamton.edu/pdf/Caryopsis.pdf.

Blumler, Mark A. "Seed Weight and Environment in Mediterranean-type Grasslands in California and Israel." Ph.D. diss., University of California, Berkeley, 1992.

Blumler, Mark A., and R. Byrne. "The Ecological Genetics of Domestication and the Origins of Agriculture." *Current Anthropology* 32 (1991): 23–53.

Bosworth, Michael L. "The Rise and Fall of 15th Century Chinese Seapower." http://www.cronab.demon.co.uk/china.htm.

Bower, B. "Domesticated Goats Show Unique Gene Mix." *Science News*, May 12, 2001. http://www.findarticles.com/cf_dls/m1200/19_159/75309403/p1/article.jhtml.

Bower, B. "Maize Domestication Grows Older in Mexico." *Science News* 159 (2001): 103. http://www.findarticles.com/cf_dls/m1200/7_159/71191553/p1/article.jhtml.

Bowen, H.V. *War and British Society 1688–1815.* Cambridge: Cambridge University Press, 1998.

Bradley, Daniel G. "Genetic Hoofprints: The DNA Trail Leading Back to the Origins of Today's Cattle Has Taken Some Surprising Turns along the Way." *Natural History*, Feb 2003. http://www.findarticles.com/cf_dls/m1134/1_112/97174195/p1/article.jhtml?term=.

Braude, Benjamin. "Ham and Noah: Race, Slavery and Exegesis in Islam, Judaism, and Christianity." *Annales: Histoire, Sciences Sociales* (2002), Department of History, Boston College. http://www.bc.edu/bc_org/research/rapl/word/braude01.doc.

Braudel, Fernand, and Sian Reynolds. *Civilization and Capitalism, 15th-18th Century.* London: Fontana, 1985.

Braudel, Fernand. *Der Alltag. Sozialgeschichte des 15.-18. Jahrhunderts.* München: Kindler, 1985.

Budiansky, Stephen. "Horse." Microsoft Encarta Online Encyclopedia. http://encarta.msn.com/encyclopedia_761562654_2/Horse.html#s6.

Cartwright, Frederick F. and Michael D. Biddiss. *Disease and History: The influence of disease in shaping the great events of history.* New York: Thomas Y. Crowell Company, 1972.

Celiac Disease & Gluten-Free Diet Support Page. http://www.celiac.com.

Childe, V. Gordon. *Social Evolution.* New York: Abelard Press, Inc., 1951.

Cipolla, Carlo M. *The Economic History of World Population.* Hassocks, Sussex: Harvester Press, 1978.

Claiborne, Robert. *Climate, Man, and History.* New York: W.W. Norton & Company, 1970.

Cohen, Joel E. *How Many People Can the Earth Support?* New York: W.W. Norton & Company, 1995.

Columbia Electronic Encyclopedia, The, Sixth Edition, "Japan." http://www.bar tleby.com/65/ja/Japan.html.

Columbia University, "Asia for Educators." http://afe.easia.columbia.edu/chinawh/web/images/sect5/0781_5_grand_canal_600x580.jpg.

Cook, Earl. *Man, Energy, Society.* San Francisco: W.H. Freeman, 1976.

Cotton Times-Understanding the Industrial Revolution. "The Unsung Thomas Highs." http://www.cottontimes.co.uk/highs.htm.

Cottrell, Fred. *Energy and Society—The Relation between Energy, Social Change, and Economic Development.* New York: McGraw-Hill Book Company, 1955.

Crosby, Alfred W. *The Columbian Exchange : Biological and Cultural Consequences of 1492.* Westport: Greenwood Press, 1972.

Davies, Glyn. *A History of Money: From Ancient Times to the Present Day.* Cardiff: University of Wales Press, 1994.

Davistown Museum, The. "Steel-and Toolmaking Strategies and Techniques before 1870." http://www.davistownmuseum.org/PDFs/Vol6_SteelToolMaking.pdf.

de Vries, J., and A. van der Woude. *The First Modern Economy: Success, Failure, and Perseverance of the Dutch Economy, 1500–1815.* Cambridge: Cambridge University Press, 1997.

de Wit, C.T. "Photosynthesis: Its Relation to Overpopulation." In *Harvesting the Sun,* ed. A. San Pietro, F. A. Greer and T.J. Army, 315–320. New York: Academic Press, 1967.

del Carmen Rodríguez Martínez, Ma., et al. "Oldest Writing in the New World." *Science* 313 (2006): 1610–1614.

DeVries, Kelly. *Medieval Military Technology.* Peterborough: Broadview Press, 1992.

DeZeeuw, J.W. "Peat and the Dutch Golden Age: The Historical Meaning of Energy Attainability." *AAG [Afdeling Agrarische Geschiedenis] Bijdragen* 21. Wageningen: Landbouwuniversiteit Wageningen, 1978, pp. 3–31.

Diamond, Jared. *Guns, Germs, and Steel: The Fates of Human Societies.* New York: W.W. Norton & Company, 1997.

Diamond, Jared. "Spacious Skies and Tilted Axes." *Natural History* (1994). http://www.mc.maricopa.edu/dept/d10/asb/anthro2003/lifeways/hg_ag/agspread.html.

Diamond, Jared. "The Worst Mistake in the History of the Human Race." *Discover,* May 1987.

Dunn, Adam. "Did the Chinese discover America?" *CNN,* January 13, 2003. http://www.cnn.com/2003/SHOWBIZ/books/01/13/1421/index.html.

Economist, The. "Cotton—A great yarn." December 18, 2003. http://www.economist.com/displaystory.cfm?story_id=2281685.

Economist, The. "Goodbye to the Mamluks, Millennium issue: The Turkish empire." December 23, 1999. http://www.economist.com/displaystory.cfm?story_id=347137.

Economist, The. "Letter from Samarkand: Soviet hangover." July 24, 1997. http://www.economist.com/displayStory.cfm?Story_id=152626.

Economist, The. "Survey: Food—Make it cheaper, and cheaper." December 11, 2003. http://www.economist.com/displayStory.cfm?Story_id=2261831.

Economist, The. "Tamerlane: Lame but never halting." August 26, 2004. http://www.economist.com/displaystory.cfm?story_id=3127398.

Encyclopædia Britannica, "China: History." http://www.britannica.com/EBchecked/topic/111803/China/214398/History#toc214398.

Encyclopædia Britannica. "Longbow." http://www.britannica.com/EBchecked/topic/347452/longbow.

Encyclopædia Britannica. "Military Technology." http://www.britannica.com/EBchecked/topic/382397/military-technology.

Encyclopædia Britannica. "Musket." http://www.britannica.com/EBchecked/topic/399353/musket.

Fagan, Brian. *The Long Summer: How Climate Changed Civilization.* New York: Basic Books, 2004.

Food and Agriculture Organization of the United Nations. *FAOSTAT.* Time-series and cross sectional data relating to food and agriculture for some 200 countries. http://faostat.fao.org/default.aspx.

Forde, C. Daryll. *Habitat, Economy and Society: A Geographical Introduction to Ethnology.* New York: E.P. Dutton & Co., Inc., 1946.

Gans, Paul J. "The Medieval Technology Pages." http://scholar.chem.nyu.edu/tekpages/Subjects.html.

Gepts, Paul. "Evolution of Crop Plants: The Origins of Agriculture and the Domestication of Plants." Department of Agronomy and Range Science, University of California, Davis. http://www.plantsciences.ucdavis.edu/gepts/pb143/pb143.htm.

Gepts, Paul. "Lecture 10: Where Did Agriculture Start? Centers of Origin and Diversity." http://agronomy.ucdavis.edu/gepts/pb143/lec10/pb143l10.htm.

Gies, Frances, and Joseph Gies. *Cathedral, Forge, and Waterwheel—Technology and Invention in the Middle Ages.* New York: HarperPerennial, 1995.

Gordon, Bryan C. "Preliminary Report on the Study of the Rise of Chinese Civilization Based on Paddy Rice Agriculture." Canadian Museum of Civilization, Hull, Quebec, February 1999. http://www.carleton.ca/~bgordon/Rice/research__resources.htm.

Güterbock, Hans G. "Hittites." Microsoft Encarta Online Encyclopedia. http://encarta.msn.com/encyclopedia_761563583/Hittites.html.

Hadingham, Evan. "Ancient Chinese Explorers." In *NOVA Online—Sultan's Lost Treasure.* http://www.pbs.org/wgbh/nova/sultan/explorers.html.

Hall, Bert. "Medieval Iron and Steel." Medieval Science and Technology: Original Essays. http://www.the-orb.net/encyclop/culture/scitech/iron_steel.html.

Haywood, John. *World Atlas of the Past, Vol. 1–4.* Oxford: Andromeda Oxford/Oxford University Press, 1999. Danish edition: John Haywood, *Historisk Verdensatlas* (Köln: Könemann, 2000).

Henderson, Donald A., and Bernard Moss. "Smallpox and Vaccinia." In *Vaccines.* Philadelphia: W.B. Saunders, 1999. http://www.ncbi.nlm.nih.gov/books/bv.fcgi?rid=vacc.chapter.3.

Hendrickson, J. "Energy Use in the U.S. Food System: A Summary of Existing Research and Analysis." Center for Integrated Agricultural Systems, College of Agricultural and Life Sciences, Wisconsin Institute for Sustainable Agriculture, University of Wisconsin-Madison, 1996. http://www.cias.wisc.edu/wp-content/uploads/2008/07/energyuse.pdf.

Henretta, James A. "American Revolution." Microsoft Encarta Online Encyclopedia. http://encarta.msn.com/encyclopedia_761569964/American_Revolution.html.

Herren, Ray V. *The Science of Animal Agriculture.* Albany: Delmar Publishers Inc., 1994.

Hevia, James Louis. *Cherishing Men from Afar: Qing Guest Ritual and the Macartney Embassy of 1793.* Durham: Duke University Press, 1995.

Higham, Charles, and Tracey L.-D. Lu, "The Origins and Dispersal of Rice Cultivation." *Antiquity* 72 (1998): 867–877.

Hillman, G.C., and M.S. Davies. "Measured Domestication Rates in Wild Wheats and Barley under Primitive Cultivation, and Their Archaeological Implications." *Journal of World Prehistory* 4 (1990): 157–222.

Hobhouse, H. *Seeds of Change: Five Plants that Transformed Mankind.* New York: Harper & Row, 1986.

Houghton, John. *Global Warming: The Complete Briefing.* Cambridge: Cambridge University Press, 1997.

Hsü, Immanuel C. Y. *The Rise of Modern China.* New York: Oxford University Press, 2000.

Illinois State Museum. "Ice Ages." http://museum.state.il.us/exhibits/ice_ages/.

International Institute for Applied Systems Analysis. "China's Population Growth." http://www.iiasa.ac.at/Research/LUC/ChinaFood/data/pop/pop_21_m.htm.

Jacobsen, T., and R. Adams, "Salt and Silt in Ancient Mesopotamian Agriculture." *Science* 126 (1958): 1251–1257.

Janick, Jules. "History of Horticulture." Department of Horticulture and Landscape Architecture, Purdue University, 2002. http://www.hort.purdue.edu/newcrop/history/default.html.

Janick, Jules. "History of Horticulture, Lecture 34: Horticulture, Politics, and World Affairs: Sugarcane and Plantation Agriculture." Department of Horticulture and Landscape Architecture, Purdue University, 2002. http://www.hort.purdue.edu/newcrop/history/lecture34/lec34.html.

Janick, Jules. "Sugar & the Slave Trade." http://www.hort.purdue.edu/newcrop/history/lecture34/r_34–1.html.

Jewish Encyclopedia. "Slave Trade." http://www.jewishencyclopedia.com/view.jsp?artid=849&letter=S.

Johnson, N.P., and J. Mueller, "Updating the Accounts: Global Mortality of the 1918–1920 "Spanish" Influenza Pandemic." *Bull Hist Med.* 76 (2002):105–15. http://www.ncbi.nlm.nih.gov/pubmed/11875246?dopt=Abstract.

Johnston, Mark. "The Sugar Trade in the West Indies and Brazil Between 1492 and 1700." James Ford Bell Library, University of Minnesota. http://www.bell.lib.umn.edu/Products/sugar.html#n6.

Kennedy, P.M. *The Rise and Fall of British Naval Mastery.* N.p.: Scribner, 1976.

Kerr, R.A. "Sea-floor Dust Shows Drought Felled Akkadian Empire." *Science* 279 (1998): 325–326.

King, John. "Bush Calls Saddam 'the guy who tried to kill my dad'." *CNN*, September 27, 2002. http://www.cnn.com/2002/ALLPOLITICS/09/27/bush.war. talk/.

Kirby, Richard Shelton, et al. *Engineering in History.* New York: Courier Dover, 1990.

Landes, David S. *The Wealth and Poverty of Nations: Why Some Are So Rich and Some So Poor.* New York: W.W. Norton & Company, 1998.

Lau, C.H., R D. Drinkwater, K. Yusoff, S.G. Tan, D.J. Hetzel, and J.S. Barker. "Genetic Diversity of Asian Water Buffalo (Bubalus bubalis): Mitochondrial DNA D-loop and Cytochrome b Sequence Variation." *Animal Genetics* 4 (1998): 253–64. http://www3.interscience.wiley.com/journal/119119543/ abstract?CRETRY=1&SRETRY=0.

Law, John. "On the Methods of Long Distance Control: Vessels, Navigation, and the Portuguese Route to India." In "Power, Action and Belief: A New Sociology of Knowledge?," ed. John Law, pp. 234–263, *Sociological Review Monograph* 32. Henley: Routledge, 1986. http://www.comp.lancs.ac.uk/sociology/papers/law-methods-of-long-distance-control.pdf.

LeConte, Joseph. *Manufacture of Saltpetre.* Columbia, S.C.: Charles P. Pelham, State Printer, 1862. http://docsouth.unc.edu/imls/lecontesalt/leconte.html.

Lester, Toby. "What Is the Koran?." *The Atlantic Monthly* 283 (1999). http://www. theatlantic.com/issues/99jan/koran2.htm.

Levathes, Louise E. *When China Ruled the Seas—The Treasure Fleet of the Dragon Throne, 1405–1433.* New York: Simon & Schuster, 1994.

Luikart, Gordon, et al. "Multiple Maternal Origins and Weak Phylogeographic Structure in Domestic Goats." *PNAS* (*Proceedings of the National Academy of Sciences of the United States of America*) 98 (2001): 5927–5932. www.pnas.orgycgiy doiy10.1073ypnas.091591198. http://www.pnas.org/cgi/reprint/98/10/5927. pdf.

MacDougall, J. D. *A Short History of Planet Earth: Mountains, Mammals, Fire, and Ice.* New York: John Wiley & Sons, 1996.

Maddison, Angus. "Growth Accounts, Technological Change, and the Role of Energy in Western Growth." In *Economia e Energia*, sec. XIII–XVIII, Istituto Internazionale di Storia Economica "F. Datini." Prato: Le Monnier, 2003. http://www. eco.rug.nl/~Maddison/ARTICLES/Role_of_energy.pdf.

Maddison, Angus. *The World Economy: Volume 1: A Millennial Perspective, Volume 2: Historical Statistics.* Paris: OECD Publishing, 2006.

Maidens, Melinda. "Horses and History." http://users.erols.com/mmaidens/index. html.

McIntyre, Charshee. "The Continuity of the International Slave Trade and Slave System." 1990. http://www.nbufront.org/html/FRONTalView/ArticlesPapers/ CMcIntyre_SlaveSystem1.html.

McNeill, John Robert, and William Hardy McNeill, *The Human Web: A Bird's-eye View of World History.* New York: W.W. Norton & Company, 2003.

McNeill, William Hardy. *Plagues and People.* Garden City: Anchor Press/Doubleday, 1976.

McNeill, William Hardy. *The Rise of the West: A History of the Human Community*. Chicago: University of Chicago Press, 1963.

Menzies, Gavin. *1421: the Year China Discovered the World*. London: Bantam Press, 2001.

Millard, A. R. "Mesopotamia." Microsoft Encarta Online Encyclopedia. http://encarta.msn.com/encyclopedia_761559228/Mesopotamia.html#p4.

Needham, Joseph, Wang Ling, Lu Gwei-djen, "Civil Engineering and Nautics." Part III of Volume 4 ("Physics and Physical Technology"). In *Science and Civilisation in China*, ed. Joseph Needham. Cambridge: Cambridge University Press, 1971.

Neiberg, Michael S. *Warfare in World History*. London & New York: Routledge, 2001.

Nesbitt, M. "Clues to Agricultural Origins in the Northern Fertile Crescent." *Diversity* 11 (1995):142–143.

North, Douglass C., and Robert Paul Thomas. *The Rise of the Western World: A New Economic History*. Cambridge: Cambridge University Press, 1973.

NOVA online, "Sultan's Lost Treasure." http://www.pbs.org/wgbh/nova/sultan/expl_01.html.

Ogino, Akifumi, Hideki Orito, Kazuhiro Shimada, and Hiroyuki Hirooka. "Evaluating Environmental Impacts of the Japanese Beef Cow-calf System by the Life Cycle Assessment Method." *Animal Science Journal* 78 (2007): 424. http://www.blackwell-synergy.com/doi/abs/10.1111/j.1740-0929.2007.00457.x.

Oklahoma State University, "Breeds of Livestock." http://www.ansi.okstate.edu/breeds/cattle/.

People for the Ethical Treatment of Animals (PETA). "Factsheets." http://peta.com/mc/facts.asp.

Phillips, William H. "Cotton Gin." EH.Net Encyclopedia, edited by Robert Whaples, February 11, 2004. http://eh.net/encyclopedia/article/phillips.cottongin.

Piperno, D.R. and K.V. Flannery. "The Earliest Archaeological Maize (Zea mays L.) from Highland Mexico: New accelerator mass spectrometry dates and their implications." *Proceedings of the National Academy of Sciences* 98 (2001): 2101.

Pocock, Tom. *Battle for Empire: The Very First World War, 1756–63*. London: Michael O'Mara Books, 1998.

Renes, J. "Urban Influences on Rural Areas: Peat-digging in the Western Part of the Dutch Province of North Brabant from the Thirteenth to the Eighteenth Century." In *The Medieval and Early-Modern Landscape of Europe under the Impact of the Commercial Economy*, ed. H. J. Nitz, 49–60. Gottingen: University of Gottingen Press, 1987.

Rosen, Edward, and Erna Hilfstein (Eds.). *Copernicus and His Successors*. London: Hambledon Press, 1995.

Roush, Wade. "Squash Seeds Yield New View of Early American Farming." *Science* 276 (1997): 894–895. http://www.sciencemag.org/cgi/content/abstract/276/5314/932.

Runyan, Timothy J. "Ship—Earliest Sailing Vessels." Microsoft Encarta Online Encyclopedia. http://encarta.msn.com/encyclopedia_761571524/Ship.html#s62.

Science Museum, The. "A Fashion Revolution." *Making the Modern World*. http://www.makingthemodernworld.org.uk/stories/manufacture_by_machine/01.ST.01/?scene=6.

Science Museum, The. "Textiles: From Domestic to Factory Production—The Industrial Revolution and the Textiles Industries." *Making the Modern World.* http://www.makingthemodernworld.org.uk/learning_modules/history/01. TU.01/.

Science News. "Early Agriculture Flowered in Mexico." June 16, 2001. http://www. findarticles.com/cf_dls/m1200/24_159/76653912/p1/article.jhtml.

Science Show, The. "Jared Diamond Lecture." Radio National, Broadcast October 28, 2000. http://www.abc.net.au/rn/science/ss/stories/s199676.htm.

Simmonds, Norman Willison, ed. *Evolution of Crop Plants.* London: Longman, 1976.

Smil, Vaclav. *Essays in World History—Energy in World History.* Boulder, CO: Westview Press, 1994.

Smith, Bruce D. "Documenting Plant Domestication: The Consilience of Biological and Archaeological Approaches." *Proceedings of the National Academy of Sciences* 98 (2001): 1324.

Smith, Bruce D. *Emergence of Agriculture.* New York: W. H. Freeman & Co, 1995.

Smith, Bruce D. "The Initial Domestication of Cucurbita pepo in the Americas 10,000 Years Ago." *Science* 276 (1997): 932–934. http://www.sciencemag.org/cgi/con tent/abstract/276/5314/932.

So, Kwan-wai. *Japanese Piracy in Ming China during the 16th Century.* East Lansing: Michigan State University Press, 1975.

South Tyrol Museum of Archaeology. "The Iceman." http://www.archaeologie museum.it/f01_ice_uk.html.

Spijker, Job. "Geochemical Patterns in the Soils of Zeeland." 2005. http://spkr.hack value.nl/research/thesisonline/node13.html.

Stark, Rodney. *The Rise of Christianity: A Sociologist Reconsiders History.* Princeton: Princeton University Press, 1996.

Steinfeld, H., et al. "Livestock's Long Shadow—Environmental Issues and Options." *LEAD-Livestock Environment and Development Initiative.* Coordinated by the UN FAO's Animal Production and Health Division. Rome: FAO, 2006. http://www. virtualcentre.org/en/library/key_pub/longshad/A0701E00.htm. http://www. virtualcentre.org/en/library/key_pub/longshad/A0701E00.pdf.

Tauber, H. "13C Evidence for Dietary Habits of Prehistoric Man in Denmark." *Nature* 292 (1981): 332–333.

Taylor, Robert E. *Scientific Farm Animal Production—An Introduction to Animal Science.* Upper Saddle River: Prentice Hall, 1995.

Thompson, Rebecca S. "Raising Emus and Ostriches." Special reference briefs 97–06, Alternative Farming Systems Information Center, Information Centers Branch, National Agricultural Library, Agricultural Research Service, U.S. Department of Agriculture, 1997. http://www.nal.usda.gov/afsic/AFSIC_pubs/srb9706. htm.

Unger, R. W. "Energy Sources for the Dutch Golden Age: Peat, Wind, and Coal." in *Research in Economic History*, ed. Paul Uselding, 221–53. Greenwich, Connecticut & London: JAI Press, 1984.

Van Cauwenberghe, Eddy H. G., ed. *Precious Metals, Coinage and the Changes of Monetary Structures in Latin-America, Europe and Asia (Late Middle Ages-Early Modern Times).* Leuven: Leuven University Press, 1989.

van Dam, Petra J. E. M. "Sinking Peat Bogs: Environmental Change in Holland, 1350–1550." *Environmental History*, January 2001. http://findarticles.com/p/articles/mi_qa3854/is_200101/ai_n8932821/pg_1?tag=artBody,col1.

Vandervoodt, Claire. *The Dasana Ostrich Guide: A Practical Handbook*. Devonport: Nova Creative Publishing, 1995.

Vegetarian Society. "Information Sheets." http://www.vegsoc.org/info.

Vogel, J.C. and N.J. van der Merwe. "Isotopic Evidence for Early Maize Cultivation in New York State." *American Antiquity* 42 (1977): 238–242.

Wallerstein, Immanuel. *The Modern World-system II. Mercantilism and the Consolidation of the European World-Economy, 1600–1750* (Studies in Social Discontinuity). New York: Academic Press, 1980.

Watson, Andrew M. *Agricultural Innovation in the Early Islamic World: The Diffusion of Crops and Farming Techniques, 700–1100*. Cambridge: Cambridge University Press, 1983.

Watts, Sheldon. *Epidemics and History: Disease, Power, and Imperialism*. New Haven: Yale University Press, 1999.

Weir, Fraser. "A Centennial History of Philippine Independence: Austronesian Expansion—Taiwan 4,000 BC." University of Alberta. http://www.ualberta.ca/~vmitchel/rev3.html.

White, Lynn, Jr. *Medieval Technology and Social Change*. New York: Oxford University Press, 1980.

Willcox, G. "Archeobotanists Sort Out Origins of Agriculture from Early Neolithic sites in the Eastern Mediterranean." *Diversity* 11 (1995):141–142.

Wilson, Samuel M. "The Emperor's Giraffe." *Natural History* 101 (1992). http://muweb.millersville.edu/~columbus/data/art/WILSON09.ART.

World Health Organization. *Cardiovascular diseases*. Geneva: WHO, 2007. http://www.who.int/mediacentre/factsheets/fs317/en/index.html.

World Health Organization. *The Global Eradication of Smallpox—Final Report of the Global Commission for the Certification of Smallpox Eradication*. Geneva: WHO, 1979. http://whqlibdoc.who.int/publications/a41438.pdf.

Zohary, D., and M. Hopf. *Domestication of Plants in the Old World*. Oxford: Clarendon, 1988.

Zohary, D., and M. Hopf. *Domestication of Plants in the Old World: The Origin and Spread of Cultivated Plants in West Asia, Europe, and the Nile Valley*. Oxford: Oxford University Press, 1993.

PART III

Coal Age

I cannot claim to be especially fond of coal. I remember all too well how much work, dirt, and inconvenience this fuel involved when I lived in a 19th-century house without central heating as a student. The coal came in large, heavy sacks that had to be carried into the basement for storage. From there small portions of coal had to be picked up at least twice a day, once in the morning and once in the evening, to feed the stoves, one in a corner of every major room. And when the coal had burned down, the ash had to be removed from the ovens and be allowed to cool before it could be carried outside and disposed in the garbage container.

Quite obviously coal offers none of the advantages of the liquid fuel systems that Oil Age people are acquainted with: it neither flows continuously from storage to burner, nor does it combust without leaving ashes. And yet coal changed the world. The most immediate advantage of this fuel was its availability. Timber takes a long time to regrow, while coal could be dug up from the ground in what often seemed unlimited quantities. For the first time in human history energy supplies seemed detached from the growth limitations restricting agricultural energy and biofuel consumption. What is more, coal is a very reliable fuel. While winds are moody, and water streams may run dry in the summer or freeze over in the winter, coal will serve as a fuel at any chosen time in the future. (Besides, water and wind power are also principally limited by the amount of solar radiation that reaches the planet's surface to lift water into the sky or heat the atmosphere asymmetrically so winds will blow to balance out regional pressure differences.) True, the agricultural energy system actually also provided work *on demand*. But even the best of storable grains to fuel people or beasts-of-burden had to be handled with care and be protected from pests and humidity. Coal, on the other hand, has a virtually unlimited shelf-life, even if stored under harsh environmental conditions. (Wood may be stored for long periods as well, but it has a substantially lower energy density, soaks up water and may be consumed by certain insects.)

The major impact of coal energy on the world arose with the development of the steam engine, an entirely new prime mover that began to mature towards the end of the 18th century. Steam engines translated coal energy into mechanical work to boost the output of factories and to revolutionize mobility on land (locomotives) and on the water (steam boats). Steam engines simply kept on running, day and night, summer and winter, during rain and in the desert, during storms and in dead calm, just as long as enough coal was supplied to them. The only domains that remained entirely dependent on agricultural energy during the Coal Age were individual land travel (horseback riding), short-distance transport of freight (horse-drawn carriages), and field work (slaves, workhorses, oxen).

As we would expect, the extreme boost in the amount of energy commanded by people had effects similar to those observed in the aftermath of the emergence of agriculture. The use of steam engines in production freed increasingly more hands and pushed more people towards jobs that required schooling and advanced engineering skill. Knowledge accumulated faster than ever before, and innovations followed suit. What is more, progress in coal-fueled transportation promoted trade, economies of scale (mass production), and thus regional division of labor and professional specialization. The overall wealth of Coal Age societies was therefore bound to increase.

To be sure, the transition from Agricultural Age to Coal Age was no smoother than the transition from Foraging Age to Agricultural Age. It involved radical social changes that advantaged some and disadvantaged others. As the energy captured on the fields was no longer the prime energy source of societies, power shifted from landowner to industrialist, and scores of farm and cottage workers had to relocate from the countryside to rapidly growing cities. Even after the initial wave of urbanization had wound down, people kept on being pushed towards new jobs, as their old ones kept being eliminated and taken over by increasingly more complex machinery. However, the elevated level of energy command, with all its feedbacks, eventually resulted in a higher quality of life, better education, and longer life expectancy for practically all members of Coal Age societies. Thus the global human population soared to over one billion.

Another effect that paralleled what had been experienced in the previous Energy Era was that increased amounts of energy became available only to some societies, and that these societies used their new energy advantage to dominate those who had not yet access to the new energy technology. Especially critical were coal-fired steam boats that allowed for swift mobility on the rivers, the hitherto secure inland lifelines of agricultural societies.

Compared to the Agricultural Age, the Coal Age was very short. It may be argued that it began in Britain in the 1770s or 1780s, when relatively efficient Boulton & Watt steam engines spread into commerce. But it really took off in 1800, when the initial Watt patent finally expired, and the steam engine was set free to undergo a rapid further evolution that made it fit for a host of applications. The Coal Age then lasted for about 11 or 12 decades. The end of this Energy Era may be given as 1912, the year the British Admiralty began to convert the Royal Navy from coal to oil, or sometime later in this second decade of the 20th century, during which the share of coal in the world's primary commercial energy mix began to sink.[1]

NOTE

1. Coal accounted for 93 percent of the world's commercial energy supply in 1900 as well as in 1910, but had passed its peak by 1920 (88%), and kept on sinking (to just under 80% in 1930) to continue its decline in favor of oil. United Nations, World Energy Requirements in 1975 and 2000, Proceedings to the International Conference on the Peaceful Uses of Atomic Energy, Geneva, 1955, Volume 1 (New York: UNO, 1956), quoted in Vaclav Smil, *Transforming the Twentieth Century: Technical Innovations and Their Consequences* (New York: Oxford University Press, 2006), 35. The figure for 1900 (93%) is mentioned in Stephen Hughes, "The International Collieries Study," ICOMOS (International Council on Monuments and Sites) and TICCIH (The International Committee for the Conservation of the Industrial Heritage), International Council on Monuments and Sites, Paris, France, page 6, http://www.international.icomos.org/centre_documentation/collieries.pdf.

CHAPTER 15

WHAT IS COAL?

Coal is quite a special material. It stores lots of energy, because it represents solar energy that has been captured during tens of millions of years of plant growth and has then been buried for hundreds of millions of years.[2] More specifically, coal was formed underground from the leftovers of plant matter that grew in vast tropical swamps about 330 million years ago. The lush vegetation of these swamps consisted of huge spore-bearing trees as well as ferns, horsetails, and club mosses, whose remains settled to the swamp bottom to be buried under increasingly more layers of sediment. The organic matter was thus cut off from air supply, which interrupted the usual decaying process and preserved much of the chemical energy stored in the biomass. The process that consequently turned the plant material into coal is very slow and involved bacterial decomposition at specific temperature and pressure conditions. Starting from fresh plant material, aging under such conditions results in products gradually denser in carbon and energy. It first yields peat, then lignite (brownish coal that contains recognizable plant matter), then subbituminous coal (which is dull black, showing very little plant matter), then bituminous coal (which is jet black, very dense, and brittle), and finally anthracite (or "hard coal," which is shiny black, contains little volatile matter, and burns cleaner than bituminous coal). The peat bogs found today are 9,000 years old, for instance, while lignite is about 135 million years old.

Because coal now found underground started to form around 300 million years ago out of plant material growing in the above mentioned swamps, the geological time period from approximately 360 million years ago to 290 million years ago has been named Carboniferous (carbon-containing) Period.[3]

Coal Miners   This picture was taken towards the end of the Coal Age, in 1905, in a mine at Starkville, Colorado. It was originally tilted "Digging coal half mile underground." (Library of Congress image LC-USZ62-72863, edited.)

It is not entirely clear what was so special about this period as to create abundant biomass and preserve it in just the right way for us to inherit such large amounts of energy. One speculation is that much wood ended up being buried in the swamps because bark-bearing trees emerged in the absence of (yet-to-emerge) bacteria (and animals) that can effectively break-down the lignin component of wood. However, climate and sea level changes associated with alternating glaciation at the southern pole during this period played an important role as well. Because of worldwide sea level fluctuations, coastal regions (such as those that were to become eastern North America) experienced alternating shifts between terrestrial and marine environments. After plant material had been turned into peat in swamp forests, the sea flooded these swamps, and marine sediments covered the peat, which in turn continued the coalification process.

Rich coal deposits occur throughout northern Europe, Asia, and Midwestern and Eastern North America. These regions were distributed entirely differently over the globe at the time of the great coal swamps. (As a result of tectonic plate movements the position of continents is not static. Europe and America, for instance, are now drifting apart at a speed of about 2 cm per year. This is roughly equivalent to the rate at which a fingernail grows.) In the Carboniferous Period a supercontinent called Gondwana existed. It consisted of present-day Africa and South America and occupied much of the Southern Hemisphere. Laurussia, a continent consisting of present-day Europe and North America, bumped straight into Gondwana, which created the Appalachian mountain belt of eastern North America and the Hercynian Mountains of Britain. A further collision, between eastern Europe and Siberia, created the Ural Mountains, while North China and South China for the time being remained separate continental entities.

Much of these continents and continental fragments must have enjoyed climates that allowed for extensive plant growth. Otherwise, the world's total coal stock would not have ended up being as huge as it (still) is. At the beginning of the 21st century, after more than two centuries of extensive coal use, known reserves are still large enough to last for another two centuries at current consumption rates. (Some mines actually have proven reserves to operate for almost a thousand years into the future if the rate of production was to remain constant.) Nevertheless coal is a non-renewable, and thus limited, energy source. As the process that yields coal takes hundreds of millions of years, coal deposits cannot be replenished within time frames meaningful to the current human civilization.

Coal is not a uniform substance. Its composition varies from deposit to deposit, depending on the type of original plant matter, the extent to which the plant matter decomposed, and other factors. (There are over 1,200 distinguishable types of coal.) However, for all practical purposes coal is a black or brownish-black, solid, stone-like, but relatively soft material that is combustible. It contains chemical energy in a more concentrated, compact form than plant material does, and hence releases more (heat) energy when burned. Burning a kilogram of coal yields some 7,000 kcal in heat energy, compared to 3,000 kcal released when a kilogram of dry wood is burned.[4] The bulk of global coal extraction consists of bituminous (black) coals. These contain slightly less energy per weight than anthracites, and slightly more energy than lignites. Peats contain less than a third of the chemical energy contained in the same weight of bituminous coal.

NOTES

2. Find an estimate of how much photosynthetic energy was captured to yield the fossil fuels we use today in the following paper: J. S. Dukes, "Burning Buried Sunshine: Human Consumption of Ancient Solar Energy," *Climatic Change* 61 (2003): 31–44, http://globalecology.stanford.edu/DGE/Dukes/Dukes_ClimChange1.pdf.

3. University of California Museum for Paleontology, "The Carboniferous: 354 to 290 Million Years Ago," http://www.ucmp.berkeley.edu/carboniferous/carboniferous. html; "The Carboniferous—354 to 290 Million Years Ago," The Paleontology Portal, http://www.paleoportal.org/time_space/period.php?period_id=12.

4. To be sure, wood, unlike coal, is a renewable fuel. One ton of coal equals the sustainable timber yield from about 1 to 1.5 acres of prime mid-latitude forest. Kenneth L. Pomeranz and R. Bin Wong, "China and Europe: 1780–1937, What Happened?," China and Europe: 1500–2000 and Beyond: What is "Modern"?, http://afe.easia.columbia. edu/chinawh/web/s6/s6_2.html.

CHAPTER 16

THE EMERGENCE
OF COAL TECHNOLOGY

Contrary to agriculture, coal technology (or "coal culture") emerged but one single time, in late 18th-century Britain. That's not especially surprising, as the world had long been open by this stage. All continents were connected and none isolated, which meant that no region had (a chance) to recreate the experience. Instead, coal technology spread rapidly around the globe and was adopted wherever technological standards and available resources allowed this to happen.

To be sure, the emergence of coal culture, like that of agriculture, was not a single strike of genius. It was a slow process, the sum of many small innovative steps in a society that had, over many generations, accumulated a lot of knowledge and engineering expertise. Most notably coal was not initially used as a source of concentrated energy to be turned into mechanical power. Rather, it served as a fuel and feedstock for iron smelting. (Similarly, the pioneers of herding did not have milking, drafting, or animal-back riding in mind when they first domesticated their prey.) And even after the first steam engines had been constructed to turn coal energy into work, these new prime movers operated exclusively at mines for nearly a century.

The chain of events that led to the emergence of coal technology can be traced very far into the past. Iron technology had spread to Britain from continental Europe, and Europe as a whole owed its technological standards to the fact that it was part of the large Eurasian knowledge and technology pool. Hence, the emergence of coal culture in western Europe is ultimately connected to the early emergence of agriculture in the Fertile Crescent and to the large population size and density on the Eurasian landmass. Not a

single plant or animal has been domesticated in the British Isles, and much
of Europe's medieval technological progress was the result of the influx of
know-how that had arrived all the way from China.

The accumulation of mining, metal-making, and machining technology in
medieval Europe was a somewhat closer starting point of the events that led
to the emergence of coal technology. Waterpowered blast furnaces, the prerequisite for cast iron production, were developed in continental Europe from
the 13th century. Horse-powered and waterpowered winding apparatus was
pioneered in nonferrous mines of German-speaking central Europe. (Nonferrous refers to minerals other than iron ore.) Many of Britain's early mining experts actually came straight from Germany. Already King Edward I
of England (1239–1307) had invited some to live in Wales, and from the
16th century German-speaking miners were brought in in large numbers
to many of England's nonferrous mines. The 1556 publication "De Re Metallica" by German scholar Georg Bauer, who is better known by his Lati-
nized name Georgius Agricola, played a significant role in spreading mining
know-how. And by the 17th century waterwheel winding had reached the
Great Northern Coalfield of Britain.[5]

COAL FOR IRON AND STEEL PRODUCTION

The initial impetus for large-scale coal mining came from the English iron
industry, which was troubled by charcoal supply problems. Charcoal was
indispensable in metallurgy for its chemical purity, but it was made from
timber, a commodity that was increasingly scarce in the island nation. Charcoal production consumed huge amounts of wood, and iron production in
turn consumed large amounts of charcoal. It took some 5 kilograms (kg)
of air-dried wood to produce 1 kg of charcoal, and between 8 kg and 20 kg
of charcoal to produce 1 kg of iron. Hence, the production of 1 kg of iron
consumed at least 40 kg of dried wood.[6] In addition, heating, cooking, house
construction, ship building, barrel production, glassmaking, and various
other industries competed for wood or charcoal. As early as the 16th century
the radical deforestation around furnace sites forced the king of England to
close down many iron mills, but the situation worsened dramatically in the
Super-Agricultural Era with the expansion of the British iron and shipbuilding industry. England had to import large amounts of iron from Sweden,
Germany and Russia, and many English furnaces were operated only every
second or third year.

Prohibitions on cutting wood prompted people to look for substitute fuels.
Coal was an obvious choice as it happened to be abundant and easily accessible in the region. The Romans had already used coal for heating purposes
in Britannia, and so did local people from the ninth century c.e. From 1228,
in order to relieve a scarcity in fuel wood, notably for lime production, coal

was shipped to the growing city of London from Northumberland.[7] These were lumps of sea coal, broken from submarine outcroppings and washed ashore, where they could be easily collected. (To be sure, the term sea coal, or *carbonem marus*, probably refers to the mode of transportation of coal, by sea.) Later in the 13th century, monks began to mine outcroppings in the north of England, but real coal mining had actually begun in Germany and Belgium a century earlier. However, these early coal mining ventures remained small: there simply was not enough demand for coal to justify the digging efforts.

During the 16th and 17th centuries, when the price of wood increased, England managed to accelerate the trend that had started in the 13th century, and switched quite a few industries to coal. Coal was used to fire bricks, tiles, and earthen ware; to heat forges; to make starch and soap; and to extract salt. From about 1610 coal was also used in glassmaking. The impurities contained in coal were actually a problem for this application, but the issue

Coal Mining in the 1880s   Coal mining operations became gradually more complex. Initially people collected sea coal on beaches, then they mined coal from open pits, wherever the coal strata extended to, or close to, the surface. With growing demand for coal, surface extraction methods evolved into to deep shaft mining. The depicted illustration is contained in Pierre Foncin's 1888 *La deuxième année de géographie* (Librairie Classique Armand Colin et Cie., Paris). It shows a pit pony supporting the miners.

was resolved by the introduction of reverberating (heat-reflecting) furnaces. These heated the glassmaker's raw material inside closed vessels to avoid direct contact with the fuel and its combustion products. Unfortunately, this solution did not work in iron production: the raw material had to be in direct contact with the fuel in order for the iron to absorb some carbon, which is necessary to turn iron into the hard carbon-iron alloy that was useable for tools. However, during such contact, the impurities contained in coal were transferred to the metal as well, which resulted in brittle iron products. Hence, iron smelters kept using charcoal.

Coke for Beer Brewing and Cast Iron Production

Interestingly, the innovation that finally made coal applicable to iron smelting was achieved by the beer brewing industry. Traditionally, the brewers had to use charcoal rather than coal because coal released gasses that resulted in a smelly, unappetizing beer. (As in iron production, the sulfur content of coal was the main problem.) But from about 1640 brewers began to use coke in malt drying. Coke is about as chemically pure as charcoal, but it is produced from pit coal rather than timber. (The processes that yield charcoal and coke are actually very similar. The main difference is the feedstock: timber versus coal.) To make coke, coal is simply heated in an airtight oven to some 1000 C, which causes some coal components to evaporate, while others melt and can be removed. The resulting product is a solid but highly porous, nearly pure, carbon product.

It took surprisingly long, about seven decades, before coke was applied in iron smelting as well. Abraham Darby of Coalbrookdale, Shropshire, took this step in 1709, when nearly all of England's accessible natural forest growth had disappeared. (Note that this is when the Little Ice Age was at its worst in England: between 1650 and 1715 the Thames froze over regularly.) Darby had been an apprentice in a Birmingham works that made brass mills for grinding malt. Hence he was familiar with coke being used to fuel the malting ovens of the time. And when Darby eventually started his own business to produce cast iron, he began using coke (which he produced himself) instead of charcoal, to fuel his blast furnace.

Given England's scarcity in timber, and the strategic importance of cast iron to produce cannons and cannon balls, this was a tremendously important break-through. However, Darby apparently did not share his innovation. He excelled in the technique of mass casting and manufactured good-quality thin iron castings that competed successfully with more expensive traditional brass pots and other hollow ware. But the general iron production in the Birmingham area stagnated. Only after 1750, when England entered the "Very First World War," were blast furnaces increasingly fed with coke.

The use of coke then allowed for major changes in the iron industry. Waterpower (to drive the bellows of blast furnaces) and iron ore were still essential, but rich forest resources were no longer a prerequisite if coal was available. What is more, the use of coke removed the size limitations imposed on blast furnaces by the friability of charcoal.[8] When charcoal-fed blast furnace stacks exceed a height of about eight meters, the charge of charcoal and iron ore would simply collapse under its own weight and the charcoal be crushed into dust. Coke, in sharp contrast, allowed for much taller stacks and larger internal volumes. Hence, furnace size gradually increased to a height of 24 meters and a width of 6 meters by 1840.

Wrought Iron and Steel

Once cast iron (with a high carbon content) had become cheap and abundant, the search was on for methods to turn it into malleable wrought iron (with a low carbon content), and strong steel (with a moderate carbon content). The first economic process to convert cast iron into wrought iron was achieved by Englishman Henry Cort in 1784. He used a reverberatory furnace in which he placed solid cast iron (pig iron) on a hearth between a coal fire and a chimney. The cast iron melted without getting in direct contact with the coal, and was stirred to bring air into the mix, which turned some of its carbon content into volatile carbon monoxide gas. This *puddling process* yielded large quantities of wrought iron that were in turn rolled into bars or sheet metal, or traditionally worked into straps, hinges, horseshoes, and such. (The Eiffel Tower of Paris is made of wrought iron, or puddle iron, as well.)

Steel, which is harder than wrought iron and less brittle than cast iron, remained scarce. It depended on a modest carbon content that was difficult to achieve. A method to produce large amounts of steel was finally introduced in 1856 by Henry Bessemer, a Huguenot born in England. Bessemer's procedure was similar to Cort's, but removed less carbon. Bessemer placed cast iron inside an egg-shaped *converter*, in which cast iron was transferred into liquid steel by blowing a controlled amount of compressed air into the melt. (The process was actually controlled through the melt's temperature, not the exact airflow, which is possible because the melting point of the various forms of iron sensitively varies with the exact carbon content.)

The Bessemer process is regarded the starting point of inexpensive, large-scale steel making. But it actually did not work very well and frequently yielded a very brittle product, at least when phosphorus-rich English iron ore was used. However, Alfred Krupp pioneered and perfected the Bessemer method in Germany in 1862. Another German, Karl Wilhelm Siemens, improved the Bessemer process in England. Siemens pre-heated the air directed into the cast iron and hence achieved high enough temperatures to burn away

Puddling Furnace   It was a relatively simple construction, but it made a major difference: the use of a reverberatory puddle furnace allowed for the conversion of cast iron (pig iron) into wrought iron on a relatively large scale. The fireplace is shown on the left, separated from the actual hearth (A) by a bridge (D) to avoid direct contact between the fuel and the iron. The chimney (C) was equipped with a damper at the summit to regulate the draught. (Image from *The Household Cyclopedia of General Information* printed in 1881.)

certain impurities. Furthermore, he used a shallow open-hearth furnace, in which he added iron scrap and iron ore to the molten cast iron, and he heated the mix with overhead gas burners. In France, Pierre and Emile Martin introduced a nearly identical process at the same time, and the method thus became known as the Siemens-Martin process. It is also referred to as open-hearth process, and largely replaced the Bessemer process towards the end of the 19th century because it allowed precise control of temperatures and resulted in better quality steel.

The application of coal in iron production had major consequences. As all forms of iron were now being mass-produced, tools (and weapons) became available in very large quantities, ship hulls were made in unprecedented size, buildings were getting a lot taller and bridges a lot longer. Brooklyn

Eiffel Tower of Paris   Made of wrought iron for the Centennial Exposition of 1889, the Eiffel Tower is a manifestation of coal energy. It remained the tallest building in the world until 1930. The photo shown here has been taken from a balloon by Alphonse Liébert (in 1889) when the tower was brand new. (Library of Congress image LC-USZ62-94571.)

Bridge, which links Brooklyn to Manhattan island over the East River, was the first bridge to use steel for cable wire. Completed in 1869, its main span of 486 meters (1,595 ft) was then the longest in the world, and its towers were then the highest structures of New York. The 300 meter (984 ft) Eiffel Tower of Paris was built of wrought iron for the Centennial Exposition of 1889 and remained the world's tallest building until the completion of New York's Chrysler Building in 1930.

Increase in Iron and Coal Production

England, where coke-fueled iron smelting was pioneered, had the lead in the world's tremendous expansion in iron production. English annual iron consumption more than doubled from 60,000 tons in 1760 to 125,000 tons in 1800. A short 15 years later, England produced as much as one million tons of iron per year, more than the rest of Europe combined. Following the introduction of the Bessemer and Siemens-Martin processes, global steel production soared as well, from 22,000 tons per year in 1867, to 500,000 tons in

1870, to one million tons in 1880, to 28 million tons in 1900, and to as much as 70 million tons per year in 1913.[9]

Obviously, this extreme growth in iron production depended on a parallel growth in coal production. Almost all of England's major coalfields had already been opened between 1540 and 1640, but a radical increase in mining output had to wait for the Super-Agricultural Era and the introduction of coke in iron smelting. After the outcropping seams had been depleted, increasingly deeper pits were being dug. These were rarely deeper than 50 meters in the late 17th century, while the deepest shafts surpassed 100 meters right after 1700, and 200 meters by 1765. By 1830 the deepest coal mining shafts surpassed 300 meters.[10] Hence it was possible for British annual coal output to grow from two million tons around 1650, to three million tons by 1715, six million tons by 1760, and 15 million tons by 1815. (British annual coal production was then over five times the combined production of continental Europe.) And yet this was only the beginning: By 1913 coal production had expanded to as much as 289 million tons annually in Britain alone, and to 1,300 million tons annually worldwide. (By this stage British coal production had peaked.[11]) To be sure, this enormous increase in mining operations and fuel consumption was by no means the result of iron production alone: a machine had been invented that turned the chemical energy contained in coal into practical work.

COAL AND THE STEAM ENGINE

During the 1760s, when British annual coal extraction soared to over 6 million tons and the deepest mining shafts surpassed 200 meters, the use of waterpower reached new heights as well. The iron smelting industry now connected the two energy sources, as it relied on coal for its coke supply and on waterpower to drive the bellows that allowed blast furnaces to reach high enough temperatures. In addition, more and more manufactories were set up along water streams. Birmingham entrepreneur Matthew Boulton in 1762 purchased Soho, an area of undeveloped land, to develop it into a major manufacturing center that delivered buttons, buckles, watch-chains, trinkets, and the like. Within two years he employed over six hundred workers, who were assisted by the power obtained from two large waterwheels that drove a variety of lathes and polishing and grinding equipment. Richard Arkwright set up the first waterpowered cotton spinning operation in Cromford in 1771, and by 1820 some 110,000 workers labored in Britain's (still mainly waterpowered) cotton-spinning mills.

But despite the obvious importance of waterpower in Britain's early industrialization phase, various sectors of industry did not benefit from it. They could not be relocated to water streams and continued to rely on muscle power, with oxen, horses or people doing all the work. Mines certainly could not be relocated. They were getting deeper and deeper, and it was increasingly more difficult to pump flood water out of them. As muscle power could hardly keep up with the task, those who were extracting coal began wondering if it was possible to use the energy source they dug out of the ground to assist their efforts directly. In short, the search was on for a prime mover that would translate coal energy into mechanical work.

Early Steam Engines

Steam engines did exactly that. They had their origins in the construction of vacuum pumps. German scientist Otto von Guericke presented the world's first vacuum pump to the Imperial Diet at Regensburg in 1654. Famously, he also evacuated a metal sphere consisting of two well-fitting hemispheres, about a foot in diameter, and showed that 16 horses, eight on each side, could not pull them apart.[12] (The hemispheres were held together only by the atmospheric pressure that surrounds us all.) Later on, Robert Hooke constructed a vacuum pump for Robert Boyle in England, which caught the interest of Dutch scientist Christiaan Huygens during a visit to London in 1661. Hence, he had a vacuum pump built for him as well. Frenchman Denis Papin, who had earlier been Huygens' assistant, built a tiny steam engine in 1690 that worked remarkably well. Papin's model was a simple cylindrical metal tube (closed at one end) with a piston inside. Inside the tube, under the piston, was a small quantity of water. When the tube was heated, the water was transformed into steam, and hence raised the piston to the edge of the cylinder. And when a stream of cold water was in turn sprayed onto the outside of the cylinder for cooling, the water steam condensed and created a vacuum inside the tube. Hence, outside air pressure could easily force the piston back down.

Papin himself never really put his engine in practice, but his work was well documented. It was therefore just a small innovative step to use a coal fire to evaporate enough water to run a full-scale steam engine. Englishman Thomas Savery patented a steam-driven pump in 1698, but his pump operated without a piston. Instead, it created a vacuum by condensing water steam inside a suction pipe that directly reached into the flood water to be removed from mines. Savery-type pumps were applied in the Cornish mining district, but it was obvious that they had serious limitations: the pumping cycle was very slow, and the maximum suction height to pump water was only about 25 to 30 feet. The pump thus had to be constructed within the mine shaft itself.

Thomas Savery worked with Thomas Newcomen, who eventually developed a true steam engine. Newcomen's engine was directly inspired by, and very similar to, Papin's engine. It was little more than a large metal cylinder, filled with a bit of water and heated by a coal fire. The moving piston was

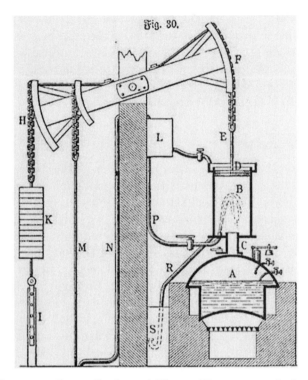

Thomas Newcomen Steam Engine   A Thomas Newcomen as depicted in *Meyers Encyclopedia* (Meyers Konversations-Lexikon, 4. Auflage) 1885–1890. Note that the steam is condensed directly in the cylinder (B) by injecting cold water into it. Thus, much of the energy employed went into reheating the cylinder in each cycle, which made the Newcomen design fuel-inefficient.

connected to a large, centrally pivoted wooden beam, whose other end was used to lift water from the mine shaft. Unfortunately, this engine was very inefficient. One major problem of Newcomen's early machines was that it took too long to condense the steam created inside the cylinder. To solve this problem Newcomen constructed a cooling jacket out of lead, which surrounded the cylinder and could be filled with cold water. This solution did not work all that well but, as the story goes, a defect in the inner cylinder wall provided Newcomen with the ideal solution to condense the steam rapidly inside the cylinder: As soon as water from the surrounding jacket leaked into the cylinder, the piston descended with considerable force.

Right after this incident, about 1712, Newcomen introduced a new, better-functioning steam engine design that purposely released a water jet into the cylinder to condense the steam as soon as the piston was at the top of its stroke. But even this improvement left the engine quite inefficient. Its power output was small, its operation slow, and its coal requirements enormous: the cylinder had to be heated, and thereafter cooled, on each and every stroke. What is more, cylinders of the time could not be produced in iron, nor could they be machined. They were cast in brass, and the inside wall was smoothed by hand with sand or a similar abrasive. Cylinders were therefore inaccurate and expensive to make. And to effect a seal, Newcomen had to fit a leather flap to the top of the piston and to cover the piston with water.

Newcomen died in 1729, but from about 1750 his type of steam engine became quite widespread in the mining districts. This was possible because of advances in engineering skills and material expertise. The Darby factory at Coalbrookdale, initially founded by the coke pioneer who achieved thin-wall iron casting, cast over 100 Newcomen cylinders by 1758. John Smeaton achieved superior steam engine efficiency in 1772 by measuring the machine parts better and by improving the sealing gaskets. And a major efficiency leap forward was finally accomplished with a radically new steam engine design presented in 1776.

Boulton & Watt

Scottish engineer James Watt (1736–1819) had learned to make mathematical instruments in London before opening a shop at the College of Glasgow at the age of 21. In 1765 he was asked to repair a Newcomen engine and in turn began improving steam engines. In 1769 Watt patented a relatively simple idea that was going to make a major difference. Watt understood that the inherent inefficiency of Newcomen engines lay in the fact that its cylinder had to be alternately heated and cooled, which takes time and consumes a lot of energy. Thus, he suggested to keep the cylinder permanently hot, and to connect it with a pipe to a separate condenser that was kept permanently cold. Allowing the hot steam to escape through the pipe instead of heating and cooling the cylinder on every stroke promised to save huge amounts of fuel, about 75 percent in comparison to the Newcomen design. This innovation was eventually going to open the way for steam engines to emerge into a generally useful prime mover that could work outside the mining districts as well. However, to get from the patented idea to a functioning model took a lot of development work.

Watt was short in cash and went into partnership with entrepreneur and inventor John Roebuck, who had been studying medicine in the Netherlands and later pioneered mass-production of sulfuric acid as well as a method to turn cast iron into malleable iron. Roebuck received two-thirds of Watt's patent rights in exchange for financial help, but he passed them on to Matthew Boulton in 1774 when he ran into financial difficulties himself. Boulton then persuaded Watt to relocate to his Soho manufacturing complex at

Birmingham to develop steam engines. Boulton provided vast financing and the two men established a partnership that lasted for 25 years, until 1800. This time frame was no coincidence. It was connected to Boulton's successful lobbying to achieve an Act of Parliament, the Steam Engine Act of 1775, which extended Watt's patent of 1769 for another 25 years.

The first Boulton & Watt engine was completed in 1776. The initial target market was the replacement of Newcomen engines at mines.[13] Some 75 Newcomen engines were operating in the mines of Cornwall, but within a short four years they were all replaced by more powerful and coal-efficient Watt engines. One Boulton & Watt engine was built in 1779 for the Birmingham Canal Navigation Company, which had taken the canal between the coal mines of the Black Country and the thriving city of Birmingham over the Smethwick Summit via a series of six locks. It struggled with water problems until Boulton & Watt installed an engine that pumped water back to the summit. (It lifted 170 gallons of water on each stroke and completed ten strokes per minute. The piston made a stroke of 8 feet, and the powering cylinder had a diameter of 32 inches.) To some extent Boulton & Watt engines, even though they were huge, house-filling structures, also found customers in the production sector, where they replaced waterwheels. A 1785 Boulton & Watt steam engine was used to grind malt in Whitbread's London brewery, for instance.

To be sure, Watt and his colleagues for about 20 years remained merely consulting engineers, rather than actual steam engine producers. Boulton & Watt fabricated ancillary gear and certain components for the steam engines, but otherwise only supervised their construction and left customers to source many parts locally. This changed when the firm in 1795 decided to build a foundry at Soho. And since Boulton & Watt could not match the quality of the cylinders produced by Matthew Murray's Round Foundry at Holbeck, Leeds, the company from 1799 targeted the Round Foundry with industrial espionage and other dubious assaults.[14] Boulton and Watt actually retired from their firm in 1800, when the patent on the separated condenser had finally expired. (They handed it over to their sons, Matthew Robinson Boulton and James Watt junior.) By this stage Boulton & Watt had produced some 450 steam engines. An average Boulton & Watt engine of 1800 had a power output of 20,000 watts, which was five times as powerful as ordinary contemporary water mills, and the largest Boulton & Watt engine matched the largest waterwheels of the time by delivering just over 100,000 watts.

James Watt's many design improvements included an insulated steam jacket around the cylinder and an air pump to maintain the vacuum in the steam condenser. Eventually he also introduced a *double-acting machine*, which admitted steam at both ends of the cylinder to achieve an equally powerful up and down stroke. Watt also adapted steam engines to provide rotary motion (for mills) and invented a centrifugal governor to maintain a constant speed with varying loads. But for all his achievements, James Watt rejected

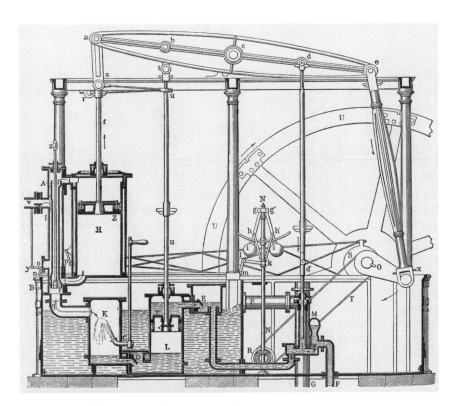

Low-Pressure Steam Engines Top: A typical Boulton & Watt low-pressure steam engine as depicted in *Meyers Encyclopedia* (Meyers Konversations-Lexikon, 4. Auflage) 1885–1890. Note that an external condenser (K) is clearly separated from the cylinder (H), which was Watt's principal advance over the Newcomen design. Bottom: The picture shows a steam engine located at the ruins of a sugar mill in Puerto Rico (Hacienda Azucarera La Esperanza). This beam engine was produced by the West Point Foundry in Cold Spring, New York, in 1861. (Library of Congress image HAER PR,55-MANA,1A-3, edited.)

high-pressure steam engines (apparently over safety concerns).[15] This was a major problem as it delayed the development of small, powerful and fuel-efficient steam engines for decades.

High-Pressure Engines

Many of the contemporary engineers who did indeed understand the potential of a high-pressure set-up were deterred by the long protection of Watt's principal patent on the separated condenser. Hence, most steam engines that existed in 1800 were operating at mines, where the fuel was basically free and the enormous size of the engines did not matter much. But as soon as Watt's patent expired, a radical proliferation of the steam engine set in. For this reason it is reasonable to claim that the Coal Age truly began in the year 1800. A whole host of (often very basic) innovations occurred within a few years, and the steam engine quickly broadened its applications to become finally a universal prime mover that could be flexibly located.

High-pressure boilers were introduced by Richard Trevithick in England in 1804, and by Oliver Evans in the United States in 1805. Arthur Woolf patented a high-pressure *compound* engine in 1804, but his design was flawed and was used widely only after William McNaught improved it in 1845. Compounding steam engines recycled their used (high-pressure) steam from the first cylinder into additional cylinders to increase the overall power output. Such engines used only two-thirds the amount of coal necessary to fuel a contemporary Watt-type low-pressure engine. By the 1890s, the largest steam engines were about 30 times as powerful as the ones of 1800, and their fuel efficiency was up to 10 times higher. Much of this gain came about through a hundred-fold rise in operating pressure.[16]

Slow Break into Manufactories

Acceptance of steam engines in (light) manufacturing was actually slow. Steam engines had been operating almost exclusively at mines for some eight decades, and then diversified mainly into other water pumping applications, for instance for the canal system and to supply water to towns. (Similar later applications were for field irrigation and fire-fighting.) In iron smelting, steam engines were eventually used to drive the bellows of blast furnaces. (Iron production was now liberated from both its traditional dependencies: timber/charcoal and rivers/waterwheels. Only iron ore and coal were prerequisites for iron production in Coal Age societies.) But only few manufactories of the day used the room-filling Watt-type engines.

With the expiration of the Watt patent, smaller, self-supporting steam engines appeared on the scene. William Murdoch's model was the first such engine to be commercially available (from 1799). It delivered merely 1,800 watts.

Freemantle's Grasshopper engine was released in 1803, and Henry Maudslay's table engine in 1807. These low-pressure engines soon provided power for many small workshops and mills. Arguably textile production was Britain's most important industry at the time, but the majority of firms in this booming sector kept relying on waterpower until the 1830s. Still in 1838 as many as 2,230 waterwheels were employed in British textile manufactures, compared to 3,053 steam engines. However, by 1850 about 85 percent of Britain's roughly 1,900 factories in the cotton industry were utilizing coal energy, and during the following years, when power weaving was firmly established, mill size and engine power increased significantly.

To be sure, it was a lot more difficult to achieve mechanized weaving than mechanized spinning, because power looms had to handle dozens of warp threads and had a tendency to break them. English inventor Edmund Cartwright in 1785 invented a power loom that was still inadequate, but established the principle for later, functioning models. Cartwright opened a weaving shed in Doncaster, using a bull as his principal power source for the first two years. Thereafter he purchased a steam engine. In 1799 a Manchester firm acquired several hundred of Cartwright's power looms, but soon thereafter its factory was burnt to the ground, possibly by weavers fearing to lose their jobs. In turn power weaving was introduced successfully only in the years after 1813, when William Horrocks had invented the variable speed batten. By 1821 some 5,732 power looms operated in 32 mills in the Manchester region, and the year after the number of power looms in operation had doubled.[17] In concert, the number of handloom weavers employed in the booming British cotton industry decreased from 250,000 in 1820 to practically none by 1850. Their work (like that of cotton spinners) was now done by the energy contained in coal. British consumption of raw cotton increased from 61 million pounds in 1802, to 1.2 billion pounds in 1875, to 2.2 billion pounds in 1913. This extreme expansion reflected the triumph of steam technology in manufacturing.

STEAM TRAINS

Steam technology also revolutionized the transport of people, food, raw materials, and finished goods. There had actually been no significant progress in overland transportation since the domestication of animals, the invention of the wheel and the improvement of saddles and harnesses. But now locomotives and steam coaches entered the scene.

As early as 1765 French engineer Nicolas-Joseph Cugnot started to experiment with steam-powered vehicles designed to haul heavy cannons for the French army. Cugnot presented his first functioning steam wagon in 1769. (Note that this was several years before first Boulton & Watt engine was released.) The heavy three-wheeled vehicle carried the steam boiler in

Nicolas-Joseph Cugnot's Steam Wagon of 1771   French artillery officer Nicolas-Joseph Cugnot built and ran a 3-wheeled carriage powered by a steam engine. Depicted here is his alleged crash into a wall, which may be considered the world's first car accident.

front of the stirring tiller and had two wheels in the back. It reached a top speed of 2 mph and could pull heavy loads. However, it had to stop about every 11 minutes to rebuild enough steam pressure. Running out of control, this vehicle crashed into a garden wall in 1771, causing the world's first automobile accident. The same year Cugnot constructed a second vehicle, but thereafter his efforts were disrupted by political turmoil.[18]

In Britain, Boulton & Watt employee William Murdoch in 1784 built a number of small steam carriages that were powered by tiny, highly-advanced high-pressure steam engines. On his way to London to apply for a patent, Murdoch happened to meet Matthew Boulton himself, who successfully discouraged him from handing in the patent application. (Boulton urged Murdoch to focus on stationary low-pressure engines for the mining industry instead.) This event practically sidelined the use of steam for transportation for a whole generation. Twenty years were going to pass before Richard Trevithick would build a steam locomotive, right after the original Watt patent had expired.

Trevithick's Steam Carriages and Railroad Locomotives

Trevithick was probably the single most instrumental person to establish the use of coal energy in overland transportation. He worked in Cornwall's tin mines from early age and improved a steam engine for a local company before developing his own small high-pressure engine in the 1790s. He then built a vehicle that was little more than a steam engine mounted on a three-wheeled platform, but it carried Trevithick and seven friends half a mile up a hill on Christmas Eve 1801. Soon thereafter the vehicle's boiler burnt dry while it was parked outside a bar. Trevithick then built another 3-wheeled steam carriage, this time equipped with seats. In 1803 he took it to London to demonstrate it to several leading scientists, including James Watt, but none of them caught on. Trevithick ran out of funds and sold the power unit of his vehicle to a mill.

Fortunately, Trevithick was immediately thereafter commissioned by the Penydarren Ironworks in Merthyr Tydfil (a center of the Welsh coal and steel industries) to build a steam locomotive for on-rail transport of loads to the closest canal. Rails have their origins in the wooden railways constructed in the mines of central Europe from the early 16th century. Rails are quite easy to lay, provide a smooth surface for wheels to roll, and distribute the vehicle's weight pressure from beneath the wheels to a long stretch of rail. Thus, rails were popular in the mining sector for the transport of heavy loads in vehicles drawn by horses to the nearest riverside or canalside. Meanwhile, England's first horse-drawn railway to offer a public service was established in 1778.[19]

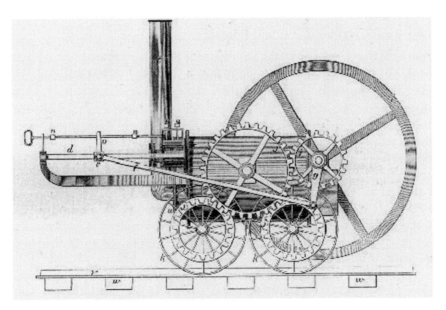

Richard Trevithick Locomotive of 1803 Richard Trevithick was perhaps the single most important person to forward and demonstrate early steam technology for mobility on land. He used a 3-wheeled vehicle by 1801, and introduced the world's first steam engine to run successfully on rails in 1804. The model shown here is depicted in his son Francis's work *Life of Trevithick* of 1872 under "Trevithick's Tramroad Locomotive, South Wales, 1803."

In February 1804 Trevithick demonstrated the world's first steam engine to run successfully on rails. The locomotive managed to haul 10 tons of iron, 70 passengers, and 5 wagons at a top speed of nearly 5 miles an hour on the 9-mile stretch between the ironworks at Penydarren and the Merthyr-Cardiff Canal. Overall, it was no faster than horses, but it clearly showed that steam locomotives could pull far greater loads. Unfortunately, Trevithick's locomotive made only three journeys because the seven-ton steam engine each time broke the brittle cast iron rails. This led investors to the conclusion that the invention was unlikely to reduce transportation costs.

Trevithick then built a five-ton locomotive for a colliery in Northumberland. Again, this locomotive was far too heavy for the existing (wooden) wagonway, and Trevithick was soon back in Cornwall, where he constructed yet another locomotive. He used it in 1808 to operate a circular railway, or steam circus, in London, and had quite a lot of customers before the rails once again broke. In 1816 Trevithick relocated to Peru, where his steam-engines were successfully employed in silver mines. Trevithick was even able to acquire his own silver mines, but was forced to escape from Peru in 1826 after a war broke out. Via Costa Rica and Columbia he returned to England, where he died a poor man.

Cogwheels, Puffing Billy, and Blücher

However, Trevithick's achievements inspired a number of small, private railroads in the mining industry. John Blenkinsop, manager of Middleton Colliery, teamed up with engineer Matthew Murray to construct a locomotive that transported coal to the nearby town of Leeds from 1812. As they rejected the idea that a steam locomotive with smooth-surface wheels would have sufficient adhesion to carry loads on a smooth rail, they added a cog-toothed driving wheel on each side of the locomotive, right between the smooth back and front wheels. (The toothed wheels, on a slightly wider axis, reached into a toothed rack-rail laid right outside the smooth rails.) This locomotive was sufficiently light not to break the cast-iron rails and became a commercial success. Three more were built, but they all suffered heavy wear (between the driving gear wheel and the horizontal rack). The system was therefore not used by other collieries.

Also in 1812, the Wylam Colliery (near Newcastle-upon-Tyne) commissioned its engineer William Hedley to produce a steam locomotive that was operational about 1813. This machine, named Puffing Billy, was improved several times and remained in use until 1862. George Stephenson, enginewright at the Killingworth Colliery, heard about Hedley's efforts and started to develop a steam-powered machine as well. In 1814 Stephenson, together with William Losh, completed his first locomotive, naming it Blücher for the

Prussian general who was instrumental in helping Britain to defeat Napoleon about this time. Stephenson kept on improving the Blücher and used it in 1821 to impress Edward Pease, an investor planning a horse-drawn railway between the West Durham collieries and the port of Stockton. Pease teamed up with George Stephenson, George's son Robert Stephenson, and Michael Longridge, to form a company that became the world's first locomotive producer. They also hired Timothy Hackworth, who had helped building Puffing Billy, and in 1825 opened the Stockton & Darlington Railway.[20]

Stockton & Darlington and Liverpool & Manchester Railway

The Stockton & Darlington Railway was organized much like a canal. It was the first public railway, allowing any carrier in exchange for a fee to take his goods along the line by using his own wagons and horses. In addition, a purpose-built horse-drawn railway coach offered transport to 18 seated passengers. The jewel of the Stockton & Darlington Railway, however, was a steam train called Locomotion, which maintained a speed of around 10 to 15 miles an hour. One of the true innovations of the Stockton & Darlington Railway was to use wrought iron rather than brittle cast iron tracks. George Stephenson had previous (negative) experience with cast iron rails and contracted the Bedlington Iron Works to produce wrought iron rails, which were going to be lighter and stronger, and could be welded together on site to make the joins smoother. (Wrought iron was available in quite large quantities at this stage as Cort's puddling process of 1784 had become wide-spread.)

By 1825, when the Stockton & Darlington Railway started to operate, coal energy had begun to make a difference in the British textile industry. Power weaving had been introduced from 1814, and Manchester turned into the center of coal-fueled cotton processing. At least 1,000 tons of freight passed between Liverpool's harbor and Manchester per day in the early 1820s, with raw cotton being shipped from Liverpool to Manchester and finished cotton goods the other way. The existing canals and roads had problems handling these loads, and local businesspeople in 1826 achieved approval from Parliament to build a railway between the two cities. George Stephenson was hired to oversee the construction. The line was going to be double track throughout, and was to transport both freight and passengers. However, the project involved serious challenges, including the crossing of Chat Moss, the erection of a bridge at Rainhill, and the construction of a nine-arch viaduct across the Sankey valley.

Curiously, the railroad construction began without making a decision in regards to the motive power to be used. The Stephensons promoted their steam locomotives, while other prominent engineers recommended building stationary steam engines at intervals along the line, which would wind endless ropes between the tracks for trains to hook on to. In order to find the best possible solution, the railroad's directors decided to organize the Rainhill Trials, offering a prize of £500 to encourage engineers to develop new ideas in locomotive construction. These were to be demonstrated on a completed section of the railroad in October 1829. Robert Stephenson's locomotive model named *Rocket* won the Rainhill Trials and was chosen from among three serious competitors to be used for the Liverpool & Manchester Railway, which opened in 1830. Stephenson in turn won contracts to equip the Bolton & Leigh Railway and in 1833 was appointed chief engineer of the London & Birmingham line.[21]

At the end of the Rainhill Trials, Stephenson ordered his machine to be freed from all loads it had shown it could carry. The *Rocket* then reached an astonishing speed of 35 miles an hour. This was the first time humans traveled at a sustained speed faster than on horseback. The *Rocket*'s main inno-

Robert Stephenson's Steam Locomotive *Rocket* of 1829   Constructed for the Rainhill Trials of 1829 that were to identify the best option for the Liverpool & Manchester Railway, Robert Stephenson's locomotive was the only contestant that did not break down. Stephenson's design defined a whole generation of steam locomotives to come. (Reproduced from the *Mechanics Magazine*, October 24, 1829.)

vation was a multi-tube boiler with a water-jacket firebox, evaporating the water much more efficiently than a single large flue. The blast-pipe exhaust induced a draught through the fire and made the engine self-regulating. The pistons were directly coupled to the wheels by cranks set at right angles to each other, providing for a certain start and smooth drive. Hence, the *Rocket* established the basic architecture for most steam locomotives to come.[22]

The success of the Liverpool & Manchester Railway and the *Rocket* locomotive triggered a railroad boom in Britain. A network of lines first emerged in industrial northwestern England, but by 1840 Birmingham, Liverpool, Manchester, Leeds and other population centers were all connected by railroad to London. A total of 2,300 miles of railroad track was laid in Britain between 1833 and 1843, and by 1860 there were some 16,000 km of British track. The rails were made from malleable wrought iron obtained from large Cort furnaces until the Bessemer process was introduced in 1856 to yield large quantities of steel. Bessemer steel was of medium quality only, but still far better than wrought iron. It was directly cast into inexpensive, durable rails.

Off Rails

Off rails steam transportation did not work very well due to the high weight of inherently massive steam engines. Nevertheless there were steam-powered coaches operating between various English towns from about 1820 to 1840. Especially well known was Dr. Church's unusually designed three-wheeled steam coach of 1833, which ran daily from Birmingham to London at an average speed of 14 miles per hour, carrying up to 44 people, 22 inside the carriage and 22 outside.

Perhaps the main reason why steam coach operators were driven out of business was opposition by the providers of horse-drawn passenger services who blocked roads and persuaded the government to impose crippling tolls. Development of road transportation in Britain was entirely stalled with the Locomotive Act of 1865, which set a speed limit of two miles per hour in towns and four miles per hour in the country, and required a person with a red flag or lantern to walk 60 yards ahead of every road locomotive to warn horse riders and horse-drawn traffic of the approach of a self-propelled machine. The Locomotive Amendment Act of 1878 made the red flag optional under local regulations (and reduced the distance of the warning to a somewhat more manageable 20 yards). But only in 1896 was the Red Flag Act finally repealed, and the speed limit increased to 12 miles per hour. In short, the Coal Age remained the era of the horse in terms of individual short-distance land travel.

Unprecedented Mobility

Overland transport of freight and people on set routes, on the other hand, was revolutionized by railroad locomotives, as it became faster, cheaper, and more reliable. Land transport had been depending exclusively on human and animal power throughout the Agricultural Age. Horse-drawn coaches of the pre-railway era reached maximum speeds of perhaps 12 miles per hour (20 kilometers per hour), but usually traveled less than 6 miles per hour (10 kilometers per hour). Heavy freight wagons moved at only half that speed. Steam trains, in contrast, reached speeds of well over 60 miles per hour (100 kilometers) per hour towards the end of the 19th century, and the streamlined German Borsig achieved a speed of 125 miles per hour (200 kilometers per hour) in 1936.[23] (August Borsig's first locomotive of 1840 outperformed a Stephenson locomotive in a direct comparison in Berlin, and his company quickly advanced to become Europe's largest locomotive producer. The firm presented its 500th steam locomotive after just 14 years, in 1854, and completed its 3,000th locomotive in 1873.) Railways thus changed the economics of transport on land entirely (which led to the downfall of coastal water-borne transport and the canal networks), and urban centers could now flourish at various in-land locations away from coasts and riversides. Best of all, steam technology did not necessarily depend on coal. Fueling steam trains with wood was relevant in the colonization of the United States as well as imperial Russia.

Steam Ships

On the water, the high weight of steam engines (and their fuel) did not matter much. Steam engines were therefore bound to revolutionize mobility on the water. As soon as small engines became available, the idea of fitting steam engines to boats occurred to many inventors. The advantage was especially obvious for propulsion on rivers and canals, where mobility still depended on agricultural energy in form of grain-fueled oarsmen or draft horses. On the oceans it was going to be far more difficult to establish coal power, simply because sailing technology had matured over many centuries and was very well developed. In fact, the British tea clippers of the 19th century reached unprecedented speeds with significant cargo. (In 1866 two British clippers managed to sail 16,000 miles from China to London fully laden with tea within just 99 days, a bit over three months.) Nevertheless, it was only a question of time before coal energy would conquer the oceans as well. The benefit of being independent of weather conditions was just too obvious. English ships sometimes had to wait up to three months for the right wind to take them into Plymouth Sound, and any military leader would immediately realize the strategic advantage of being mobile at sea in the absence of wind.

Paddlewheels

The main technological challenge was how to translate the energy derived from a steam engine into ship propulsion. All early steam ships were propelled by paddlewheels, which were either mounted amidships (on both sides of the hull) or astern (at the end of the hull). In essence, these wheels were copies of the waterwheels used by factories alongside rivers. In 1770, Jacques Perrier, an early pioneer in steam application, ordered a British engine for the waterworks of Paris. As the steam engines of the time filled up a whole house, Perrier in turn built his own small engine to fit a boat. (It was, however, too underpowered to move upstream on the Seine River.) Another Frenchman, the Marquis de Jouffroy d'Abbans, experimented with steam boats for years, placing one on the Doubs River at Baum-des-Dames in the Franche-Comté in 1776, incidentally the same year the first Boulton & Watt engine was completed. He demonstrated the "Pyroscaphe," the world's first successful steamer, to representatives of the scientific community and thousands of spectators in July 1783. This 45-meter (140-foot) steam vessel traveled upstream on the Saône River near Lyon at a speed of six miles per hour. It operated for sixteen months, but authorities refused to grant Jouffroy a patent (or to provide public financing), and he later had to leave due to the outbreak of the French Revolution.

In America, John Fitch demonstrated a well-functioning 45-foot steamboat on the Delaware River in August 1787 and later built a larger vessel that carried passengers and freight between Philadelphia and Burlington, New Jersey. Fitch was granted a U.S. patent for a steamboat in 1791 after fighting off rivaling claims by James Rumsey, another American steam boat pioneer experimenting on the Potomac River. In England, where experiments with steam engines were curbed by the 25-year Watt patent, Patrick Miller's *Charlotte Dundas* operated in 1802 and inspired American Robert Fulton, whose *Clermont* traveled on the Hudson River between New York and Albany in August 1807, completing the 150-mile trip in 32 hours at an average speed of about 5 miles per hour. Encouraged by this truly successful paddle-steamboat operation, Fulton by 1814 (together with the Livingston brothers) also offered regular steamboat passenger and freight service between New Orleans, Louisiana, and Natchez, Mississippi. His steam boats traveled on the Mississippi River at rates of eight miles per hour downstream and three miles per hour upstream.

On open oceans, steam power was initially installed as an additional means of propulsion on fully-rigged sailing ships. The first known Atlantic crossing using only steam power was completed in 1833 by the *Royal William*. This sailing/steam ship, with paddlewheels amidships, reached London coming from the Province of Quebec, where it had been built with government funding. The first coal-fueled westward run across the Atlantic was completed five years later.

Steamship *Washington*, 1847   Early oceangoing steamers were hybrids, relying on wind power as well as coal energy. Though ship propellers were by this stage already in use, the *Washington* of 1847 employed well-established paddlewheels amidships to utilize the energy derived from two coal-fired steam engines for propulsion. This ship's Atlantic crossings took about 17 days. (Library of Congress image LC-USZC2-570, edited.)

Screw Propellers and Iron Hulls

The ultimate solution for engine-driven ship propulsion was the screw propeller rather than paddlewheels. The development of practical ship screws took surprisingly long, given that early concepts date to the late 18th century. A screw propeller is basically nothing else than a short Archimedes' screw, the ancient water-lifting device. Austrian inventor Josef Ressel designed an operational ship screw and tested his innovation on a ship to which he had himself fitted a steam engine. The *Civetta* became the world's first steamer with a screw propeller, sailing over the Gulf of Trieste (then part of the Austrian monarchy, now part of Italy) in 1829 with about 40 passengers on board. The cruise demonstrated the feasibility of the ship screw as a means of propulsion, but a breakdown of the steam engine at open sea prompted Austrian authorities to declare the use of steam engines on board of ships too dangerous. The use of screw propellers was therefore delayed. Several years later Swedish John Ericsson and British Francis Smith independently obtained patents on the ship screw. In 1837 Ericsson demonstrated a well-functioning screw-propelled ship to British navy representatives, who did not appreciate his invention. However, a U.S. navy officer realized the potential and convinced Ericsson that he should relocate to the United States, where Ericsson made a fortune. (Earlier on, Ericsson had actually built one of the few steam locomotives competing in the 1829 Rainhill Trials.)

The ship screw started to spread after the SS *Great Britain*, the first big iron steamer equipped with screw propellers (in addition to six masts for traditional sailing), was launched on July 19, 1843. This was the time when iron hulls began to replace wooden hulls in ship construction. The shift towards iron hulls was stimulated by the scarcity of timber and the availability of cheap wrought iron (and later steel), and it removed the size restrictions that limited wooden hulls to a length of about 100 meters. The shortest Atlantic crossing was cut to less than 10 days in the 1840s, and by 1890 steel-hulled steamers normally crossed the Atlantic in less than six days, shipping more than half a million passengers per year between Europe and New York. The *Normannia*, launched in 1890 to serve the Hamburg-New York line, was 156 meters long and reached a speed of nearly 25 miles per hour (40 kilometers per hour). By the 1920s, annual North Atlantic traffic surpassed one million people. Some 60 million emigrants left Europe between 1815

Russian Cruiser *Aurora* Toward the end of the Coal Age war ships were full-metal constructions. The picture shows the Russian cruiser *Aurora*, commissioned in 1903. Equipped with 24 boilers and three steam engines, the *Aurora* operated in the Pacific, but, unlike many other Russian ships, managed to escape destruction in the 1905 Battle of Tsushima against the Japanese. Back in St. Petersburg, the *Aurora* became famous for its role in the 1917 October Revolution, as it is said that the ship fired a blank shot to signal the start of the storming of the Winter Palace by the Bolsheviks. (The *Aurora* is now exhibited in Saint Petersburg, Russia. Photograph by Manfred Weissenbacher.)

and 1930 for overseas destinations, and most were shipped to North America on iron-hulled screw-propelled steamers. In terms of cargo shipping, the world's fleet of steamers enjoyed a boost following the completion of the Suez Canal in 1869 and the introduction of effective refrigeration during the 1880s. (The latter allowed meat and other spoilable products to be transported from overseas to European population centers.) Later on, growth was stimulated by the opening of the Panama Canal in 1914.

COAL GAS

Although steam engines in most respects define what the Coal Age was all about, an alternative way to utilize coal energy was quite literally much more visible to many people: This was the production of coal gas, an energy-rich medium that entirely changed the life of people by illuminating houses, factories, streets, and entire cities. (Previously only oil lamps and candles provided for indoor illumination, offering inefficient, weak, and smoky, but portable and safe, lighting.) Coal gas was also used for cooking and heating in households, replacing the solid fuels wood and charcoal. Furthermore, coal gas emerged as a critical industrial fuel. It heated up water boilers in factories, for instance, and fueled the overhead gas burners in the open-hearth (Siemens-Martin) process to mass-produce high-quality steel.

Coal gas is also known as town gas or water gas. It was initially obtained as by-product of coke production. Since only limited amounts of air (oxygen, O_2) come into contact with coal (carbon, C) during the coking process, some of the coal's carbon is turned into combustible gaseous carbon monoxide (CO) rather than being burned all the way into (inert and non-flammable) carbon dioxide (CO_2). ($3C + O_2 + H_2O \rightarrow H_2 + 3CO$). Coal gas also contains other volatile combustible compounds released during coking. Nevertheless it tends to have a quite low energy content if it is obtained from ordinary coking. A much better coal gas was achieved when water steam was directed over glowing coal/coke. This procedure was introduced in the 1850s and is accredited to Karl Wilhelm Siemens, the inventor of the open-hearth steel furnace. (The use of water steam enriched the obtained gas with combustible carbon monoxide (CO), hydrogen (H_2) and methane ($C + 2H_2 \rightarrow CH_4$), but the relatively high content of carbon monoxide was also a bit of a problem as this component is toxic when inhaled.)

Gas Plants for Factories and Cities

The roots of practical coal gas application go back to the work of Frenchman Phillipe Lebon, who used gas obtained from the destructive distillation of wood (that is, heating wood in the absence of excess air). Lebon demonstrated gas lighting to the public in Paris in 1801, but he was soon killed in a street robbery, and no one in France followed up on his work at the time. However, Lebon's demonstration had been attended by James Watt's second son, Gregory, who made sure that Boulton & Watt employee William Murdoch would get support from the firm for experiments with the gases released during coke production. (Yes, this was the same Murdoch who had previously been discouraged by Boulton from applying for a steam carriage patent.) Murdoch had been running a gas pipe into his front room at Cornwall as early as 1792, and in 1802 he installed gas burners in a Boulton & Watt factory near Manchester. From 1804 the firm offered commercial gas plants to illuminate English cotton mills and other factories, which improved working conditions and extended production hours.

Meanwhile a German, Frederic Winsor, had different ideas. He had also seen Lebon's work in Paris and independently demonstrated coal gas lighting in London in 1804. Winsor's vision was to supply gas by mains from central gas generating plants, which was principally different from the independent house or mill gas systems sold by Boulton & Watt. The Gas Light and Coke Company was chartered in 1812 to provide such centralized gas supply for London, and piping immediately began to spread throughout the city. Other towns soon followed suit. Coal was shipped into the cities, and town gas was produced and stored locally at gasworks. The huge cylindrical structures in which coal gas was stored are still now landmark signs in many Western cities. From these tanks the gas was directed through wooden or cast-iron pipes to street lamps, households, and factories. (Meanwhile, the rural households of Europe and North America relied on whale oil, and later kerosene, for illumination.)

In 1885 two inventions were advanced that both further stimulated the use of gas for lighting purposes. One was the Bunsen burner by German chemist Robert Wilhelm Bunsen, the other the Welsbach mantle by Austrian

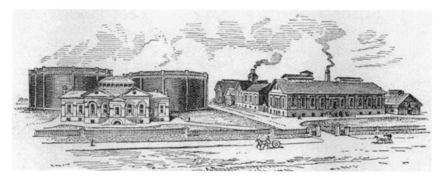

Illuminating Cities with Coal Gas   The coal gas (town gas) works of the Coal Age have left landmarks in many Western cities in the form of huge cylindrical structures once used to store coal gas. The illustration shows Baltimore's Bayard Street Station as depicted in *Progressive Age Magazine* in 1889. (The image is shown in Housewerks, "The History of Bayard Station," http://www.housewerksalvage.com/history.html , and in Allen W. Hatheway, "History and Chronology of Manufactured Gas," http://www.hatheway.net/01_history.htm.)

scientist Carl Auer von Welsbach. The Bunsen burner mixes gas with air, thus allowing the flame to be regulated, while the Welsbach mantle radiates off bright light when placed over a flame. (The Welsbach mantle is a cylindrical framework of gauze impregnated with oxides of thorium and cerium. When heated by a gas flame, it produces a very bright light because of the incandescence of these oxides. It is nowadays used in outdoor and camping lamps.) However, around this time the development of incandescent electric lamps was well on its way to replace gas lamps. Coal gas illuminated European and American towns for about eight decades; thereafter it was mainly used for heating and cooking purposes. (In the second half of the 20th century coal gas was almost entirely replaced by natural gas, which became available from the 1930s and 1940s. Natural gas is chiefly methane and does not contain toxic carbon monoxide.)

ELECTRICITY

Yet another way to utilize coal energy is to generate electricity. Electricity is a phenomenon that involves the separation of charges at the atomic level. Static electricity may occur when two different materials are rubbed against one another, because the surface atoms of one material may lose electrons (negatively charged particles) to the other material. This is sometimes observed when people brush their hair: A nylon comb may become negatively charged while the hair becomes positively charged. In turn hair tends to stick to the comb because conversely charged objects attract one another. Such attraction between contrary charges can be utilized by connecting two conversely charged objects with an electric conductor such as metal wire. An electric current will then flow through the wire as the two objects attempt to redistribute their charges. This flow of charges can be turned into mechanical work by electric motors. It can also be used to generate light or heat.

To be sure, energy has to be invested first to separate charges. Only thereafter can electricity be put to work. Muscles or any other prime mover may rub two materials against each other to achieve this separation, but the best way to produce electricity is to exploit its relationship with magnetism. Magnetism is closely associated with electricity because electric fields (as generated by electric charges) induce magnetic fields, and vice versa. Danish physicist Hans Christian Ørsted observed in 1819 that an electric current flowing in a wire deflects a nearby magnetic needle. Before long it was speculated that the system might work both ways, that is, that a moving (rotating) magnet may induce an electric current in a nearby wire. In 1831, Englishman Michael Faraday showed exactly that, demonstrating the first primitive electricity generator, a machine that converts mechanical energy into electricity.

Generators typically consist of stationary magnets that induce an electric current in a wire or coil rotating between them. But there were problems involved. One was that permanent (or static) magnets produced weak magnetic fields, and hence weak flows of electricity. Another was that large permanent magnets are very heavy, and they lose their magnetism due to the vibrations created by the rotation of an electricity generator. The alternative was electromagnets, which consist of an insulated wire coiled onto an iron core. However, electromagnets depended on electricity flowing through their wire, which means they consumed electricity in a setup that was supposed to produce electricity. Werner Siemens, brother of open-hearth steel furnace inventor Karl Wilhelm Siemens, is credited with helping the Electricity Age on its way. He invented the self-excited generator that could do without a permanent magnet, being set in motion merely by the weak residual magnetism of its powerful electromagnet. Siemens demonstrated this *dynamoelectric principle* in Berlin in 1866, and thereafter large generators without external electricity supply were constructed.

Belgian Zénobe-Théophile Gramme around 1870 built the first ring-wound armature generator that could run continuously, and from 1873 gen-

Edison Dynamo   The first power plant completed by Thomas Edison's company was the one installed in 1882 at London's Holborn Viaduct, a bridge initially illuminated by electric carbon arc lamps rather than Edison-type light bulbs. This was a little after a waterpowered Siemens alternating current generator had been set up to deliver electricity for arc lamps and Swan filament light bulbs in the town of Godalming in Surrey (southeastern England) in September 1881. The Edison electricity generator (dynamo) shown here was located at New York's Pearl Street power station that opened the same year as the one at Holborn Viaduct. Driven by steam engines, such direct current generators could supply little electricity only for a nearby area of one square mile or less. But even though Edison's direct current system ultimately failed, it deserves credit for advancing the public acceptance of the idea of electric lighting and central power stations. (U.S. Department of the Interior, National Park Service, Edison National Historic Site, Image Number 15.400/8, edited.)

erators became available that were capable of prolonged operation. With this technology established, the first electricity production plants (power plants) could be set up. These were initially little more than sheds hosting steam engines that turned electricity generators. In San Francisco the California Electric Light Company Inc., organized by George H. Roe, started to produce and sell electricity on a small scale in 1879, while the first larger power plants were built by American inventor Thomas Alva Edison. His London company completed its first electricity plant at Holborn Viaduct in 1882, and his American company built the New York Pearl Street station that started to transmit power later that same year. These plants supplied areas of about one square mile, and within two years similar power stations were set up by entrepreneurs in various Western countries.[24]

Electric Lighting

The electricity generated by early power plants was mostly used to power incandescent electric lamps. The first development in electric lighting was the arc lamp, which evolved from the carbon-arc lamp demonstrated by Englishman Humphry Davy in 1801. (In carbon-arc lamps an electric current bridges a gap between two carbon rods and forms a bright discharge called an arc.) In 1858, American Moses G. Farmer experimented with a different kind of incandescent electric lamp. His had an electric current passing through a resistance filament enclosed in a vacuum tube. In this set-up the filament becomes hotter and hotter until it starts glowing. In 1879 practical lamps of this kind were forwarded independently by Joseph Swan in England and Thomas Edison in the United States, while Siegfried Marcus (who is now best known for the construction of the first practical gasoline-powered automobile) patented an electric lamp in Austria two years earlier. Also in 1879, carbon-arc lamps enclosed in glass were used in Cleveland and San Francisco to illuminate streets.

Out of all incandescent lamps available at the time, Edison's became most widely marketed. It was quite inefficient, converting less than 0.2 percent of the electric energy that passed through it into light. (What is more, Edison's first electricity generating plants converted less than 10 percent of the energy in coal into electricity.) Nevertheless, these light bulbs were ten times brighter than Welsbach gas mantles and 100 times brighter than candles. This translated directly into a major improvement of quality of life and work conditions.

While Edison's lamp was a bulb with a carbon filament (that is, carbonized cotton thread), Auer von Welsbach introduced the first metal filament light bulb in Austria in 1898. Welsbach's lamps were more than three times as efficient as Edison's, but the osmium metal used by Welsbach turned out too rare for commercial use. Instead tungsten proved to be the filament metal of choice for the modern light bulb. Another type of electric lamp, the mercury-vapor electric lamp, was devised by American inventor Peter Cooper Hewitt in 1903. This type of lamp makes use of a small pool of liquid mercury in a vacuum. When an electric current passes through the mercury it produces ionized vapor, which gives off a blue-green light. Modern improvements have given this lamp a much greater efficiency. The neon lamp, a type of gas tube lamp in which neon gas is charged by electricity to emit light, was developed by several inventors, including French physicist Georges Claude (1911). It is now commonly used in commercial signs.

Electric Motors

As soon as electricity became reliably available to factories, electric motors were used to drive machinery. Electric motors are the counterpart to electricity generators, working exactly the opposite way. Werner Siemens realized this opportunity and in 1867 pointed out that electricity can be turned into mechanical work by enunciating the reversibility of the electricity generator. Frenchman M. Hippolite Fontaine demonstrated this on a practical scale at the Vienna International Exhibition of 1870. He used a waterwheel to turn a generator that was wired to a second generator that served as electric motor and pumped water to above the waterwheel. (To be sure, these, like all, energy conversions involved energy losses. Less water can be pumped than drives the waterwheel of the first generator.)

Outside factories, electric motors found their niche in transportation to power electric tramways in cities. Werner Siemens demonstrated an electric train in Germany at the Berlin Industrial Exhibition of 1879. His circular railway was 1,000 feet long with a one meter gauge. The necessary electricity was produced by turning a generator with a steam engine. This train was subsequently exhibited in many European metropolises, and in 1881 Siemens built the world's first electric tramway line in Berlin. Clean and silent, tramways had clear advantages over steam trains when used in cities, but electric trains did not become a practical proposition for use outside urban areas before the 20th century. For one thing, it was very expensive to string overhead wires along the entire length of long-distance tracks; and even more prohibitive, the necessary electricity capacities had not yet been built up.

Alternating Current

As the electricity grid expanded, the cost of electricity lost during transportation through wires became a major concern. The problem is associated with direct current, the kind of continuous flow of electric current as delivered by the early power plants. Thomas Edison fought hard to keep his direct current (DC) system established as the general standard, but it was outperformed and replaced by the alternating current (AC) system developed by Nikola Tesla.

Tesla was of Serbian background, born in the Austrian monarchy in the region of present-day Croatia. He studied engineering at the Technical University of Graz, Austria, and worked in Paris and Budapest before coming to New York City in 1884 at age 28. Tesla had four cents in his pocket and a letter of recommendation from one of Edison's business partners in Europe. Apparently it read: "My Dear Edison: I know two great men and you are one of them. The other is this young man!" Edison had little interest in Tesla's plans to develop an AC motor, but he hired Tesla, reputedly promising him $50,000 if he succeeded in making certain improvements in DC generation plants. When Tesla delivered within just a few months, Edison refused to pay. He explained that the offer of $50,000 had been made in jest: "When you become a full-fledged American you will appreciate an American joke." Shocked and disgusted, Tesla immediately resigned and temporarily made his living as a construction worker.[25]

Fortunately A. K. Brown of Western Union Company decided to invest in Tesla's idea for an AC motor and sponsored a small lab in which Tesla quickly developed all the components of the system of AC power generation and transmission as it is now used universally throughout the world. In the end of 1887 Tesla filed for seven U.S. patents covering a full system of AC generators, transformers, transmission lines, motors, and lighting. George Westinghouse, a Pittsburgh industrialist who had invented a railroad air brake, heard about Tesla's patents and purchased them for $60,000 worth of cash plus shares in the Westinghouse Corporation and royalties of electrical capacity sold.

Alternating current flows in alternately reversed directions through or around a circuit. Its advantage over direct current is that its voltage can be raised or lowered economically by a transformer, which is useful because high voltage allows for the transmission of electricity over long distances, while safe, low voltage is desired for many applications. Hence, energy losses during transmission were significantly reduced compared to the DC system. What is more, Tesla's AC induction motor was simple in concept and could be made sufficiently small to power individual machines in factories.

AC Beats DC

With the publication of Tesla's breakthrough patents a full-scale industrial war erupted between the Westinghouse Electric Company and General Electric Company, which had taken over the Edison Company. Edison launched

Triumph for Nikola Tesla: The 1893 Chicago World's Fair   Nikola Tesla's polyphase alternating current (AC) system by far outperformed Thomas Edison's direct current (DC) system. After the demonstration of alternating current (AC) at the Chicago World's Fair in 1893, it became the general standard in the United States and internationally. The picture shows the Statue of the Republic overlooking the Court of Honor at the World's Columbian Exposition, Chicago, 1893.

a propaganda campaign against alternating current, but Westinghouse won the bid for illuminating the 1893 Chicago World's Fair, the first all-electric fair in history, by undercutting General Electric's million-dollar bid by half. This was the beginning of the triumph of Westinghouse and the efficient and cost-effective AC system. Twenty-seven million people attended the fair, and henceforth more than 80 percent of all the electrical devices ordered in the United States were for alternating current.

As AC electricity could be conveniently transmitted through power lines over long distances, it allowed for electricity plants to be built far away from city centers. This brought into play the possibility of turning electricity generators with waterpower (rather than coal energy). George Westinghouse's company installed a hydroelectric plant at the Niagara Falls in 1895. It was named for Tesla, who designed it and oversaw its construction. The plant went into operation in 1902 and supplied electricity to local industries, delivering alternating current to the town of Buffalo, New York, 35 km (22 miles) distant. As planned, the number of generators at Niagara Falls reached ten within a few years and soon electrified New York City via long-distance power lines.

Water Turbines

By the time hydropower was used for electricity production, waterpowered prime movers had undergone a radical evolution. In the earlier Coal Age enormous waterwheels with diameters of over 20 meters were constructed, but even those could not at all match the power of mid- 19th-century steam engines. However, eventually a new type of waterpowered prime mover entered the scene that was even stronger than the most powerful high-pressure steam engines.

Water turbines emerged gradually out of waterwheels. The main difference between the two is that in water turbines water is flowing parallel to the wheel's axis, and the wheel is encased in a housing. All water flowing into the housing is being efficiently utilized to turn the wheel. Generally defined, turbines are rotary engines driven by the impulse of a current of fluid. They are usually made with a series of vanes on a central rotating spindle. Thin and curved vanes had previously been difficult to produce, but were now made from metal that had become available in large quantities at low cost.

The early development work on water turbines was done in France, where coal was scarcer than in Britain and Germany, and the government funded innovations that would improve the utilization of the country's vast water resources. In 1832 Benoit Fourneyron built an efficient water turbine that was used to power forge hammers. With a rotor diameter of only 2.4 meters his turbine delivered as much as 38,000 watts in power. (For comparison, a giant contemporary waterwheel with a diameter of 21.9 meters yielded 200,000 watts of useful power.) In America, England-born James B. Francis in 1847 invented a turbine that worked even better. It replaced traditional waterwheels and competed with steam engines for use in factories along riversides. (In New England steam power was about three times more expensive than waterpower in the mid-1850s. Waterpowered machines gained even more importance during the following two decades, but by 1880 large-scale coal mining and more efficient engines had made steam cheaper than waterpower virtually everywhere in the United States.)

Water turbines finally had their full breakthrough when capable electricity generators became available in the 1870s, and various countries began setting up large hydropower schemes from the mid-1880s. Three types of water turbines gained permanent world-wide importance: the Francis turbine, for water heads between 10 and 700 meters; the Pelton turbine (patented by American Lester A. Pelton in 1880), for heads of 200 to 2,000 meters; and the Kaplan turbine (invented by Austrian Viktor Kaplan in 1912), for heads between zero and 30 meters, that is, for direct utilization of the flow of (undammed) rivers. Water turbines were the strongest of all prime movers between about 1850 and 1910.[26] A large water turbine of 1900 delivered as much as 10 million watts, which is 100 times stronger than the strongest steam engines of the year 1800.

Francis Runner Water turbines were the strongest of all prime movers between about 1850 and 1910, and are now being used for hydro-electricity production. The turbine designs by Francis, Pelton, and Kaplan turned out to be the most important ones during the Coal Age and beyond. The picture shows a Francis turbine being installed at the Grand Coulee Dam in Washington in 1947. (U.S. Department of the Interior, Bureau of Reclamation image.)

Steam Turbines

When the electricity generators installed at hydropower plants were growing in size in the early 20th century, those installed at coal-fired power stations followed suit. However, the generators soon became so enormous that even the strongest contemporary steam engines were not powerful enough to turn them. This promoted the emergence of a new coal-fueled prime mover that would fully re-establish the role of coal energy in electricity production.

The problem that became apparent with steam engines in electricity production was that their reciprocating (up-and-down) motion had to be translated into rotating motion. This resulted in relatively slow rotation while electricity generators need to rotate at high speeds to be efficient. Engineers hence realized the need for a rotating machine that would convert the energy of expanding steam more directly into electricity. The result was a new prime mover that represented a cross-over between steam engines and water turbines: the steam turbine.

Steam turbines actually look a lot like water turbines. They produce fast rotary motion and, despite their much smaller engine size, deliver higher

Steam Turbine Power: Coal Energy Utilization at Its Best   To this day, the steam turbine remains the most powerful prime mover of all. The top picture shows the *Turbinia*, powered by a Parsons turbine, at the 1897 Fleet Review at Spithead. This opened a new era of steam propulsion, as most newly built steamers were equipped with steam turbines rather than steam engines. What is more, steam turbines were almost immediately used for electricity production, which they dominate to this day. The bottom picture shows a steam turbine with the typical blades. (Photograph of *Turbinia* by Alfred John West, 1897. Photo of turbine: Library of Congress image LC-USW33-034627-ZC, edited).

power than steam engines. (They are also more thermally efficient. In a turbine the steam expands continuously through the engine so that each stage has a constant temperature and is not affected by its own exhaust.) The best known steam turbine design was patented by British engineer Charles A. Parsons in 1884. Another turbine design, invented by Swedish engineer Gustaf de Laval around 1887, was simpler and less expensive, but also less efficient. Parsons' design maximized the pressure of steam, while Laval's design maximized the speed of the steam passing through the turbine.

Parsons used his steam turbine almost immediately to drive an electricity generator. His 1888 model rotated at 4,800 rounds per minute, yielding 75,000 watts. Just 12 years later, in 1900, Parsons installed two steam turbines with alternators (that is, electric generators that produce alternating current) at the German city of Elberfeld. Each of these turbines delivered 1 million watts. Soon thereafter, in 1904, Parsons turbines were installed at the Carville power station near Newcastle upon Tyne. This plant set the pattern for power station design for most of the 20th century: coal has been the most important energy source for the generation of electricity ever since.

Electricity is clean and silent at the point of consumption and a very versatile form of energy. In Europe and the United States it made fast inroads in industries and households (and on railroads) when a complete electricity infrastructure was built up during the first decades of the 20th century. Germany was supplied with electricity in the 1920s and Russia in the 1930s. The British electric grid covered the entire country by 1936. In the United States the capacity of industrial electrical motors grew from less than 5 percent to over 80 percent of all installed mechanical power between 1899 and 1929. (Nowadays electricity accounts for about one third of the world's primary energy use, and in the United States for a bit over 40 percent. Modern steam turbine plants convert about 40 percent of the chemical energy contained in coal into electricity.)

Steam Turbines for Marine Propulsion

While electricity generation was the most important use for steam turbines, they also replaced steam engines in just about every other application. Steam turbines simply are a lot more efficient than steam engines. (Their only practical disadvantage is their lack of low speed torque: A turbine is best used in applications where continuous, unvarying power is needed.) Most significantly, steam turbines rapidly replaced steam engines for ship propulsion. Parsons himself designed the *Turbinia*, a small ship (just 100 ft. long and 9 ft. wide) that convincingly demonstrated the potential of turbines on the ocean. With a power of 1.5 million watts and a top-speed of 34 knots the *Turbinia* was easily the fastest ship in the world. Right after it had distanced the most powerful battleships of the day at Queen Victoria's Diamond Jubilee

Fleet Review at Spithead in June 1897, all the world's major navies quickly adopted turbine engines for their warships. The commercial fleet followed suit, and the new prime mover soon enabled passenger vessels to achieve speeds in excess of 20 knots.

By 1910 steam turbines had evolved into the most powerful prime mover of all, a position they maintain to this day. (Even the gas turbines developed in the Oil Age for jet plane propulsion are actually less powerful. They are, however, lighter.) Both large water turbines and steam turbines delivered an astounding 10 million watts by 1910. But by 1920 steam turbines delivered as much as 100 million watts. (What seems to be a final power plateau was in turn reached around 1960, when the strongest steam turbines began to deliver 1 billion watts.[27])

WHY DID COAL TECHNOLOGY EMERGE IN BRITAIN?

Since the emergence of coal culture transformed the world, quite a bit of effort has gone into attempts to answer the question why coal-based steam technology initially emerged in Britain. But given the fact that other regions did not get a chance, or need, to reinvent coal technology in a tightly connected world, the question may also be formulated as, "Why was Britain first?" We have seen that there are obvious answers explaining this, including that Britain was part of the European and Eurasian technology pool; that the island nation was especially stressed by timber scarcity; that easily accessible, good-quality coal was abundant; and that coking technology had been established by other industries before it was adopted by iron makers, which in turn increased demand for coal enough to promote the construction of engines to pump water from ever deeper coal mines. However, in addition to these, a variety of other factors were in place, including efficient agriculture, colonialism, slave work, markets, capital, inventiveness, security from invasion, and so on. Several of these factors were connected to the previous onset of the waterpowered industrialization of the textile industry, and some cannot be shown to have been truly essential for coal culture to emerge.

Infrastructure

Infrastructure is one factor that was undoubtedly special in Britain, at least in comparison to most of continental Europe. In times when water-borne transport was a lot cheaper and faster than road transport, the disposition as island nation in itself was an advantage, as it allowed for water-borne transport along coastlines. However, Britain also enjoyed many naturally navigable rivers and eventually built a vast canal system, notably to serve the booming waterpowered textile industry. Yet the first important canal, the

Bridgewater canal of 1761, was actually built for the very reason of transporting coal, from the Worsley mines to Manchester.[28] (To be sure, Manchester did not have a coal-fired textile industry by this stage. This coal served household use and such industries as beer brewing and iron smelting. The Duke of Bridgewater, the financier of this first truly important English canal, was cheered by the general population because his initiative cut the price of coal in Manchester by half.) Much of the canal-building technology came from the mining industry. One early coal-fueled Boulton & Watt steam engine pumped water for the Birmingham canal, and James Brindley, the engineer behind the Bridgewater canal and the first Birmingham canal, had earlier improved the performance of the steam engine and draining of the Clifton coal mines.[29] The waterborne transport infrastructure in its entirety provided for mass transport before the advent of steam locomotion, hence facilitating mass production that depended on high-volume shipments of raw materials to production sites and finished goods to markets.

Warfare

Another factor that helped the rise of coal as a fuel by increasing the demand for metal was warfare. The "Very first World War," or Seven Years' War, fought globally by Britain between 1754 and 1763, was a boom period for the British mining and iron-making industry. It promoted both the use of coke in iron-smelting and the spread of Newcomen steam engines in the mining districts. (It was just two years after the war, in 1765, that James Watt took on a repair job for a Newcomen engine.) Britain then fought the American Revolutionary War between 1775 and 1781, and the Napoleonic Wars between 1799 and 1815. Britain was the principal winner of these wars that left continental Europe devastated. (Ravaging French troops proceeded all the way to Russia in this conflict). In addition, British troops fought the three Maratha Wars against native forces in India between 1775 and 1818.

Britain enjoyed a safe island position, and remained unharmed during this whole belligerent period. (A century later, when the Oil Age had begun and people had started to take the skies, the British Isles were no longer secure in the same way. Britain was attacked from the air already during World War I.) The Napoleonic wars were actually a period of rapid industrial growth for Britain, which rushed ahead of continental Europe for decades to come. Notably, the outbreak of the war coincided with the expiration of the original Watt patent in 1799, which was followed by the diffusion of steam engines into all kinds of industries and the development of efficient high-pressure steam engines. Britain's secure island position also allowed it to attract capital as well as brain power (a position later held by America). Skilled engineers and experienced bankers (especially from the Netherlands), for instance, emigrated to England to escape from the war. And by the end of the Napoleonic wars, in 1815, Britain was the only country that still had a merchant fleet of useful size.

Exports

Though the British military and the growing domestic industrial base accounted for much demand during Britain's transfer into the Coal Age, export markets were critical as well. Coal technology itself positioned Britain well in terms of exports. Many western European regions had supply problems in terms of charcoal or iron, while Englishmen produced cast iron (goods) in decent quality at a low price by using coke. Steam technology was exported as well. As soon as Watt's patent was granted in 1769, he was approached by Matthew Boulton, who conceived of global markets and wrote: "It is not worth my while to manufacture [your engine] for three countries alone; but I find it very well worth my while to make it for all the world." Besides, Britain had so much coal (and modern means to extract it) that it could afford to sell it overseas to become the world's largest coal exporter. In contrast, the export of textile machinery was prohibited by the British government until 1843. In the period 1760 to 1830, when Britain's industry was, first, waterpowered, and then, coal-powered, no other country was nearly as mechanized as Britain. In this period the British government prohibited the export of machinery, skilled workers, and manufacturing techniques.

As far as consumer goods, and especially textiles, are concerned, much demand came from the soaring population inside Britain. The British population more than doubled from 5.7 million in 1750 to 11.5 million in 1820. In addition, demand came from captive markets (that is, markets that are not left a choice to buy any other products) in the British colonies and from voluntary international trading partners who appreciated inexpensive British manufactures. The latter included the southern states of the United States, while the former included the booming British sugar islands. Generally, Britain enjoyed vast colonial markets due to the rapidly growing size of the global British Empire. (At the end of the Seven Years' War Britain gained Canada, India, Australia, New Zealand, and much of Africa became British colonies during the Coal Age.) Without those markets, at least some mass production might not have been economical.

Economic System

Colonial trade had provided the tobacco merchants of Glasgow, tea merchants of London and Bristol, and various sugar and slave traders, with plenty of money to finance the early British factories. Absolute individual property rights had been guaranteed to land and commercial owners in England after the Restoration settlement of the 1680s, and the economic system of capitalism allowed businesspeople to invest their private capital freely wherever they wanted, with relatively few government regulations. Britain also had a patent system in place, which between 1660 and 1800 evolved from an instrument of royal patronage into one of commercial competition among inventors and manufacturers. A patent system is generally considered crucial for societies to secure growth and development. It provides inventors with an incentive to devote time, capital, and energy toward developing new technology, products, and processes, or to improve established ones. (In essence, a patent is a contract between society and inventor, which allows the inventor to reap profits during a limited time period, usually about 20 years. Thereafter, patent protection ends and the invention becomes a common good of society.) On the down side, patents may stifle progress if they are not specific enough and cover too wide a field. Hence the British patent system of the time may not have been all that helpful. Most importantly, the Watt patent of 1769 to 1799 delayed the development of small efficient high-pressure steam engines. What is more, many important innovations of this period were not patented at all. Many inventions were freely shared within Britain as well as with continental Europe and America.

Cotton and Slaves

As far as steam technology for the textile industry is concerned, the emergence of coal culture is also connected to slave-produced cotton. The initial British textile production boom was waterpowered, but from about the 1830s, coal energy was used to turn imported cotton into textiles. Slave-based cotton agriculture expanded in the newly independent United States following the invention of the cotton gin in 1793, and the British turned into the most notorious slave traders of all. Slave trade was abolished in British colonies in 1807, but slave work remained legal until 1834, and thereafter the British had little scruples importing cotton from the United States, where slavery was still thriving. (The cotton industry was then truly operating at the interface between Agricultural Age and Coal Age: raw cotton was produced by slave labor, that is, grain-fueled muscle power, but was processed by coal-fired machinery in England.)

The availability of cheap American slave-grown cotton, as opposed to more expensive Egyptian and Indian cotton, seems to have made a significant difference to a British economy that was dominated by the textile industry. In fact, U.S. slave cotton may have saved Britain as much as four percent of national product in peak years and two percent in average years in reduced foreign materials prices in the first half of the 19th century. Economic growth would thus have been somewhat slower in the absence of cheap raw cotton.[30] Note, however, that cotton was really the exception in this context:

Western Europe was almost entirely self-sufficient in terms of raw materials during its industrialization period.[31]

NOTES

5. Stephen Hughes, "The International Collieries Study," ICOMOS (International Council on Monuments and Sites) and TICCIH (The International Committee for the Conservation of the Industrial Heritage), International Council on Monuments and Sites, Paris, France, http://www.international.icomos.org/centre_documentation/collieries.pdf.

6. Vaclav Smil, *Essays in World History: Energy in World History* (Boulder, CO: Westview Press, 1994).

7. Peter Brimblecombe explains where the widely cited year 1228 C.E. comes from: The quantities of lime produced were very large, as lime was extremely important to medieval society, being used in mortar and for agricultural application. Lime was produced by heating limestone ($CaCO_3$) to a high temperature in a kiln. This drove off the carbon dioxide to produce lime (CaO). When mixed with water to form cement it was converted into slaked lime [$Ca(OH)_2$]. Traditionally, limestone was burnt with oak brushwood. In requisition orders, such as those issued by Henry III in the building operations at Westminster in 1235, oak brushwood was specified as the fuel. However, change was so rapid that only 11 years later we find that a similar requisition order specified a different fuel:

> To the Sheriffs of London, 23 July, 1264
> Contrabreve to purvey for the King in the City of London without delay and without fail a boat-load of sea-coal and four millstones for the King's mills in Windsor Castle and convey them thither by water for delivery to the constable of the castle.

Sea coal, or carbonem marus, seems to have been so called because it was brought by sea to coastal centers in England during the 13th century. It must have appeared in London by early that century, because by 1228 there was a street in London called Sacoles Lane (Sacoles ≈ sea coal). Its approximate location is still evident to the pedestrian in London today—both Old Seacoal Lane and Seacoal Lane are to be found near Ludgate Circus. Despite the difficulties in establishing the way in which Seacoal Lane came to be named, it does signify a very early beginning for the importation of coal into London. Peter Brimblecombe and László Makra, "Selections from the History of Environmental Pollution, with Special Attention to Air Pollution. Part 2: From medieval times to the 19th century," *International Journal of Environment and Pollution* 23 (2005): 354, http://www.sci.u-szeged.hu/eghajlattan/makracikk/Brimblecombe%20Makra%20IJEP.pdf.

8. Vaclav Smil, *Essays in World History: Energy in World History* (Boulder, CO: Westview Press, 1994).

9. Paul Bairoch, *Economics and World History. Myths and Paradoxes* (Chicago: University of Chicago Press, 1993).

10. Ibid.

11. Stephen Hughes, "The International Collieries Study."

12. J. B. Calvert, "Hydrostatics," University of Denver, http://www.du.edu/~jcalvert/tech/fluids/hydstat.htm.

13. Alessandro Nuvolari, "Collective Invention during the British Industrial Revolution: The Case of the Cornish Pumping Engine," Eindhoven Centre for Innovation Studies, Faculty of Technology Management, Technische Universiteit Eindhoven, Working Paper 01.04, May 2001, http://www.tm.tue.nl/ecis/Working%20Papers/eciswp37.pdf.

14. "Centres of Excellence: Engineering Pioneers," *Making the Modern World*, The Science Museum, http://www.makingthemodernworld.org.uk/stories/manufacture_by_machine/03.ST.01/?scene=5.

15. Ben Russel, "Powered by Steam: The Steam Engine 1780–1830," Science Museum, London, http://www.fathom.com/course/21701780/session4.html.

16. "Power for Production," *Making the Modern World*, The Science Museum, http://www.makingthemodernworld.org.uk/stories/the_age_of_the_engineer/03.ST.02/?scene=7.

17. William E. A. Axon, ed., *The Annals of Manchester—A chronological record from the earliest times to the end of 1885* (Manchester: Manchester Central Library, Salford Local History Library, 1886).

18. "Cugnot," The Online A-Z of 3-Wheeled Cars, http://www.3wheelers.com/cugnot.html.

19. "Constructing the Railway System," *Making the Modern World*, The Science Museum, http://www.makingthemodernworld.org.uk/stories/the_age_of_the_engineer/01.ST.04/.

20. "George Stephenson" (and related links on early trains and lines), Spartacus Schoolnet, http://www.spartacus.schoolnet.co.uk/RAstephensonG.htm.

21. "The Rainhill Trials, October 1829," *Making the Modern World*, The Science Museum, http://www.makingthemodernworld.org.uk/stories/the_age_of_the_engineer/01.ST.04/?scene=5.

22. "Stephenson's Rocket Locomotive, 1829," *Making the Modern World*, The Science Museum, http://www.makingthemodernworld.org.uk/icons_of_invention/technology/1820-1880/IC.007/.

23. Vaclav Smil, *Essays in World History: Energy in World History* (Boulder, CO: Westview Press, 1994).

24. "New Power," *Making the Modern World*, The Science Museum, http://www.makingthemodernworld.org.uk/stories/the_second_industrial_revolution/05.ST.01/?scene=7.

25. "Coming to America," *Tesla: Life and Legacy*, Public Broadcasting Service (PBS), Arlington, Virginia, http://www.pbs.org/tesla/ll/ll_america.html; "War of the Currents," *Tesla: Life and Legacy*, Public Broadcasting Service (PBS), Arlington, Virginia, http://www.pbs.org/tesla/ll/ll_warcur.html

26. Vaclav Smil, *Essays in World History*, 226.

27. Ibid., 229.

28. "The Bridgewater Canal," Canal Archive: Bridging the Years, http://www.canalarchive.org.uk/stories/pages.php?enum=TE133&pnum=0&maxp=7.

29. Roslyn Tappenden, "Birmingham's Canal Network—In Brindley's Footsteps," 24 Hour Museum—official guide to UK museums, galleries, exhibitions and heritage, http://www.24hourmuseum.org.uk/trlout_txo_en/TRA23386.html.

30. J. Bradford DeLong, "Slouching Towards Utopia?: The Economic History of the Twentieth Century, Part B: The Path from the Pre-Industrial World, VIII. The First Global Economy: Production and Trade," University of California at Berkeley and NBER (National Bureau of Economic Research), January 1997, http://econ161. berkeley.edu/TCEH/2000/eight/html/Slouching2_8preWWI.html.

31. Paul Bairoch, *Economics and World History. Myths and Paradoxes* (Chicago: University of Chicago Press, 1993).

CHAPTER 17

HOW COAL TECHNOLOGY SPREAD

Those regions where the new coal-based technologies first emerged, and those regions where coal technology was rapidly adopted, soon became extremely wealthy and powerful. Like the agrarian energy experts in the previous Energy Era, the coal technologists expanded their power to other regions and dominated those who were slower or unable to adopt coal technology. The pattern of how coal technology spread therefore critically influenced the political developments of the 19th and 20th centuries, and we can still now observe its consequences.

The spread of coal technology is actually quite easy to follow. While agriculture emerged in natural environments and spread somewhat freely to various other natural environments (of similar climate), coal culture emerged in a complex human-made environment and could at first only be adopted in those few regions that enjoyed similar technological standards. Building a steam engine was difficult and even to copy an imported one required substantial engineering skills and a well-established iron industry. In addition, entering the Coal Age required coal, which (in contrast to the solar radiation that was available everywhere) occurred in relatively limited, concentrated underground deposits. In short, coal technology was bound to spread less easily than agricultural technology.

On the upside, the spread of coal technology was a lot less cruel than the spread of agriculture. Agrarians had been ruthlessly expanding into gatherer-hunter regions by killing, enslaving, or expelling foragers. They were looking for an outlet for their growing populations and for ever more fertile land, the source of their energy, wealth, and power. Coal culture, on the other hand, spread mainly through technology transfer, not through population migration and expansion. The fate of North American natives and Australian aborigines during the Coal Age suggests otherwise, but coal technologists were generally looking for access to foreign markets and resources rather than attempting to use their energy advantage to remove or enslave the dense agrarian populations that occupied much of the world by this stage. Hence the coal-fueled expansion of empires in the Coal Age was relatively non-violent: The coal powers had little interest in slaying their potential customers. And since the world was already connected by a tight network of trade and cultural exchange, coal technology emerged but one single time: all subsequent world-wide spread can ultimately be traced to Britain, the original center of coal technology.

SPREAD TO WESTERN EUROPE AND NORTH AMERICA

Separated from continental Europe only by the narrow English Channel, Britain was an integrated part of the western European knowledge pool and engineering culture. British achievements could therefore easily be adopted and further developed in continental Europe. The same was true for the United States (and Canada), where many people from Britain and other European regions were settling during the Coal Age. For western Europe and North America it was thus merely a question of how well the countries of these regions could mimic the conditions, and happened to enjoy the resources, that seem to have, to various degrees, helped the emergence of coal culture on its way in Britain: abundant coal supply; access to European engineering skill; sufficient investment capital; efficient governance; protected markets for industrial produce; and secure grain supply, usually from within their own national borders. Britain's advantage of a superior water-borne infrastructure plus a close proximity of coal and iron ore deposits became less relevant as soon as railroad technology evolved and spread. Coal technology also changed the prerequisites for large-scale iron manufacturing, as coal-fired steam engines replaced waterpower for driving the bellows of blast furnaces, and coke replaced charcoal, thus freeing iron smelters from their ancient dependence on abundant forest resources.

The fact that early coal powers tended to be self-sufficient in grain production shows that the new energy system was built on top of a well-established agricultural energy system. Efficient agriculture ensured that industrializing societies could urbanize and sustain their hordes of centralized factory workers. The early coal powers did not rely on raw materials from non-industrialized countries either. They all engaged in imperialism, though they were generally not seeking raw materials, but access to colonial markets that made mass-production in coal-powered factories worthwhile. (And to some extent, they were also attracted to the prestige attributed to empire-building at the time.)

Coal Resources in Different Countries

Obviously, the most basic ingredient to adopt coal technology was access to coal. In analogy to the Fertile Crescent, Britain was really part of something of a *Carboniferous Crescent.* The Northwest European coal belt stretches from the Scottish lowlands through England to northern France and Belgium, on to the Ruhr region of Germany, and all the way to Silesia (present-day north-western Czech Republic and southern Poland).[32] Coal mining expanded very rapidly in Belgium and western Germany, regions that (like Britain) even became coal exporters. The Belgian coalfields of Wallonia served the Belgian as well as the northern French industrialization. In Germany, the Saarland and the Ruhrgebiet were principal coal producers. The latter, based on the Westphalian coal field, turned into Europe's greatest industrial complex. France sourced coal from its south-central Saint-Etienne coal basin (Massif Central), while the Austrian monarchy had coal deposits in Bohemia and Moravia to supplement those in Silesia (over which Austria and Prussia were battling). (In 1913 Upper Silesian coalfields accounted for 21 percent of German coal production, second only to the Ruhr basin in Imperial Germany.) The entire Carboniferous Crescent region turned into the industrial heartland of Europe, but both the quantity and quality of the coal within it varied widely. Neither France nor Germany had coal that compared to Britain's abundant smokeless coal of high energy-density.

The United States turned out to be very rich in coal as well. The coal districts of eastern North America, though they are now located far away from Europe, may actually be considered part of northwestern Europe's Carboniferous Crescent: During the Carboniferous Period (360 million years ago to 286 million years ago), when the biomass that eventually ended up as coal was growing in vast swamps, North America was still connected to Western Europe. Only about 200 million years ago did the continent of North America change course and begin drifting away from Eurasia, or more correctly, from the supercontinent Pangaea.

Best of all, the vast U.S. coal reserves were located in the Appalachian Basin, very close to the eastern U.S. population centers. (The Appalachian mountain range stretches parallel to the Atlantic coast, mostly less than 150 miles from the shoreline, from the state of New York all the way south to Alabama.) Pennsylvania, with its high-quality anthracites, and Ohio, with its excellent bituminous coal, became the leading industrial centers when U.S. coal-based industrialization gained momentum after the Civil War. (Still today, the United States has the largest coal reserves in the world. Of the roughly 60 billion tons of coal produced in the United States during

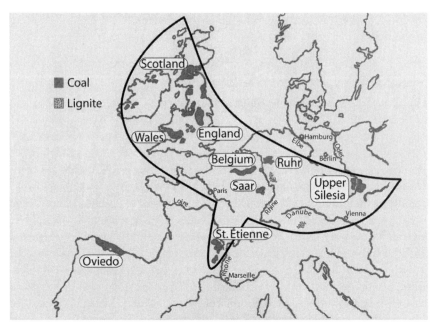

Western Europe's Carboniferous Crescent   The Northwest European coal belt stretches from the Scottish lowlands through England to northern France and Belgium, on to the Ruhr region of Germany, and all the way to Silesia (present-day northwestern Czech Republic and southern Poland). In analogy to the Agricultural Age's Fertile Crescent, Britain's rich coal deposits were thus part of a larger European Carboniferous Crescent, on which the region's coal-based industrialization depended. France also sourced coal from its south-central Saint-Etienne coal basin at the Massif Central.

the 19th and 20th centuries, about 40 billion tons have been mined from the Appalachian basin and 10 billion tons from the Illinois basin.)

Despite the large coal deposits in various countries, Britain actually maintained its position as the world's leading coal producer throughout the 19th century. Britain produced some 15 million tons annually by 1815, and over 50 million tons annually in 1850. (For comparison, the United States produced eight million tons of coal in 1850, Germany six million tons, and France five million tons.) Still in 1870 Britain produced half of the world's coal, but in 1900 U.S. annual coal production passed that of Britain. In 1912 the United States produced 485 million tons of coal, Britain 264 million tons, and Germany 172 million tons. France was now a distant fourth at 39 million tons per year. Not too far behind came Russia with 26 million tons, as well as the small country of Belgium, which produced also 26 million tons per year. (In Asia the only country to industrialize during the Coal Age was Japan, producing 17 million tons of coal annually by 1912.)[33]

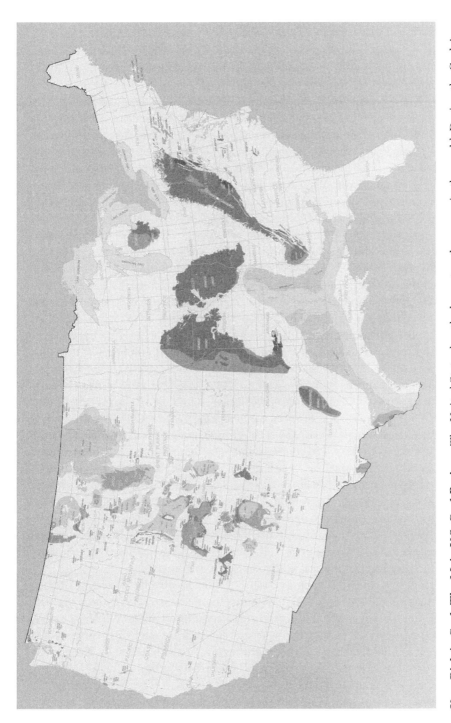

Very Rich in Coal: The Main U.S. Coal Basins The United States has the largest coal reserves in the world. During the Coal Age, the coal of Pennsylvania and of the Appalachian mountain range, close to the eastern population centers, was especially important. (Based on illustrations compiled by John Tully, U.S. Geological Survey, USGS OF 96-92.)

Engineering Skill and Technology Transfer

Next to coal, countries that attempted to adopt coal technology needed skilled engineers. In central-western Europe this was no problem at all. Much of Britain's engineering skill had its roots in continental Europe anyway. The English metal-working and mining sector had imported much of its earlier technology from continental Europe, and even the original steam and vacuum technology (as pioneered by such figures as Otto von Guericke and Denis Papin) came from regions south of the English Channel. Besides, much of the production machinery that was now driven by steam power was mainly an adaptation of general European technology developed to apply mechanical power received from waterwheels and windmills. Continental European engineers could therefore easily adopt, copy, and improve the British high technology of the day. Steam vehicles, steam boats, and rotary steam engines were built in France earlier than in Britain, and Germans pioneered many aspects of coal-fired steel production. Gas lighting, with gas obtained from the destructive distillation of wood, was first demonstrated in Paris in 1801. Coal gas lighting was demonstrated by a German in London in 1804. In turn, coal gas illuminated London in 1812, Paris in 1816, and Berlin in 1826. In short, a lot of coal technology actually emerged directly in continental Europe or spread there very fast. This situation was similar in North America.

The mechanisms by which coal technology spread were in principle the same as those that had diffused agricultural technology. They included trade, theft, migration (hiring) of specialists, espionage, and so on. A relatively new way of technology spread was through information from printed scientific journals and patents. Periodical technology publications were first printed in the 1790s. Informal philosophical societies such as the Lunar Society of Birmingham provided forums to discuss science and technology, and some of these societies published volumes of proceedings and transactions. Encyclopedias such as Harris's *Lexicon technicum* (1704), Dr. Abraham Rees's *Cyclopaedia* (1802–1819), and the French *Descriptions des Arts et Metiers* contained much valuable information as well. *Annales des Mines* published accounts by French engineers who visited Britain on study tours.

However, much technology transfer was directly facilitated by British companies selling steam engines and other industrial equipment abroad. The British government intervened by creating artificial export barriers, but it was difficult to keep hardware producers from realizing overseas sales opportunities. These producers were trying to maximize their income and were relatively ignorant of the fate of the British textile sector, for instance, which was going to suffer as soon as coal-powered, mechanized spinning and weaving machinery was sold and copied in other countries. (Shortly after the British government in 1843 lifted the restrictions on the export of textile machinery, such equipment proliferated in countries such as India, Brazil, and Japan.)

The Coal Age was also a great era of industrial espionage. Boulton & Watt, which engaged in spying on competitors itself, was a prime target for industrial espionage. Young mechanic Georg Reichenbach in 1791 bribed a Boulton & Watt night watchman to let him view a steam engine. (Reichenbach in turn designed similar machines in Germany.) Englishman Samuel Slater, initially an apprentice of official water frame inventor Richard Arkwright, learned all about mechanized cotton spinning and held an important position in a leading British textile firm before sailing to the United States. (He was masqueraded as a farmer because textile workers were prohibited from leaving Britain.) After his arrival in the United States, Slater in 1790 opened the country's first important spinning mill in Pawtucket, Rhode Island. His firm inspired many imitators and gave birth to a vast spinning industry in New England. American Francis Cabot Lowell traveled from the United States to Britain for the sole purpose of stealing designs that served for the construction of American power looms from 1813. This established the mechanized weaving technology that allowed the American textile industry to expand rapidly.

Belgium enjoyed a head-start in industrialization long before it existed as an independent country. The Catholic region of Wallonia (in east Belgium) was a traditional iron-making center and had Newcomen-type steam engines operating as early as the 1720s. Belgium was then part of the Austrian Empire, but was invaded by the French in 1794, and became part of the Kingdom of the Netherlands at the end of the Napoleonic Wars in 1815. The English Cockerill family, which initially produced spinning machines in Lancashire, broke the English laws against technology transfer and began producing textile machines in the Wallonian municipality of Liège in 1798. William Cockerill's three sons eventually followed him to continental Europe to expand the flourishing enterprise, and all three of them married into textile families of nearby Germany. They started to produce machinery in the German city of Aachen and purchased a coal and zinc mine. In Wallonia, at Seraing near the city of Liège, John Cockerill in 1817 founded what was to become one of Europe's largest iron-making and machinery complexes. King Leopold I, inaugurated when Belgium achieved independence in 1831, invited Robert Stephenson to plan the Belgian railroad network. In turn the Belgian Cockerill works produced the rails for continental Europe's first steam-operated railway (between Brussels and Mechelen, opened 1835), and established one of the first steam locomotive works on the European mainland: John Cockerill built *Le Belge*, modeled on the latest Stephenson locomotive, in 1835, and by 1844 the Cockerill firm had delivered its 100th locomotive. In short, Belgium early on achieved the highest level of coal-based industrialization of all of continental Europe. Coal-rich Belgium's wealth thus surpassed that of the neighboring coal-less Netherlands, which had been the world's most wealthy region for centuries.

Canal and Railroad Infrastructure

The spread of railroad technology was instrumental to the diffusion of general coal-based industrialization. After all, no other nation came even close to Britain's exceptional water-borne infrastructure. In the early Coal Age on-land transport was still slow and expensive, and the street networks remained inadequate. Britain was an island nation that enjoyed inexpensive water-borne coastline transport, in addition to many naturally navigable rivers. What is more, Britain built up a vast network of artificial canals during its early (waterpowered) industrialization phase.

The United States witnessed a short canal boom as well. It was generally a problem for westward expansion that there were no major east-west waterways (rivers) in the 13 colonies to allow for water-borne transport into the continent's interiors from the East coast. Only few short canals existed before the devastating War of 1812, in which the British captured Washington and burned down the White House and the Capitol, but right after this conflict, New York City's mayor De Witt Clinton promoted the construction of what was to become known as the Erie Canal, an artificial waterway that would link New York City (and more generally the northeastern U.S. population center) to the Great Lakes and the surrounding mid-western heartland. The Hudson river served as natural waterway from the Atlantic ocean up to Albany, but from there substantial investment (and English canal building expertise) was required to construct the artificial waterway that led through mountainous areas to Buffalo at the shore of Lake Erie. The 340-mile-long Erie Canal was, however, immediately profitable, and right from its completion in 1825 contributed greatly to the settlement of the Midwest. It inspired other Atlantic port cities to construct canals as well, but the canal developers of Pennsylvania, Maryland, and Virginia, soon faced funding shortfalls in light of the start of the American railroad age.

Railroad networks were especially important for inland regions located away from any coastline or river. Initially, trains were horse-drawn. In France's south-central city of Saint-Etienne, for instance, a 10-mile (16-km) railway employing horses opened in 1828 to transport coal to the Loire. A few years later, Saint-Etienne was also connected by rail to Lyon on the Rhône. The coal-rich Saint-Etienne basin in turn evolved into an important industrial center that served the more northern parts of France. In Germany, an imported Stephenson steam locomotive, the Adler, operated on the Bavarian Nürnberg and Furth line from 1835. (This line was second only to the Belgian railway in terms of steam operation in continental Europe.) In the Austrian monarchy, continental Europe's first horse-operated railway connected

Budweis to Leopoldschlag in 1827, and to Linz by 1832. The first Austrian steam railroad line, the Nordbahn, opened in 1837. The Nordbahn employed six English locomotives to connect Vienna with the northern provinces. The Semmering railway, which crossed the mountains southwest of Vienna, was constructed between 1848 and 1854 by Carl von Ghega to connect Vienna with Austria's Mediterranean port of Triest (in present-day Italy).

In the United States, steam trains operated successfully from the 1830s. By 1833 the main line of the South Carolina Railroad stretched 135 miles from Charleston to Hamburg. Steam locomotives as well as rails were imported from Britain, but machine building skills progressed sufficiently in Massachusetts and Pennsylvania for locomotives to be made in America from the 1840s. U.S. railway mileage increased from 2,800 miles in 1840 to 9,000 miles in 1850 to 30,000 miles in 1860. The railroad then helped the northern states to win the Civil War, and by 1865 more than half the railroad tracks in the world were located in the United States. And yet this was just the beginning. Another 35,000 miles of new track were laid between 1866 and 1873, and the railroad industry became the nation's largest employer outside of agriculture. The first trans-continental link was established in 1869, and by the end of the century there were four more such lines. As U.S. railroad mileage kept on growing (from 93,000 miles in 1880 to 164,000 miles in 1890 to a record high of 254,000 miles in 1916), it allowed the nation to integrate into one enormous common market that stretched from the Atlantic coast through the mid-Western grain belt to the Pacific Ocean. (Total rail freight grew from 10 billion ton-miles in 1865 to 366 billion ton-miles in 1916.)[34]

Government Involvement and Economic System

The U.S. rail system was built up with substantial public funds. Many state and city governments helped financing early railroads, and many of the western lines received federal land grants to aid their construction. This sort of government involvement indicates that the spread of coal technology was often facilitated by centralized efforts rather than independent entrepreneurs and firms. Nevertheless the role of government in the emergence and spread of coal culture was relatively minor in the United States, just as it had been in Britain: The economic system of capitalism allowed businesspeople to invest their private capital freely wherever they wanted.

Government involvement was generally more pronounced in continental Europe. One major issue all governments had to address was the hardships associated with the social transitions triggered by changing the energy base from agriculture to coal. In France, the weavers revolted in 1831 and 1834, for instance, but social reforms remained slow: 12-hour workdays were maintained, and autonomous workers' coalitions were prohibited until 1864. In

Lower Silesia weavers rioted in 1844, destroying factories after their wages for contract work had sunk below subsistence level. Entire families starved to death, while government authorities failed to react. More and more people moved towards urban centers, and cities such as Berlin began to grow rapidly. In addition, many Germans emigrated. A lot of them left for the United States, while others sought opportunities in the Russian Empire. In the United States, the percentage of the total labor force working in agriculture decreased from around 75 percent in 1790, to 60 percent in 1850, to some 40 percent in 1900. The United States then had 38 cities with over 100,000 inhabitants. The growing industrial centers were perceived as dark, dirty, and crowded places, and children were laboring in some American factories still in the beginning of the 20th century. The depressions of the 1870s and 1890s were accompanied by strikes, but both the Railroad Strike of 1877 and the Pullman Strike of 1894 were unsuccessful, as workers were generally not well organized.

Governments also sheltered their maturing industries from foreign (mainly British) competition, closing national borders to imports of manufactured goods. The German states in 1834 founded the Zollverein, a customs union that removed tariffs between the German states while restricting foreign imports. To varying degrees governments also took other measures to promote industrialization. Some involved themselves in building up the required infrastructure, others provided investment capital. Britain's Industrial Revolution had been a trial-and-error industrious evolution, but from about 1850 things turned more scientific and structured, and governments began setting up formal training programs and centralized research institutes. The taxes paid by corporations were used by state authorities to finance public technical schools from which corporations could source their personnel. The German model of combining research and teaching in one institution turned out to be tremendously successful and was generally adopted as the standard of Western universities. (Much of this success is associated with the fact that modern chemistry emerged from German universities and corporate laboratories during the later Coal Age.) In America, the first such institution was Johns Hopkins University, founded in Baltimore in 1876.

Warfare

Warfare may also be considered a kind of government involvement that actually had a major impact on industrialization and the spread of coal technology. Most importantly, wars increased the demand for iron, which was now produced with coke. From 1754, Britain fought the Seven Years' War, the American Revolutionary War, and the Napoleonic Wars. The latter delayed the usually swift technology transfer across the English channel, sheltering Britain's substantial head start. Only after Napoleon was finally defeated at

Waterloo, and peace restored at the Congress of Vienna in 1815, was continental Europe able to engage fully in all facets of the coal-fueled industrialization process. Thereafter, conflicts that involved continental European nations actually promoted all aspects of coal technology that proved a military advantage, including the laying of railroad tracks.

In the United States, the Civil War period is associated with the adoption of coal culture, and not only because it promoted railroad construction and U.S. iron production. At the core of the conflict was an industrializing North that disliked the trade relations between Britain and the U.S. South. (The South delivered raw cotton to Britain and preferred inexpensive British manufactures over products from the industrially immature U.S. North.) Hence, the South promoted a free trading regime, while the North demanded trade restrictions (and unrivaled access to cotton from the South). Eventually, the conflict over these contrary economic interests escalated into the American Civil War, which was fought between 1861 and 1865, and left 600,000 dead.

Warfare was also the means to create colonial empires that would serve as captive markets for goods mass-produced with coal energy in the mother country. Britain was ideally positioned in this respect, as its empire was growing rapidly. At the end of the Seven Years' War, Britain gained Canada, while India, Australia, New Zealand, and large parts of Africa became British colonies during the Coal Age. The other western European Coal Powers joined in on the conquest of Africa, and all of them used their military power to enforce trade relations with the population centers of the Far East.

The United States expanded coast-to-coast during the Coal Age by warring Mexico and native Americans (Indians). This created an empire that stretched from the Atlantic to the Pacific Ocean, a huge internal market with free movement of goods, people, capital, and ideas, united by a single common language. What is more, the United States established its (economic) interests in Latin America, acquired Hawaii (as a state) and the Philippines (as a colony), and used cannon steamers to force Japan out of seclusion to make it a trading partner.

Agricultural Self-Sufficiency

One of the most critical factors in place to allow Britain to industrialize was a solid domestic agricultural base that delivered plenty of traditional energy in the form of grain. Britain did actually not just lead the world in terms of industrialization at the beginning of the Coal Age; it also enjoyed the world's most efficient agricultural system. Continental Europe, too, had plenty of domestic agricultural energy available to fuel its urbanization and industrialization. France industrialized in northern regions, while southern regions delivered agricultural goods. (France also developed Algeria as its granary, encouraging the emigration of French farmers to this North

African region.) The territory covered by the not-yet-unified German states extended quite far into the east into regions located in what is now Poland. Generally, the German-speaking area industrialized in more western regions, while more eastern regions delivered grain. The multilingual Austrian monarchy was the largest empire in Europe except for Russia, covering much of northern Italy and wide parts of eastern Europe. The Austrian monarchy industrialized mainly in the west, in the regions of present-day Austria and Czech Republic, while grain was supplied from the monarchy's (south)eastern regions such as those now located in the Ukraine, Hungary, Romania, and Serbia. (The Bukovina, or Buchenland—Land of Beech Trees, a region now divided between Romania and the Ukraine, was the easternmost crown land of the Austrian Empire and one of its most important granaries.) However, western Europe, as defined after World War II, actually became a major net importer of cereals from the 1860s in order to feed its rapidly urbanizing population.

The United States had no problems at all providing food for its growing cities and factory population in the northeast, as the expanding nation took more and more highly fertile land in temperate zones from native Americans. U.S. farmers could actually afford to be far more inefficient than their western European colleagues in terms of agricultural energy captured per acre. They were also able to set much land aside to fuel horses. This led to the development of large horse-drawn harvesting machines, which made U.S. farming very efficient in terms of output per farm worker. (The largest Californian combines, that is, harvester-threshers, were drawn by up to 40 horses.)

Britain Losing Ground (against Germany and the United States)

One way of looking at how fast coal technology spread to western Europe and North America is to observe how quickly Britain lost its technological edge. Britain was perfectly positioned to take the step from the Agricultural Age into the Coal Age. All the features of the previous Energy Era were well developed (efficient agriculture, water-borne transportation, waterpower), and the prerequisites of the upcoming era were readily available (coal deposits, iron ore). So it happened that, up until 1830, most up-to-date coal technology was confined to Britain. However, during the following decades Germany and the United States rivaled Britain's position and eventually emerged as the most productive nations in the world. Germany did especially well in terms of its iron and steel, machinery, and chemical industries. As early as 1810 Friedrich Krupp established a small steel plant in Essen in Germany's Ruhr valley (Nordrhein-Westfalen), which grew into a world-famous firm under the leadership of his son Alfred, the Cannon King, who introduced new methods for producing large quantities of cast steel. By 1870 the German textile and metal industries had surpassed those of Britain in terms of productivity. In turn Britain imported a lot of steel, machinery and manufactures from Germany.

The German chemical industry became especially dominant. Starting in the 1860s, it supplied the world with pharmaceuticals, fertilizers and dyes. However, German chemists advanced highly critical innovations even before the chemical industry had emerged as a large segment of industry: Andreas Sigismund Marggraf, for instance, published a method of extracting sugar from beets in 1747, and F.G. Achard developed the corresponding large-scale industrial process in 1793. This practically put an end to slave-based sugar production in the Caribbean and South America. (Sugar from beets, *Beta vulgaris*, is identical to sugar from cane, but beets grow in temperate climate. The first sugar beet factory went into production in Cunern, Silesia, in 1802, and henceforth Europeans grew and produced their own sugar.)

The United States and Canada were geographically further away from Britain than continental Europe, but culturally very close. The shared language and continuing immigration made technology transfer from Britain to North America very easy. The United States of America emerged as a nation from the Revolutionary Wars (ending in 1781), covering the territory of the former 13 colonies. Canada, on the other hand, remained a British colony and achieved status as a self-governing dominion only in 1867. Even thereafter, Canada remained formally under the British Crown. Hence, Canada had to obey the British laws agitating against technology transfer, while the United States continuously stole British know-how and actively tried to hire British experts, often by offering substantial bounties. These realities probably contributed to the fact that Canada industrialized only after the turn of the 20th century, while the United States achieved this transition already in the mid- to late-19th century. Moreover, Canada lacked the vast agricultural base the United States was enjoying.

One critical feature that helped the United States in its early (waterpowered) industrialization phase was the fact that nearly unlimited amounts of cotton were grown and harvested (by slaves) within its own national borders. Before long Americans managed to acquire British textile production expertise, and by the 1840s the United States had 1,240 cotton factories, half the number in Britain. (Nevertheless, the United States in 1861 exported some 900 thousand tons of raw cotton, accounting for two-thirds of all the raw cotton exported around the world.) But U.S. luck in terms of resources did not end here. Vast deposits of coal and iron allowed the United States to build up a prominent iron and steel industry. Scotland-born Andrew Carnegie came to the United States as a teenager and operated his first blast furnace in 1870. The Pennsylvanian iron makers eventually took the technological lead from Europe, and by 1900 the Carnegie Steel Company's mills produced more steel than all of England (and one third of all American steel). American coal-fueled production of goods increased rapidly as well. In 1900 the United States had 30 companies manufacturing 5,000 steam engines per year, and U.S. systems of mass-production made America the only rival of Germany in terms of productivity. (Nevertheless the United States did not fully urbanize during the Coal Age: more than one out of three Americans still worked in agriculture on the eve of World War I.)

As Britain lost its edge as the world's most efficient producer, the country faced serious troubles during the Long Depression of 1873 to 1896. Britain was losing market shares in newly industrialized Western countries as well as in less-developed regions (such as Latin America, India, China and coastal Africa): its share of world trade fell from a quarter in 1880 to a sixth in 1913. What saved Britain economically during the Long Depression was the export of services, such as banking, insurance, and shipping services. (Capital investments outside Europe, including Africa, the Middle East, the Indian subcontinent, Southeast Asia, and the South Pacific, were especially important.)

SPREAD TO THE RUSSIAN EMPIRE

Russia was geographically closer to western Europe than North America, but it was culturally a lot further away. Peter I, the Great, who reigned from 1689 to 1725, attempted to modernize Russia towards the late Agricultural Age. He frequently traveled to western Europe to hire experts who in turn set up manufactories in Russia and improved its mining operations. However, Peter's coerced top-down efforts failed to trigger a general industrialization process and mainly focused on supplies for the army. What is more, Peter practically extended serfdom from farm workers to industrial laborers to create a workforce for the new manufactories. He also added to a trend of increased autocracy and enserfment on the countryside, which in effect degraded serfs to slaves. Russian serfdom then persisted for another one and a half centuries to be abolished only in the later Coal Age, in 1861, and under terms very unfavorable to peasants. Russia's rural masses therefore kept living in poverty, and the transition into the Coal Age was accompanied by pronounced social problems and political unrest. (The Coal Age's working-class and revolutionary movement in Russia eventually culminated in the assassination of Czar Alexander II in 1881.)

In other respects Russia was very well equipped to undergo the transition towards the Coal Age. As a matter of fact Russia shared quite a few features with the United States of America. Both nations were endowed with abundant iron and coal resources, both kept on expanding their territories, and both remained overwhelmingly rural and wood-fueled for much of the 19th century. The Russian Empire's extreme coal riches were mainly located in the eastern Ukraine, an area gained from Poland in the 17th century. The Donets basin, between the Dnieper and Don rivers (in parts of present-day southeastern Ukraine and southwestern Russia) is still now the most extensive coalfield in eastern Europe. Coal was mined in this region from the early 19th century, and luckily the coalfields were also adjoined by vast iron deposits. Welshman John Hughes in 1872 set up an ironworks in the city of Donetsk to deliver iron rails for the growing Russian railroad network. (Donetsk was initially named Yuzovka for Hughes.) By 1913 this plant delivered 74 percent of all Russian pig iron, and the Donets basin provided for 87 percent of all coal mined in the Russian Empire. (The empire's second major coal region, the Kuznetsk Basin, was located in southern Siberian Russia, between the Altay Mountains and the Salair Ridge. Here coal was mined from 1851, and the region also turned into a major industrial center.) Russia's coal output was as small as 0.3 million metric tons annually still in 1860, but increased to 3.3 million tons in 1880. (By this stage Britain mined 149 million tons, and Germany 59.1 million tons of coal per year). Russian coal output then doubled to 6.9 million tons annually in 1892, and nearly tripled to 18.7 million tons annually in 1905. Russian pig iron production increased accordingly, from 449 thousand tons per year in 1880 to 1.1 million tons in 1892, to 2.7 million tons in 1905. But both iron and coal production remained low in comparison to that of the major Western Coal Powers.

Growth in Russian coal and iron production was driven mainly by the construction of a railroad network throughout Russia's vast and expanding inland territory. (This improvement of infrastructure was as important for the Russian Empire as it was in the United States around the same time.) Russia's first railroad opened as early as 1837 to connect St. Petersburg and Tsarskoe Selo (Pushkin), but still in 1860 Russia had merely around 1,000 miles (1,600 km) of railroad tracks. Thereafter the Russian railroad network grew rapidly to reach 6,600 miles (10,700 km) in 1870, about 11,800 miles (19,000 km) in 1875, and 19,900 miles (32,000 km) in 1890. During the following decade, between 1890 and 1900, railroad mileage nearly doubled again, to give Russia the most track of any nation in the world except for the United States. The Trans-Siberian Railroad, stretching all the way from Moscow in Europe to Vladivostok at East Asia's Pacific coast, finally opened its full length in 1904.

The construction of the Trans-Siberian Railroad was promoted by Sergey Witte, a former railway executive, who entered the Russian government as Director of the Department of Railway Affairs in 1889. In 1893 Witte became Minister of Finance, and under his patronage Russia experienced industrial growth that was faster than anywhere else in Europe at the time. However, the merits of Witte's ambitious overall industrialization plan are still debated. Industrial development remained restricted to a few centers (such as St. Petersburg, Moscow and Baku), and much of the industry was foreign-owned. Witte was dismissed as Minister of Finance in 1903, and full and general industrial transition was not experienced in Russia until well after the turn of the 20th century. The railroad network, despite its enormous mileage, was actually not especially dense, and much of the empire's vast territory remained quite scarcely settled, while the population was concentrated in the European areas. Russia's overall population grew from 71.1 million in 1855 to 146 million in 1904, and it generally remained poor. The rapid urbanization of the 1890s was accompanied by social unrest, with abysmal working conditions, high taxes, and rural hunger continuing into the 20th century. In short, Russia neither fully industrialized nor stabilized during the Coal Age.

SPREAD TO JAPAN

In Asia, the only country to truly industrialize during the Coal Age was Japan. The way coal technology spread to Japan was altogether different from the way it diffused from western Europe to the United States or Russia. In fact, Japan secluded itself from any outside influence until the Western Coal Powers pressured it into trade relationships. In response, Japan decided to rapidly and radically modernize itself.

Still in the mid-19th century Japan was characterized by the kind of feudal organization that had dominated the country since ancient times. Shoguns (that is, military leaders) held power from the 12th century, while the role of the emperor was reduced to that of a figurehead. In the 17th century Japan entered a period of isolation and seclusion that was going to last for more than two centuries. Foreigners were expelled, Japanese citizens were prohibited from leaving the country, and overseas trade was eliminated. However, Japan's internal economy was flourishing during this period.

Japan's long seclusion ended quite abruptly when four U.S. cannon steamers appeared in Edo Bay in July 1853. The American intruders forced Japan to end its isolation and demanded access to coal, the energy source Americans used to expand their presence and power in the Asia-Pacific region. Except for coal, Japan actually did not have resources that were of immediate interest to the Western powers, which, however, were eager to sell their goods to the booming Japanese economy. Already in 1846 Americans had shown up with two warships, but Japan had turned down their request for diplomatic relationships. The Japanese simply wanted to mind their own business, but they had no means to defend their coast-line against coal-fueled iron steamers. Following U.S. navy commander Matthew Perry's forceful foray of 1853, Japan was compelled to open three (and later five) ports to international trade; to admit foreign merchants; and to sign unequal trade treaties. (After signing treaties with the United States, Japan was forced to sign similar ones with several European powers, including

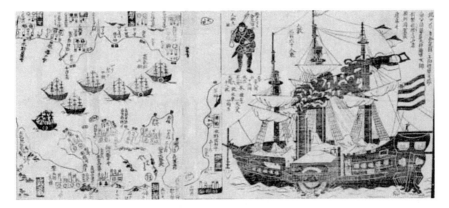

American Intruders at Edo Bay in July 1853   The U.S. Navy's coal-fired cannon ships forced Japan out of seclusion. This Japanese print of 1854 depicts one of Commodore Matthew Perry's side-wheel steamers.

Britain, Russia, and the Netherlands.) In consequence, it did not take long before unwelcome Westerners were attacked in Japan, but these assaults were answered by the bombardment of the coastal cities of Kagoshima and Shimonoseki.

Meiji Restoration

Japan was thrown into turmoil. There was no precedent to such threat to national security, and the Japanese public blamed the shogun for the humiliation. In 1867 the shogun was forced to resign, the capital was moved from Kyoto to Edo (which was renamed Tokyo), and the boy emperor Meiji was *restored* to power. (Meiji means enlightened rule. Prince Mutsuhito, born 1852, assumed the name in 1867 when he succeeded his father, emperor Komei.) In reality, a small group of senior politicians, termed genro, exercised authority and power during the following years. These leaders realized that it would be impossible to get rid of the foreigners. Even though the Meiji restoration was originally inspired by antiforeign sentiment, they opted for a radical turnaround whose key element was to strengthen Japan by copying and adopting Western technology. Scholars were sent abroad to learn as much as possible about Western technology (and political institutions), while American, French, British, German, and other Western engineers and consultants were called in to help building harbors, railroads and factories. The government invested heavily in infrastructure and set up a number of model factories. The first railroad was opened between Tokyo and Yokohama in 1872, and thereafter the railroad network quickly grew across the nation. Western ideas gained wide acceptance. The feudal system was abolished; compulsory education was extended to all; a new currency and banking system was established; a uniform code of law was introduced; and a constitution based on the imperial German model was instituted in 1889.

Though Japan evidently rushed into the Coal Age, much of the government's initial efforts focused on reforming the agricultural sector. Farmers were entirely freed from restrictions on land use and could grow whatever they wanted. (From 1872 people had the right to buy and sell land.) The government established agricultural colleges and experimental farms, and provided farmers with technical advice. More irrigation systems were installed, more fertilizer was applied, and pest control strategies were improved. By 1890 agricultural production was almost double that of 1873. There was much focus on the production of tea and raw silk, which, combined, soon accounted for about 80 percent of Japan's export revenue. (This included the export of silkworm eggs to the European silk industry.) Green tea was exported mainly to the United States to replace some Chinese green tea, as well as black tea, cocoa, and coffee. Silk was exported to Britain, France and the United States to capture significant market share from Chinese and Indian exporters.

Tax revenues from the agricultural sector were then used to modernize Japan's traditional light industry. Here, the textile sector was in focus. Japanese weavers began to use the flying shuttle from 1874, and the cotton industry was soon able to compete against British and Indian cotton textiles on the Chinese market and elsewhere in South and Southeast Asia. From about 1890, some 20 years after the Meiji period had first begun, Japan's industrial economy began to expand rapidly. Foreign investment and foreign entrepreneurs played an important role in this Japanese industrial expansion. Americans set up a flourishing beer brewing and paper industry, for instance.

Revenues from the agricultural and light manufacturing sector were used to import heavy industrial equipment. Japan received from the West everything from metal tools to machinery, steam engines, ocean-going vessels, and battleships. The government organized these strategic imports and eventually invested into the buildup of heavy industries that were important for the military: shipbuilding, munitions, steel. But the development of heavy industry was relatively slow, as Japan lacked iron ore as well as other critical resources. (This shortcoming was a main reason why Japan was soon going to follow the path of Western Coal Powers all too closely to become an imperialistic nation itself.) On the upside, Japan had energy resources. The country's many rivers allowed for the utilization of waterpower and the installation of hydroelectric capacity. Japan's coal resources were modest, but they were sufficient to fuel the country's steam engines and iron production.

Using imported iron ore, Japan's first coal-fueled blast furnace operated from 1901 at Yawata ironworks, the predecessor of Nippon Steel. And by 1910 Japan mined between 10 and 15 million tons of coal per year, more than any other Asian country.

The general population suffered during Japan's rapid industrialization, much as the populations of Europe and the United States had earlier on during the transition from Agricultural to Coal Age. Japanese miners worked 12 hours a day under close supervision, and the government outlawed strikes. Japan's population grew from 35 million in 1873 to 43.8 million in 1900, but it was still not very urbanized at the beginning of the 20th century. In part this was because Japan kept much of its focus on decentralized light manufacturing. For the time being, and up until the early 1930s, Japan's main export products were raw silk, tea, cotton textiles, and inexpensive light industrial and consumer goods of moderate quality.

WHERE THE COAL AGE
REMAINED AGRICULTURAL

In the global perspective, only a few regions industrialized during the Coal Age. Only western Europe, North America, Japan and parts of the Russian Empire entered the new Energy Age, while most of the world kept on relying principally or solely on agricultural energy. (As of 1900, Europeans and their overseas descendants accounted for 30 percent of the world's population, but consumed 95 percent of fossil fuel energy.) Given that Western firms were eager to sell their technology internationally, and coal was available on world markets, this may sound somewhat surprising. Why did most of eastern Europe, much of southern Europe, the Middle East, central Asia, southern Asia, nearly all of eastern Asia, plus Latin America and Africa, remain overwhelmingly rural during the Coal Age?

One of several answers to this question is vested interests. In regions such as eastern Europe, southern Europe, and Latin America, landlords remained powerful enough to maintain the medieval feudal structures that associated wealth and prestige with land ownership. The more power held by landlords, the weaker the central government whose coordinated efforts would have been essential to manage the transition from agricultural to coal energy. The Polish landlords took political power from their king and forced rural workers into a new serfdom. (Poland then turned into a chief exporter of wheat to western Europe.) Spain and Portugal had lost their technological edge (much of which had been inherited from the former Muslim inhabitants of the Iberian peninsula) as they failed to turn income from colonial exploitation into domestic self-sustaining economic growth. Instead, they used their sudden riches to purchase goods from other western European regions, promoting foreign rather than domestic industries.

Agricultural efficiency did not improve much in Iberia either. Southern Italy, which for long remained under Spanish control, had a similar fate, while northern Italy, which was part of the Austrian monarchy until the 1860s, began to industrialize relatively early. Southeastern Europe, including Greece, was part of the Turkish Ottoman Empire, which enjoyed limited technological exchange with western Europe and was feudally structured as well. (Greece gained independence from the Ottomans with the help of the European Coal Powers in 1829, though within borders that were substantially smaller than those of modern Greece.) Latin America remained rural, with the temperate Southern Cone of Argentina, Chile, and Uruguay becoming wealthy as a supplier of wheat and other agricultural products to international markets.

Another factor impeding industrialization was lack of sizeable domestic markets. The countries that rapidly followed Britain's path of industrialization closed their borders for foreign competition to give their infant industries time to develop. However, these countries were already relatively wealthy and enjoyed domestic markets to which their maturing industries could sell their products. What is more, these economies had already experienced some degree of waterpowered industrialization and the workforce had over centuries become acquainted with increasingly more complex machinery. Overwhelmingly rural societies, in contrast, may have found the step into the Coal Age too steep. Even if a purchase of foreign machinery was accompanied by extensive training of local people, a traditional society would have had problems copying a steam engine or a locomotive. And some societies may also have been unwilling to face the high social cost that industrializing economies had to pay during the transitional period. (Specializing in plantation agriculture for export involved far fewer social changes.)

Increasing domestic labor costs should have been an incentive for Western business owners to transfer their industries to less developed countries. However, even in the later Coal Age, when Western wages kept rising and transportation costs began to tumble, production was not generally moved to other countries. Factory owners often preferred to produce close to their domestic markets, probably to hedge against the risk of political instability, either in host countries or internationally. (Conflicts between globally operating rivaling Coal Powers would have threatened to interrupt shipments at any time.) Besides, the countries that were already industrialized offered a superior infrastructure and a trained workforce that made it easier to keep the machines running at home than in less-developed regions.

The Coal Powers also had reasons to deliberately slow down the spread of their technology. Coal-fueled productivity and mobility directly translated into wealth and political power. Industrialized nations therefore restricted the export of machinery. They were primarily interested in selling manufactured goods overseas, not in helping others to become efficient producers. None of Asia's three largest economies, present-day China, India, and Indonesia, industrialized during the Coal Age. Indonesia was a Dutch colony delivering agricultural goods, timber, and mineral resources. India became a British colony, delivering raw cotton, indigo, and spices. Meanwhile China was forced to trade under Western terms. Africa was partitioned between the European powers for plantation agriculture and mineral exploitation as much as for reasons of prestige. In short, the Coal Powers built international empires to serve their own interests rather than to promote global industrialization.

Nevertheless Coal Powers to some extent introduced technology to their colonies. India had its own mechanized textile mills within 10 years after the British government in 1843 lifted the restrictions on the export of textile machinery. Railway networks were built all over the world to serve colonial interests, but they also benefited the local population. India's first railroad opened in 1853, and the rail network spread steadily over the sub-continent to reach 9,000 miles by 1880 and 43,000 miles by 1935. Besides improving a country's infrastructure, railroad construction is generally associated with development and employment. However, in India disease and accidents accounted for a heavy death toll among local rail workers, and India kept on importing steam locomotives from Britain, producing only wooden passenger carriages locally.

NOTES

32. The term Carboniferous Crescent has been used by J. R. McNeill and William H. McNeill in *The Human Web*:

> Britain has abundant coal deposits, part of a "carboniferous crescent" that stretched from the Scottish lowlands through England to northern France and Belgium and the Ruhr region of Germany. This would become the industrial heartland of Europe, a region as important for modern history as the fertile crescent was for ancient history.

John Robert McNeill and William Hardy McNeill, *The Human Web: A Bird's-eye View of World History* (New York: W. W. Norton & Company, 2003), 231. What the McNeills are overlooking is the role of Silesia, which was part of this Carboniferous Crescent. The Silesian coal deposits were important for the coal-based industrialization of Germany and the Austrian Empire. At the end of the First Silesian War in 1742, Austria kept small parts of Upper Silesia (which, however, hosted substantial coal mining operations in the 19th century), though most of Silesia was then ceded to Prussia. In 1913 Upper Silesian coalfields accounted for 21 percent of imperial German coal production. F. Gregory Campbell, "The Struggle for Upper Silesia, 1919–1922," *The Journal of Modern History* 42 (1970): 361–385, http://www.jstor.org/pss/1905870.

33. Stephen Hughes, "The International Collieries Study," ICOMOS (International Council on Monuments and Sites) and TICCIH (The International Committee for the Conservation of the Industrial Heritage), International Council on Monuments and Sites, Paris, France, http://www.international.icomos.org/centre_documenta tion/collieries.pdf

34. Eric Foner and John Arthur Garraty, *The Reader's Companion to American History* (New York: Houghton Mifflin Books, 1991), 908. Find more on this topic in: Alfred D. Chandler, *The Railroads: The Nation's First Big Business* (New York: Harcourt, Brace & World, 1965); John F. Stover, *The Life and Decline of the American Railroad* (Oxford: Oxford University Press, 1970); Oliver Jensen, *Railroads in America* (New York: Bonanza Books, 1975).

CHAPTER 18

TRANSITIONAL DOWNTURN: POLLUTION, POVERTY, AND HARD WORK

The step into the Coal Age caused transitions that were almost as radical as those following the emergence of agriculture millennia earlier. The unprecedented boost in command of energy entirely changed the way people lived their lives and the way societies were organized. It affected families, nations, and international relations. Paralleling the effects observed following the onset of agriculture, the new energy transition created winners and losers within society; it caused enormous hardship during the transitional period; it resulted in societies that were more powerful than traditional societies; and it created positive feedbacks that enhanced technology, productivity, mobility, agricultural production, and population growth. But while the emergence of agriculture initially only had regional effects, the onset of the Coal Age was embedded in a globally open infrastructure and had immediate international consequences.

Most of the social changes occurring in the Coal Age are associated with increased urbanization and the complete reorganization of work. At the wealthy end of society, power shifted from landowners to industrialists as the principal source of energy behind society shifted from agriculture to coal culture (or coal-fueled industry). The rural poor migrated into the cities to evolve into a large class of very poor urban factory workers. In Britain, rural exodus, urbanization, and the emergence of an urban working class began with the land enclosure laws. Seventy percent of England's population lived on the countryside in the beginning of the 19th century, but by 1851 manufacturing employed more workers than agriculture, and by 1913 merely 12 percent of the British workforce labored on the countryside.

Working in the early factories was no fun. Everyone was closely overseen; people had to work with complete strangers (which was highly unusual for people born into a family-based agrarian economy); and skilled artisans often found themselves engaged in simple, repetitive, routine tasks. Wages remained low as there was an oversupply in labor, sustained by the continuous stream of people arriving from the countryside. Besides, the population grew rapidly and jobs kept being eliminated as machines took over more complex tasks. Women and children worked the same 12-hour days as men, but were paid less and hence drove the male workforce out of certain industries. (Tens of thousands of children, often recruited from orphanages, labored in England's cotton mills.) Working conditions were often extremely dangerous. A 20-year-old starting to work at U.S. Steel in Homestead, Pennsylvania, in the end of the 19th century had almost one chance in three that the factory would disable him and one chance in seven that the factory would kill him before the age of 50.[35] The work in the coal mines was hazardous as well. Roofs fell, mines were flooded, toxic methane and carbon monoxide gas accumulated, and methane or coal dust exploded. (Coal is still now considered the most dangerous source of energy in terms of people being killed while producing it.)

As people walked to work, they basically lived right next to polluting factories, and the cities were unsanitary and overcrowded. In Liverpool thousands of people lived in basements without light or heat. In parts of Manchester just one toilet existed per 200 inhabitants. Deadly diseases such as cholera, tuberculosis, measles, and pneumonia all spread quickly and easily. (Britain experienced devastating cholera epidemics in 1831–1832, 1848, and 1854–1855.)[36] Nevertheless, the population of the cities grew rapidly. All of Europe combined had barely two dozen cities with a population of 100,000 people in 1800, but Britain alone had nine such cities in 1850, and all of Europe had 150 such cities by 1900.

The new energy source had a direct negative impact on the health of Coal Age people as well: the burning of coal severely polluted the air, as it releases quite large amounts of solid particulates during combustion. By 1900, air pollution caused or exacerbated respiratory diseases that killed hundreds of thousands annually in coal-burning cities around the world. Coal also releases sulfur dioxide (SO_2) during its combustion. (Sulfur is contained by all living organisms, including those plants that grew in the coal swamps.) Once SO_2 gas is released into the atmosphere it interacts with moisture to form tiny droplets of sulfuric acid (H_2SO_4) that fall down as acid rain to damage forests, lakes, and rivers. Air pollution, acidic rain, and mining accidents were all direct effects associated with the new energy source, just as malnutrition and hard field work had been for early agrarians.

NOTES

35. J. Bradford DeLong, "Slouching Towards Utopia?: The Economic History of the Twentieth Century, Part B: The Path from the Pre-Industrial World, VIII. The First Global Economy: Production and Trade," University of California at Berkeley and NBER (National Bureau of Economic Research), January 1997, http://econ161. berkeley.edu/TCEH/2000/eight/html/Slouching2_8preWWI.html.

36. Graham Thomas, "Population Growth and Dynamics in Britain and Gloucestershire," http://www.grahamthomas.com/population.html.

CHAPTER 19

LONGER LIVES, BETTER EDUCATION,
MORE RIGHTS

However, living conditions soon improved for those living in energy-rich Coal Age societies. Despite local and temporary surges in mortality, Britain's longer-term overall death rate actually declined during the time the cities were rapidly growing. The nation's mortality rate peaked in 1710 at around 35 deaths per 1,000 population per year, and then decreased to 22 deaths per 1,000 people in 1850. (It is now about 11 deaths per 1,000 population annually.) Improved sanitary conditions and better nutrition, combined with the development of pharmaceuticals, radically decreased infant mortality and allowed for people to live longer, healthier lives. Life expectancy in Britain increased from 35 years in 1780, to 39 years in 1820, to 41 years in 1870, and to 53 years in 1913.[37] Meanwhile birth rates in Britain grew from annually 30 births per 1,000 people per year in 1700 to 40 births per 1,000 population during the 19th century. As a consequence of these changes in birth and death rates, the British population surged from some 7.5 million in 1750 to 10.8 million in 1801, to 20.9 million in 1851, and to 37.1 million in 1901.

GLOBAL POPULATION GROWTH

The dynamics of population growth were similar in all other regions that entered the Coal Age, and even in many areas that remained agrarian. (World population growth was about 0.4 percent annually between 1650 and 1750 as well as between 1750 and 1850, and then increased to about 0.6 percent per year between 1850 and 1900.) To a great extent, worldwide population growth simply reflected the effects of the global agricultural revolution observed in the Super-Agricultural Era. Industrialization and coal energy as such do not seem to have been a prerequisite for decreasing death rates and accelerated population growth: in southern China the population began to grow faster than in western Europe in the absence of industrialization, and Russia experienced the same effect in the presence of very limited industrialization. In the Americas indigenous societies were displaced by Europeans, who efficiently farmed Fertile Crescent domesticates. There was plenty of fertile virgin land, and Europeans expanded rapidly into the vacant space created by either Old World diseases or the outright killing and displacement of Native Americans. In consequence, European population growth in North America contributed significantly to the rapid overall global population growth: In 1750, just about 2 million people lived in the United States and Canada (0.3 percent of the global population), while it was as many as 82 million in 1900 (5 percent of the global population). Meanwhile in Asia the population nearly doubled from roughly 500 million in 1750 to about 950 million in 1900. During the same period Europe's population increased two and a half fold from about 160 million to 400 million, while Africa's population remained relatively stable, growing from 100 million people in 1750 to 134 million in 1900. Adding these figures together shows that the global human population increased from perhaps 780 million in 1750 to about 1,650 million (or 1.65 billion) in 1900. The one-billion mark was reached sometime between 1800 and 1850. Significantly, the number of people of European origin and descent increased from about a fifth of the global population in 1750 to a third in 1900.

One truly revolutionary trend in population development began in the Coal Age when people started to consciously decrease their reproduction rate. From ancient times, many civilizations (including medieval Christian Europe) had attempted to reduce population size by infanticide (that is, the killing of infants), but the Coal Age saw the first wave of large-scale non-infanticidal birth control. Apparently it all began among the French nobility, as well as in Japan, in the 17th century. According to a 1776 French account by the Baron de Montyon, "they" had started to "cheat nature even in the villages!"[38] By the 19th century the movement had spread to all classes in most European countries, as well as to New England. (The principal technique is assumed to have been withdrawal before ejaculation, coitus interruptus, a method that was probably known since ancient times. It is described in the biblical story of Onan, for instance.) Perhaps this trend was caused by parents' perception that their children had better chances of survival, but in some places fertility fell before mortality did. Literacy, a change in the status of women, or industrialization were not prerequisites of a fertility transition either. (In Japan the deliberate control of family and population size may have been the response to obstacles to further agricultural expansion.) Anyway, this trend towards birth control was not pronounced enough to keep the overall global population growth in check. The human population was now already very large, and even a small growth rate was bound to create a serious avalanche problem. (A population of one billion, growing at 0.5 percent per year adds five million people to the planet in the first year.) Within just one century, between 1830 and 1930, another one billion people were added to the planet, the same number that had been reached during all the centuries and millennia prior to the Coal Age.[39]

EDUCATION AND POLITICAL RIGHTS

It was not people's health and life expectancy alone that took a major turn to the better during the Coal Age. Energy-rich societies also achieved a remarkable boost in general education and political rights. Urbanized populations are easier to school and to organize politically than dispersed rural populations. In Britain literacy increased from 50 percent (1780) to 54 percent (1820) to 76 percent (1870) to 96 percent (1913). Elementary school education for practically all children became the norm in Western nations in the late Coal Age, and increases in higher education helped improve the quality of the workforce to further improve productivity. In 1910 about one million students were enrolled in post-elementary education in Germany, and some 355,000 (or five percent of their age cohort) were attending college in the United States.[40]

The political and social rights situation improved drastically as well. In Britain income levels for most workers began to increase from the 1820s, and people began to adjust to the new circumstances and conditions. The urban masses began organizing themselves and soon achieved social reforms. Workers started to form trade unions, and governments enacted regulations to protect workers as the balance of power between employee and employer tends to favor the latter. The Factory Act of 1833 was the first law against child labor in Britain. It limited the workday for those between 9 and 13 years to eight hours, and that of the 14- to 18-year-olds to 12 hours. About a decade later children and women were excluded from the work in mines.

Such legislation became possible because of the rise of democracy. In this system of governance each person has one vote, and the elected government represents the interests of the majority of the people, not those of a wealthy minority. To be sure, the fact that Britain and other western European nations were not democracies towards the end of the Agricultural Age likely helped them to enter the new Energy Era: the radical changes experienced in the transitional period adversely affected a majority of the population and would have been unthinkable in a democratic environment. But eventually political power shifted to the people. The transition towards democracy varied from radical and fast to gentle and slow. The French achieved democracy early and violently. After storming the Bastille fortress on 14 July 1789, they executed King Louis XVI in 1793. The French Revolution seemed a failure by 1799, and appeared nullified by 1815, but it established the precedents of such democratic institutions as elections, representative government, and constitutions. The British took a very different approach. They moved slowly towards democracy and were almost alone, across Europe and the Americas, in achieving democracy without civil war or revolution. (Power passed from Crown to Parliament in the 17th and 18th centuries as Britain became a constitutional monarchy. In the 19th and early 20th centuries Parliament itself became more democratic, and the House of Lords was subordinated to the House of Commons.)

As democracies matured, the right to vote was extended to all of society. In Britain, this process took about 100 years. In 1832 the vote was extended to middle class men, then to working class men, and finally to women. In 1918 British women from age 30, and British men from age 21 had full voting rights. Only in 1928 did British women gain the vote on the same terms as men. New Zealand, in contrast, granted equal voting rights to all women, including native Maori women, in 1893. In the United States women's suffrage was won in 1920.

ABOLITION OF SLAVERY

With civil and political rights spreading to all members of coal-fueled societies, voices were raised for the abolition of slavery throughout the global empires of the major Coal Powers. Unfortunately, coal energy could not be applied to field work, which aided the persistence of agricultural slavery. Vested interests remained strong, and progress slow, even though many realized that slave work was often uneconomical, as unfree workers were unmotivated and had to be closely overseen. But with time the humanitarian antislavery movement gained momentum, first achieving prohibitions to the shipping of slaves, then a ban to trade humans, then gradual manumission, and finally the total abolition of slavery.

Britain was the most active slave-trading nation in the world. Sugar merchants opposed change, and slave-produced U.S. cotton became critically important for the British textile industry. Nevertheless Britain in 1807 banned the shipping of slaves on British ships and the trading of slaves in British colonies. (Slave labor and slave breeding were still allowed.) In France Napoleon reintroduced slavery in 1802, eight years after it had been abolished by the revolutionary movement. However, in 1820 the European Coal Powers jointly closed the West African slave trade, and in the mid-1830s Britain finally abolished slavery in its huge colonial empire (though India was excluded for nearly 10 years).[41]

On the other hand, Britain kept importing slave-produced raw cotton from the United States of America, where slavery persisted much longer than in other Western nations. Already in 1776 Englishman Thomas Day had written, "If there is an object truly ridiculous in nature, it is an American patriot signing resolutions of independence with the one hand, and with the other, brandishing a whip over his affrighted slaves"; however, with the introduction of the cotton gin, U.S. slavery expanded even more. (The number of slave states increased from six in 1790 to 15 in 1860.) Mexico's 1829 decision to abolish slavery was one of the major reasons that led to the U.S. invasion of its southern neighbor. When shipments of African slaves to the United States were stopped, outright slave-breeding was encouraged. Slavery was finally abolished nationwide only at the end of the Civil War in 1865, but the subjugation of the black population remained. (Civil rights for African Americans in the South were not granted for another century, until the Civil Rights struggles of the 1960s.)

Portugal, the first European nation to trade African slaves, was even more hesitant. Following transitional legislation only those slaves who lived beyond 1878 could enjoy being liberated. In Portugal's colony of Mozambique slave trade continued even into the early 20th century. Otherwise the East African slave trade was organized mainly by Muslims who sold Africans to the Middle East. However, the European Coal Powers shut down much of this Arab slave trade from the mid-19th century. (Scottish explorer David Livingstone is a known figure in these efforts.) In the Spanish colonies the abolition of slavery was not enforced until 1820, that is, until after the South American countries began declaring their independence. Slavery was abolished in Argentina in 1813, in Colombia in 1821, and in Peru in 1854. Brazil declared its independence from Portugal in 1822, and received the last legal importations of African slaves in 1852. However, Brazilian planters strongly resisted the abolition of slavery, which was enacted only in 1888.

In South and Southeast Asia a form of European slave trade continued long after the African slave trade had been stopped. To make up for the shortage of African slaves the British (and Spanish) from around 1840 shipped coolies from such countries as India, China, and the Philippines to the Caribbean islands, South America, Africa, and various southeast Asian regions. Coolies were usually illiterate, unskilled laborers who signed contracts agreeing to heavy workloads and inadequate pay. To be sure, this was indenture, which is in principle different from real, involuntary enslavement, but it is well documented that the British on the largest scale misled Indian coolies, promising them a better life, and often leaving them without any idea of how far they were going to be shipped. The result was a mass shipment of Indian workers, organized by the British government to ease the chronic labor shortage in the sugar colonies, rubber plantations, and tea gardens of the British Empire. The Dutch used slaves in Indonesia until 1860, and forced labor thereafter. The French used forced labor in their colonies until 1860, and the Spanish did the same in the Philippines until these islands were ceded to the United States in 1898.

In the western United States, thousands of Chinese coolies were hired to construct the railroad, but this trend was opposed by Americans of European descent. In 1882 Congress passed the Chinese Exclusion Act, suspending further entry of most Chinese immigrants into the United States.[42] To be sure, there was also a sort of white slavery in the United States through informed, voluntary indenture. Many Europeans, especially Irish and Scottish, were living under such horrifying conditions in their home regions that they gave up their families and liberty and put themselves into voluntary servitude for the price of the fare to a better world. They usually agreed to labor for five or six years in exchange for the price of their passage to the New World, a practice that continued long after slavery was abolished.

NOTES

37. J. Bradford DeLong, "Slouching Towards Utopia?: The Economic History of the Twentieth Century, Part B: The Path from the Pre-Industrial World, VIII. The First Global Economy: Production and Trade," University of California at Berkeley and NBER (National Bureau of Economic Research), January 1997, http://econ161. berkeley.edu/TCEH/2000/eight/html/Slouching2_8preWWI.html.

38. Joel E. Cohen, *How Many People Can the Earth Support?* (New York: W. W. Norton & Company, 1995), 57. Presumably infanticide was still going on at this stage, and a large share of babies were left to foundling hospitals in Paris, for instance, in the late 18th century. Well over half of these supposedly died.

39. Find more on the mentioned population issues in: Joel E. Cohen, *How Many People Can the Earth Support?*; The World at Six Billion, United Nations, http://www. un.org/esa/population/publications/sixbillion/sixbilpart1.pdf.

40. J. Bradford DeLong, "Slouching Towards Utopia

41. Marjie Bloy, "The Anti-Slavery Campaign in Britain," http://www.victorian web.org/history/antislavery.html.

42. Twenty years earlier California passed the Anti-Coolie Act of 1862. "California Imposes a Tax on Chinese Laborers," Digital History, http://www.digitalhistory. uh.edu/asian_voices/voices_display.cfm?id=16.

CHAPTER 20

KNOWLEDGE AND TECHNOLOGY

The pace at which knowledge accumulated and technology was developed in the Coal Age was even faster than in the Super-Agricultural Era. The technology that defined the Coal Age was the steam engine. This prime mover radically increased the output of factories and mines, and revolutionized transport on land (railroad), as well as on the rivers and oceans (steam ships). Steam technology also fundamentally changed earthmoving capabilities. (William S. Otis of Philadelphia in 1835 invented the steam shovel, initially used as mechanical excavator for railroad construction.) On the other hand, steam technology did not have a lot of impact on individual short-range mobility. Steam engines are inherently heavy, and steam coaches were never produced in large numbers. The Coal Age therefore became the age of the horse. The unprecedented freight and passenger volume leaving and arriving in railroad stations had to be delivered or picked up locally by horse-drawn vans, wagons, and carts. London alone had about 300,000 horses in the beginning of the 20th century. What is more, horses remained the chief prime mover for field work in temperate climate agriculture. The world's population of (domesticated) horses therefore soared to record levels. The United States alone had some 25 million horses and mules during the first two decades of the 20th century. (For those who could not afford buying and keeping a horse, individual mobility increased when bicycles were invented. Modern bicycles with equal-sized wheels emerged during the 1880s. Pneumatic tires and back-pedal brakes were introduced in 1889, and bicycles were frequently equipped with various load carriers.)

COMMUNICATIONS

In communications, coal energy drastically increased the output of printed material. German inventor Frederick Koenig developed the steam-operated printing press in 1810, which gave rise to the production of mass editions such as newspapers. By the 1820s runs of 2,000 sheets per hour were possible, and rotary presses raised the rate tenfold by the late 1850s. In addition, several inventors produced typewriters. German Franz Xaver Wagner is credited with constructing the first modern typewriter, selling his patent rights to Englishman John Underwood in the United States.

Long-distance communication became critical with the rise of the railway. Telegraph communication was introduced to avoid collisions between trains going opposite directions on single-track rails. The idea of sending and receiving a signal through a wire was actually quite straightforward. In 1819 Danish Hans Christian Ørsted published his observation that a magnetic compass needle is deflected when an electric current flows through a nearby cable. Many inventors realized immediately that this could be used for communications. They telegraphed signals by orderly disruption of the electricity flow in cables, using power sources such as the battery presented by Italian Alessandro Volta in 1800. On the receiving end, they simply observed the movement of a magnetic needle indicating that a current was flowing in the wire. The first practical and successful telegraph models were patented jointly in England in 1837 by Charles Wheatstone and William Cooke, who had seen a needle telegraph while studying in Germany. Telegraph landlines expanded rapidly alongside railroads, and American Samuel Morse's coding system of 1838 was generally adopted for transmissions. (It uses sequences of long and short signals and breaks to encode letters.) In Germany, Siemens & Halske manufactured telegraph systems from 1847. Siemens developed the pointer telegraph (with a needle pointing to letters, thus eliminating the necessity to learn the Morse code) as well as automatic fast-speed writers for sending messages. Siemens & Halske installed the German and the Russian national telegraph network. The company also constructed a machine for covering copper wire with melted gum for cable production. The firm laid a telegraph cable across the Mediterranean in 1860; completed the 11,000 km telegraph line between London and India by 1870; and laid the first direct transatlantic line in 1875. By 1900 multiplex wires with automatic coding carried millions of words every day, and by 1919 13 transatlantic cables were in operation.

Telephones, that is, devices that can transform the human voice into an electronic signal and vice versa, were developed by several inventors. In Germany Johann Philipp Reis started to construct telephones during the 1850s and presented a model to Frankfurt's Physics Association in a 1861 lecture, "Telephony Using Galvanic Current." In the United States Italian immigrant Antonio Meucci demonstrated a telephone in 1860.[43] In 1871 he filed for a caveat, that is, a one year renewable notice of an impending patent, but he could not afford to renew it after 1874. Meucci handed some telephone models to the vice president of Western Union Telegraphs, and Alexander Graham Bell, who is no longer considered the inventor of the telephone, conducted experiments in the very laboratory where Meucci's materials were stored. Bell then applied for a patent in his own name, and even though the U.S. government annulled Bell's patent on the grounds of fraud and misrepresentation in 1887, the Bell patent did not expire until 1893. What is more, American inventor Elisha Gray independently filed a caveat describing a functioning telephone apparatus in 1876, not knowing that Bell had applied for the actual patent just a few hours earlier. The telephone enjoyed even faster acceptance than the telegraph, at least for local and regional service. The development of long-distance telephony, on the other hand, was quite slow. The first trans-American link was established only in 1915, and the first transatlantic telephone cable was laid only in 1956. Earlier transatlantic telephony had been established in 1927 based on radio transmission.

Radio communication had its starting point with German physicist Heinrich Hertz generating radio waves in 1887. Like light, radio waves are electromagnetic waves, but their wavelength is too long to be seen by the human eye. (Hence Hertz confirmed the ideas of Scottish physicist James Clerk Maxwell, who developed the mathematical equations describing electric and magnetic forces and fields.) Again, it was relatively straightforward to conclude that adequate interruption of such radiation would allow for the transmission of messages using the Morse (or any other) code. Nikola Tesla is considered the initial inventor of radio communications.[44] After he had sold his alternating current (AC) patent rights to Westinghouse in 1888, he engaged in the exploration of high frequency electricity and patented the Tesla coil in 1891, using it to transmit energy through the air. He illuminated a vacuum tube wirelessly in November 1890, and demonstrated the world's first public radio communication in 1893. He filed his basic radio patent applications in 1897 and in 1900 began to construct an ambitious broadcasting station on Long Island to establish a global wireless communication system. However, Tesla's financial backers pulled out of the expensive project after Italian Guglielmo Marconi on 12 December 1901 successfully signaled the letter "S" across the Atlantic. Tesla noted that the Italian had used 17 Tesla patents to accomplish the transmission, and Marconi himself admitted to borrowing the patents of German physicist Karl Ferdinand Braun, who worked on wireless telegraphy as well. (Braun is best known for his 1897 invention of the cathode-ray tube, the device on which television cameras and receivers are based.) Nevertheless Marconi, who had family connections with English aristocracy, received patent protection and a Nobel price. (Only in 1943, few months after Tesla had died at age 87, did the U.S. Supreme Court rule in Tesla's favor in the radio matter.) The first wireless transmission of language and music was achieved in June 1904 by Otto Nußbaumer at the Technical University of Graz, Austria. Ships were using (non-verbal) radio for communications with shore stations from 1905. Transatlantic wireless telegraph service was established for public use in 1907, and transatlantic radio-telephone service was available from 1927. Radio broadcasting of music and news programs began in the 1920s as well.

NATURAL SCIENCE

In natural science, Coal Age scholars gained a better understanding of their planet and the nature of nature. During the Super-Agricultural Era and the Coal Age several European powers launched global exploratory expeditions to develop maritime commerce and to discover overseas natural resources. These were frequently accompanied by scientists who thoroughly cataloged the landscapes and species they encountered. The number of registered plant species, for instance, doubled from about 20,000 to 40,000 during the 18th century. (Now it is nearly 400,000.) In a highly influential work, French scientist Georges-Louis Leclerc de Buffon between 1749 and 1788 summarized in 36 volumes everything known about the natural world. He observed the similarities between humans and apes (considering the possibility of a common ancestry), and (correctly) suggested that the age of the Earth was much greater than the 6,000 years proclaimed by the Christian church. (Buffon actually also observed the relative scarcity in large and powerful animals in the Americas and argued that nature in the New World was therefore inferior to that of Eurasia. However, he had no plausible explanation why this situation might have come about.) Scottish scientist James Hutton in 1788 published his theory of uniformitarianism, which explained the Earth's geologic processes without reference to the bible and emphasized that slow cyclical processes such as erosion, deposition, sedimentation, and volcanic upthrust sufficiently account for all geologic change of the past. Uniformitarianism held that all these processes operate on Earth with general uniformity, that is, everything observed in the present has been working in a similar manner in the past. German naturalist Alexander von Humboldt began the study of the relationship between living organisms and their environment. He explored South America between 1799 and 1804 and published *Idea for Plant Geography* in 1805. In this work he exposed relationships between plants and climate, and explained the geographic distribution of plant species according to varying environmental conditions. In 1809 French scientist Jeane Baptiste Lamarck published his ideas on evolution. He theorized that individuals adapt to their environment during their own lifetime (increasing specific capabilities by exercising them, while losing others through disease). Acquired traits are then passed on to offspring, which adapts from where the parents left off, enabling evolution to advance. (One of his examples was that giraffes may have developed long necks as a result of stretching to reach the higher branches of trees.) Scotsman Robert Chambers, who criticized both Lamarckism and (Christian) Creationism, published *Vestiges of the Natural History of Creation* in 1844. This examination of the notion of biological evolution influenced British naturalist Alfred Russel Wallace, who explored Brazil's Amazon rain forest as well as the East Indies (Indonesia). Wallace also read Scottish geologist Charles Lyell's three-volume *Principles of Geology* (1830–33), which promoted uniformitarianism and described how long-term change can be affected through the operation of slow, ongoing processes.

In 1855 Wallace published the essay "On the Law Which Has Regulated the Introduction of New Species", a paper that Charles Lyell brought to the attention of Charles Darwin.[45] While still in the East Indies, Wallace in 1858 wrote another paper, "On the Tendency of Varieties to Depart Indefinitely From the Original Type," this time sending it directly to Darwin, who he knew was interested in the topic. In this work, Wallace added to his theory the ideas of Thomas Malthus, who had published his "Essay on the Principle of Population" in 1798. By connecting the limits to population growth to a mechanism that causes long-term physiological change, Wallace in effect formulated the concept of the survival of the fittest, which holds that those individuals that are best adapted to their environmental surroundings have the best chance of surviving, thus leaving the most offspring (which inherit the beneficial traits).

As the story goes, Darwin had very similar ideas on his own, and both Darwin's yet unpublished work and Wallace's paper on evolutionary theory were presented to the Linnean Society of London the same day. One and a half years later, in November 1859, Darwin published his famous book *On the Origin of Species*, firmly establishing the idea of natural selection, which held that random variations occur among offspring and that (changing) environmental conditions select for those individuals that are best adapted to prevalent conditions, which changes the overall makeup of a population and eventually gives rise to new species.

UNDERSTANDING DISEASES

The better understanding of how nature works also began to improve medical science during the Coal Age.[46] Smallpox (variola) was a major killer in Europe during the Super-Agricultural Era. While it was still unknown that the disease was caused by a virus, or what a virus actually was, Europeans in the early 18th century began to adopt the method of variolation, which had been practiced in China since ancient times and had spread across Asia to the Ottoman Empire. The technique involved deliberate self-infection with a mild form of smallpox to gain immunity against contracting the full-blown disease. (This was initially done by blowing dried smallpox scabs into the nose of a person.) In England variolation was tested on prisoners and abandoned children by inoculating smallpox pus under the skin. The treatment killed many, but it became common practice anyway to avert full-blown epidemics. In America variolation was introduced by African slaves and was first applied among European Americans during a smallpox epidemic in Boston in 1721.[47]

Edward Jenner of Gloucestershire is credited with the development of real vaccination. He observed that people who had contracted cowpox, the cattle equivalent of smallpox, rarely caught the deadly human version. In 1796 he deliberately infected an eight-year-old boy with the pus from a cowpox sore. After the boy had recovered from cowpox Jenner infected him with smallpox, which the boy never caught. Following many more successful vaccinations, Jenner published his results in 1798. Unfortunately, it took half a century before the method was made compulsory in Britain in 1853.

To be sure, it was still unclear in the early Coal Age what caused diseases. The classic idea was that disease spread by miasmas, a poisonous atmosphere thought to rise from swamps and putrid matter. In 1840, German professor Jacob Henle of the University of Göttingen published the theory that diseases were transmitted by small, living, parasitic organisms: This might be regarded the starting point of modern medicine. In the Austrian monarchy, Hungarian physician Ignaz Semmelweis advocated the use of antiseptics from the 1850s. He recommended that doctors should wash their hands in chlorinated lime before any contact with patients, which sharply decreased mortality in hospitals. Florence Nightingale, who received training as a nurse at the German Kaiserwerth Institution near Düsseldorf, promoted general high standards of hygiene and sanitation in hospitals. In Britain, surgeon Joseph Lister pioneered the use of carbolic acid as the first antiseptic to clean wounds and surgical instruments. His procedures were published in 1867. (Besides infections, the second main cause of death during and after surgery was blood loss. Blood transfusions became a lot safer following the work of Austrian immunologist Karl Landsteiner, who in 1901 documented the A, B, and O blood groups that determine the compatibility of blood of donor and recipient.)

Robert Koch, one of Jacob Henle's students in Göttingen, is considered the founder of modern medical bacteriology. In the early 1870s, Koch once and for all eliminated all ideas of mysterious substances rather than specific microorganisms causing infectious disease: He showed that anthrax was caused by the bacterium Bacillus anthracis, the first microorganism identified as the causative agent of an infectious disease. Koch's research methods were adopted all over the world and helped scientists to identify further causative agents and strategies against them. Koch himself discovered and isolated the bacterium that causes tuberculosis (1882); traveled to cholera-plagued India to identify the cholera bacillus (showing that it was transmitted to humans primarily through water); and studied insect-borne diseases in Africa. Meanwhile French chemist Louis Pasteur developed methods to produce vaccines against such viral diseases as anthrax in animals and rabies in people. Pasteur also invented the process of pasteurization, a heat treatment to kill all bacteria and molds present in such liquids as milk. Koch's student Paul Ehrlich is recognized as the founder of modern immunology. Studying blood properties and the immune system, Ehrlich developed a method of stimulating production of antitoxins by injecting increasing amounts of toxins into animals. In 1909 Ehrlich found the first modern drug, the arsenic-containing compound salvarsan, which effectively treated syphilis.

APPLIED CHEMISTRY: FROM DYES TO PHARMACEUTICALS

The development of modern pharmaceuticals in the Coal Age was based on coal as the raw material. More specifically, the chemicals initially used to produce artificial dyes, and thereafter medicines, were contained in coal tar. At the beginning of the Coal Age the understanding of chemistry and the nature of matter was still very limited. It was only in the years after 1800 that European scientists through innumerable experiments established the idea that matter is not a structureless continuum, but consists of tiny indestructible particles (atoms) teaming up into larger particles (molecules). (As a philosophical concept, the idea of a universe consisting of invisible, indestructible, indivisible, and incompressible particles had already been proposed in Greece in classical times, chiefly by Democritus.) In terms of applied chemistry, demand for chemicals came from the metal and textile industries, but there were only few human-made chemicals in use for practical purposes at the beginning of the Coal Age. John Roebuck, James Watt's early partner, introduced mass-produced sulfuric acid in 1746 (together with Samuel Garbett) to serve, first, the metal refiners, and then, the papermakers and tanners. In turn, the textile industry substituted (diluted) sulfuric acid for sour milk in the last step of the traditional bleaching process; however, French chemist Claude Berthollet in 1785 suggested using chlorine for commercial bleaching, which soon became dominant.

The textile industry also demanded dyes. Synthetic colors for fabrics were the main product behind the tremendous expansion of the chemical industry. Up until the 1850s all colors for clothing, paint, and printing came from natural sources such as flowers, berries, roots, insects, mollusks, and so on. Thereafter, synthetic dyes were created from by-products of the coal and iron industry: the coal tar obtained as waste product in coking plants. The critical compound that chemists extracted from coal tar was aniline oil, a substance that eventually gave rise to a wide range of synthetic coal tar dyes.

Germans held the lead in chemical science. Wilhelm von Hofmann was therefore chosen to head the new Royal College of Chemistry in London, which was modeled on the laboratory of Hofmann's former teacher in Germany, and most famous chemist of his time, Justus von Liebig. In 1856, 18-year-old student William Perkin worked in Hofmann's lab (attempting to synthesize quinine, a cure for malaria that was traditionally derived from the bark of a South American madder plant) when he incidentally obtained a purple substance with excellent dyeing properties. The juvenile Perkin had in fact discovered mauveine, the first aniline dye. Perkin set up a factory to produce mauveine on large scale for the textile industry, and his huge commercial success motivated other chemists to found companies that eventually produced a rainbow of colored dyes, including aniline red, blue, and violet. In Germany, BASF (Badische Anilin–& Soda-Fabrik) was founded in 1865 to produce coal tar dyes and in 1869 patented a process to synthetically produce alizarin, the red color of madder plants. Thereafter the chemical industry rushed towards finding a way to synthesize indigo, a dye nowadays best known from blue jeans. German chemist Adolf von Baeyer achieved this task in 1880, and BASF financed the development of a process to mass-produce synthetic indigo, launching Indigo Pure in 1897. This had dramatic global consequences, as the European Coal Powers had established vast indigo plantations in their tropical colonies. Britain alone had set up over 3,000 (natural) indigo factories in India, which were now put out of business. (China immediately became the most important export market for BASF's synthetic indigo.)[48]

In the 1880s the large German dye companies began to diversify into pharmaceuticals to fully exploit the potential of their staff of highly-trained chemists. Adolf von Baeyer, the later artificial-indigo developer and (1905) Nobel Prize laureate, founded Bayer Chemical Company. In 1864 he synthesized a novel substance that he called barbituric acid, making it out of urea, an animal waste product, and malonic acid, a chemical found in apples. A whole variety of barbituric acid derivatives turned out pharmacologically active. One barbiturate helped people fall asleep, while others acted as hypnotics, sedatives, and anesthetics. Bayer employee Felix Hoffmann in 1897 combined salicylic acid, a natural painkiller obtained from willow bark, with acetic acid to obtain the substance that was to become known as aspirin, the world's most widely used pain-reliever. Hoechst chemical company, founded as a dyeworks in 1863, produced pharmaceuticals from 1883. In 1905 Hoechst employee Alfred Einhorn synthesized novocaine, which became a popular local anesthetic to replace cocaine. Hoechst also mass-produced Paul Ehrlich's 1909 arsephenamine, a treatment for syphilis, under the brand name salvarsan. This marked the beginning of modern chemotherapy.[49]

RUBBER

One strategically important application of chemical technology was the production of rubber. It all began with the import of water-resistant shoes, coats, and capes fabricated by natives in South America. The secret behind these products was rubber, an elastic and water repellent substance obtained from a milky white fluid called latex. This fluid was harvested by cutting the bark of Hevea brasiliensis, a tree of the spurge family.

As soon as Europeans got their hands on this material they began experimenting with it. In 1791 English manufacturer Samuel Peal patented a method of waterproofing cloth by treating it with a solution of rubber in turpentine, but both European and South American rubber products had the major drawback that they turned tacky under warm conditions, and brittle under cold conditions. In 1834 German chemist Friedrich Ludersdorf and American chemist Nathaniel Hayward independently discovered that adding sulfur to raw rubber greatly improved its properties. As the story goes, American inventor Charles Goodyear, in 1839, followed their procedure, but accidentally dropped a piece of sulfur-treated rubber onto a hot stove, thus inventing the process of vulcanization: This heat treatment turned sulfur-added rubber into the resistant elastic material that was soon going to be in highest demand for a whole host of applications. Machinists needed rubber for sealing gaskets, for instance, and eventually the market for rubber tires took off. (Solid-rubber tires were replaced by air-filled pneumatic tires after Scotsman John Dunlop had started putting them on bicycles in 1888.)

Brazil had a monopoly on crude rubber production and began making a fortune. But eventually Brazil was pushed off the rubber throne in a scheme that may be regarded one of the most consequential industrial thefts of all times: British explorer Henry Wickham collected some 70,000 seeds of Hevea brasiliensis, smuggling them out of Brazil in 1876 despite the strict laws that the government had enacted to prohibit seed exports. The British used the seeds to build up rubber plantations first in Ceylon and then on the Malay peninsula and the Malay archipelago. The Dutch did the same in the East Indies (Indonesia). The enormous rubber plantations established by the European Coal Powers in their colonies of tropical Southeast Asia (and West Africa) then replaced South America's wild rubber trees as the principal source of crude rubber for Western markets. (Total rubber plantation acreage eventually peaked at about 3.64 million hectares (9 million acres) right before World War II.)

NOTES

43. "107th CONGRESS, 1st session, H. RES. 269, Expressing the sense of the House of Representatives to honor the life and achievements of 19th Century Italian-American inventor Antonio Meucci, and his work in the invention of the telephone," House of Representatives, September 25, 2001, http://www.popular-science.net/his tory/meucci_congress_resolution.html.

44. "High Frequency," *Tesla—Life and Legacy*, Public Broadcasting Service (PBS), Arlington, Virginia, http://www.pbs.org/tesla/ll/ll_hifreq.html; "Who Invented Radio?" *Tesla—Life and Legacy*, Public Broadcasting Service (PBS), Arlington, Virginia, http://www.pbs.org/tesla/ll/ll_whoradio.html.

45. Charles H. Smith, "The Alfred Russel Wallace Page," Western Kentucky University, http://www.wku.edu/%7Esmithch/index1.htm.

46. "History of Medicine, 1700–1900: 18th and 19th Centuries," The Association for Science Education, http://www.schoolscience.co.uk/content/4/biology/abpi/history/history8.html.

47. David A. Koplow, *Smallpox: The Fight to Eradicate a Global Scourge* (Berkeley: University of California Press, 2003). Read Chapter 1, "The Rise and Fall of Small-pox" at this address: http://www.ucpress.edu/books/pages/9968/9968.ch01.php.

48. "New Dyes," *Making the Modern World*, The Science Museum, http://www.makingthemodernworld.org.uk/stories/the_second_industrial_revolution/05.ST.01/?scene=2&tv=true.

49. "New Drugs: The Industrialisation of Pharmaceuticals," *Making the Modern World*, The Science Museum, http://www.makingthemodernworld.org.uk/stories/the_second_industrial_revolution/05.ST.01/?scene=4&tv=true.

CHAPTER 21

DEADLIER WEAPONS

Advances in weapons technology were closely associated with progress in chemical science as well. The main breakthrough was the development of a new class of chemicals that allowed for the production of high explosives. These were prepared by nitration of such organic compounds as cellulose (wood fibers) and glycerin (obtained from animal and plant fats), and later phenol and toluene (obtained from coal tar). Nitration was at first a simple treatment with nitric acid (HNO_3), which is obtained by dissolving potassium nitrate (KNO_3) in sulfuric acid (H_2SO_4). The new high explosives delivered a far more powerful blast than traditional gunpowder. In combination with better-quality steel, they moved the effective range of large field guns to unprecedented levels: from less than 2 kilometers during the 1860s to over 30 kilometers by 1900. Nobody had ever been able to kill from such long distance.

The whole story started in the 1830s with the discovery that cotton can be turned into an explosive by dipping it in concentrated nitric acid. German chemists Schönbein and Böttger quickly realized that the obtained nitrocellulose (cellulose nitrate) had the potential to replace traditional gunpowder, because it exploded without developing smoke (which fouled guns, blackened the gunners, and obscured vision). Guncotton came into use from the 1860s, but it exploded too rapidly and violently, and thus had to be moderated for safe handling and application in common guns. Austrian Frederick Volkmann, around 1870, treated a blend of nitrated wood with a mix of ether and alcohol, solvents known to dissolve nitrocellulose. The resulting paste was then dried under controlled conditions to adjust the explosion rate.

The essentially same product, known as Poudre B, was produced in 1884 by Frenchman Paul Vieille and quickly became the smokeless gunpowder used by practically all Western armies.

Another explosive, nitroglycerin (glyceryl trinitrate), was first prepared in 1847 by Paris-educated Italian chemist Ascanio Sobrero. Nitroglycerin is a colorless oily liquid formed by nitric acid and sulfuric acid on glycerin. In a journal article Sobrero warned against the use of the substance, because it was too unstable and exploded extremely violently. Nevertheless, Sobrero was later mortified when the Swedish Nobel family made a fortune on the commercial exploitation of nitroglycerin. Alfred Nobel was born the third of four sons of entrepreneur Immanuel Nobel.[50] When their father's business went into bankruptcy, the Nobel brothers were looking for new business ideas, and Alfred began experimenting with nitroglycerin, trying to make it safer to handle. After his younger brother Emil and four others had been killed in a nitroglycerin blast in 1864, Alfred achieved a breakthrough in Germany: He discovered that nitroglycerin was absorbed to dryness by a type of porous sand or diatomaceous earth known in German as Kieselguhr. The resulting paste was safe to handle and became known as dynamite. It was used for both military and industrial applications (such as blasting tunnels, cutting canals and building railways and roads).

In 1887 Alfred Nobel also invented and patented a smokeless gunpowder called Ballistite. It was made of 45 percent nitroglycerin, 45 percent nitrocellulose, and 10 percent camphor. However, as Ballistite, like Poudre B, tended to become unstable over time, the British in 1889 developed their own smokeless powder, Cordite, which they made from 58 percent nitroglycerin, 37 percent nitrocellulose, and 5 percent vaseline. (As this was extremely similar to Nobel's Ballistite, Nobel sued the British inventors for patent infringement, but the House of Lords in 1895 decided against Nobel.) Notably, the British used acetone as a solvent, and extruded Cordite as rods initially called cord powder. (During World War I Britain lacked acetone until Chaim Weizmann, a Jewish chemist educated in Switzerland, helped by developing a biotechnological process to make acetone from starch with help of a bacterium. In 1948 Weizmann became the first president of Israel.)

Another well-known explosive, TNT (trinitrotoluene), was first prepared in 1863 by German chemist Joseph Wilbrand, who mixed toluene, available from coal tar, with nitric and sulfuric acids. Large-scale production of TNT started in Germany in 1891, and from 1902 the German army adopted it as standard filling for shells. Other countries followed suit, and TNT became the most widely used military explosive. TNT is very stable and can be stored under all temperature conditions for a long time. Besides, TNT in the form of demolition blocks is virtually bullet safe.

Sprengel explosives became wide-spread as well. First patented by Hermann Sprengel in 1871, these explosives are special as they are shipped and stored as two relatively harmless separate components. Yet another German chemist, Hans Henning, introduced the world's most powerful (non-nuclear) explosive in 1899. Like nitroglycerin, it was initially developed as a pharmaceutical, but eventually cyclonite (cyclotrimethylenetrinitramine) became best known as RDX, Research Department Explosive. (This was much later: it took four decades until an efficient method to produce RDX was developed.) The detonation velocity of dynamite is nearly four times, and that of cyclonite more than six times, that of gunpowder.

NEW TYPES OF GUNS

Guns underwent a rapid evolution during the Coal Age as well. Muzzle-loading was replaced by breech-loading, smoothbore barrels by rifled barrels, and flintlocks by percussion locks. Accelerating still, the evolution of firearms then proceeded towards fully automatic machine guns, which were available a short 20 years after the 1866 Battle of Königgrätz, in which the Prussians' use of breech-loading guns was still a novelty.

The first widely used breech-loader was the 1836 needle-gun (Zündnadelgewehr) by Johann Nicholas Dreyse, which the Prussian army adopted in 1841. The Dreyse gun actually had a very short effective range, as much gas pressure was escaping at the rear of the gun, but this weakness was eventually overcome through the introduction of metal cartridges: these effectively sealed the breech when they were fired and directed the entire explosion boost toward the bullet. The earliest such metal cartridges were patented in Paris by Houiller (1847) and Lefaucheux (1850).

The Dryse gun also featured a rifled barrel. It was known at least from the 15th century that imparting a spin to a projectile improved the range and accuracy of guns. Barrels with spiral grooves cut in their inside wall had been produced ever since, but never came into favor because it is much harder to muzzle-load a rifle than a smoothbore musket. Only with the development of breech-loading systems did rifled barrels make sense.

Wilhelm and Paul Mauser of southern Germany in 1867 invented a rotating bolt system and built a popular rifle that became the standard infantry weapon of the German army in 1871. The ascent of powerful smokeless gunpowder started with the French Lebel of 1886 (accurate up to 1,000 yards), and soon thereafter Friedrich Vetterli adapted a four-round box magazine repeating system (designed by Italian Artillery Captain G. Vitali) to the formerly single-shot Model 1870 rifle. Smokeless gunpowder and the magazine were immediately adopted by arms manufacturers in several countries. In Austria Ferdinand von Mannlicher's repeating rifle of 1895 soon became part of the army's standard equipment, and was still used by several European countries in World War II. (The Mannlicher-Carcano rifle, Model 1891, was made (in)famous by its use in the assassination of U.S. President

John F. Kennedy in 1963.) However, perhaps the most famous weapon of this time was the Mauser Model 1893, which was extremely popular. It was used by Spain in the Spanish-American War of 1898, when 700 Spanish regulars held off an attack by 15,000 U.S. troops for 12 hours during the Battle of San Juan Hill in Cuba. This led the United States to develop their own version of the clip-loaded Mauser design, which would become the Springfield 1903 rifle. (The slightly modified Springfield 1906 remained the principal U.S. infantry weapon until 1936.) Since the San Juan Hill incident became

The Battle of San Juan Hill   Due to rapidly improving engineering skills, the Coal Age witnessed a fast evolution of deadly hand-held guns, which appeared with breech-loading, rifled barrels and percussion locks. Repeating rifles and machine guns in turn rendered the world's cavalries useless. In the Spanish-American War of 1898, some 700 Spanish soldiers with German Mauser (Model 1893) rifles held off an attack by 15,000 U.S. troops on Cuba during the Battle of San Juan for many hours and inflicted 1,400 casualties on the attacking U.S. forces. Dubbed Rough Riders by the American press, the Cavalry Regiment that Colonel Theodore Roosevelt (reputedly) helped command during the battle was actually fighting on foot. (Celebrated as a war hero, Roosevelt was then immediately elected governor of New York and made U.S. vice president within two years.) As a result of the battle, the Springfield of 1903 was developed to replace the older Springfield and the Norwegian Krag-Jørgensen rifles carried by United States troops. (The picture shows the Sixteenth Infantry in San Juan Creek Bottom, under Spanish fire from San Juan Hill. Photograph by William Dinwiddie. Library of Congress image LC-USZ62-534, edited.)

known around the world, the Mauser company was able to sell the Model 1893/94 to numerous countries, including the Ottoman Empire, Sweden, Brazil, Mexico, Chile, Uruguay, China, Iran, and the South African Republic (Boer Transvaal and Orange Free State). The British were confronted with this weapon during the Boer War and in turn designed their own Mauser-type rifle as well. (The SMLE became the standard British infantry weapon until the 1950s.)

Repeating rifles evolved into fully automatic machine guns. The early years of the American Civil War (1861–1865) were still dominated by flint-locks, but the Confederate Army eventually introduced a hand-cranked machine gun that could fire 65 rounds per minute from one barrel. This design was still prone to overheating, but towards the end of the war Richard Jordan Gatling invented a machine gun capable of (sustainably and accurately) firing 200 rounds per minute. Used by the U.S. Army and Navy, the Gatling gun was gradually improved and widely exported after it became known for its notoriety during the Spanish-American War of 1898. It could now fire at a rate of approximately 1,000 rounds per minute when hand-cranked, and 3,000 rounds per minute when an electric motor was added. The world's first truly automatic machine gun was invented in 1884 by American Hiram Stevens Maxim, who resided in England. The British army failed to realize the strategic importance of this weapon and 30 years later, at the outbreak of World War I, possessed only few of them, while many other countries had adopted Maxim-type machine guns. The development of, first, repeating rifles and, then, machine guns in the later Coal Age put to an end the millennia-long importance of mounted warriors: by World War I a cavalry charge against a line of entrenched troops with rapid-firing small arms was suicidal.[51]

NOTES

50. "Alfred Nobel—The Man," Britannica Guide to the Nobel Prizes, http://www.britannica.com/nobel/micro/427_33.html; Tore Frängsmyr, "Life and Philosophy of Alfred Nobel—A memorial address by The Royal Swedish Academy of Sciences," http://nobelprize.org/nobel/alfred-nobel/biographical/frangsmyr/.
51. Encyclopædia Britannica, "Cavalry," http://www.britannica.com/EBchecked/topic/100566/cavalry.

CHAPTER 22

IMPROVED AGRICULTURE

Agriculture profited as much from advances in chemistry as weapons technology did. German chemist Justus von Liebig (1803–1873) formulated the Law of the Minimum, which states that plant growth is controlled not by the total of resources available, but by the scarcest resource. This implied that the growth of a plant or crop could be enhanced by increasing the availability of the limiting nutrient. Liebig analyzed many tissues and body fluids to comprehend the processes in living organisms. He understood that plants get carbon (C) for their growth by capturing carbon dioxide (CO_2) from the air that surrounds them. He also showed that water (H_2O) supplies the plant with the critical elements hydrogen (H) and Oxygen (O). And from about the 1840s it became clear that plants need another three elements in quite large quantities for their growth: potassium (K), phosphorus (P), and nitrogen (N). (Later it was found that a range of about 13 elements, most in minute quantities, are absolutely necessary for healthy plant growth.)

Once Liebig had shown that nitrogen-based fertilizer can help plant growth, various mineral fertilizers were applied to the fields to improve agricultural output over the yields achieved with traditional organic waste fertilizers. Bone meal (that is, ground up bones) was one such traditional fertilizer. It provided phosphorus to the soil, but it released it only very slowly. James Murray in Ireland and John Bennett Lawes in England used sulfuric acid to turn bone meal into a water soluble product that was readily absorbed by plants. Lawes received his original patent in 1842. A few years later he also treated mineral phosphates (which are insoluble in water) with diluted sulfuric acid to produce soluble superphosphates. This procedure became widespread after 1870 and had a significant impact on field fertility. It also led to the discovery of several new phosphate deposits, such as those in Florida in 1888 and in Morocco in 1913. The mining of potash deposits, as a source of potassium, began to expand in Europe and North America about the same time.

CHILE SALTPETER

While potassium and phosphate supplies could quite easily be managed and sustained, mineral deposits of nitrogen turned out to be extremely rare. In fact, the only large reservoir was discovered in 1809 in the rainless desert of Atacama, at the South American Pacific shore-line. These deposits consisted of sodium nitrate, $NaNO_3$, which became known as Chile saltpeter or Chilean nitrate. However, the Atacama nitrate fields were actually located in Bolivia. Chile then provoked the War of the Pacific (1879 to 1884) and defeated Bolivia and its ally Peru to take control of the saltpeter districts. Backed by British (and U.S.) capital investments, Chilean saltpeter production rose to as much as 1.5 million tons annually by 1900. Practically all of it was exported to serve two purposes: one was field fertilizing, the other production of the new class of high explosives such as Poudre B, dynamite, and TNT. (Besides, classic black gunpowder production depended on nitrate as well.)

Alternative sources of nitrate were rare. Peru's guano deposits, created by droppings of huge bird colonies and preserved in extremely dry coastal climate, were highly valued as fertilizer for their phosphorus content but also contained nitrogen. Similarly, deposits of bat guano (bat excrement) were preserved in caves all over the world, but all of them were rapidly depleted following their discovery. The only alternative inorganic source for nitrogen salts were small quantities of ammonium sulfate recovered from coking ovens. Energy-intensive processes to make nitrogen salt from other sources were developed towards the end of the 19th century, but these were not at all economical. Hence Chilean nitrates remained the principal source of nitrogen throughout the Coal Age. (A breakthrough came only in 1913, when German scientists figured out how to fix nitrogen straight from the atmosphere, a technology that was going to have a profound impact in the Oil Age.)

DRAMATIC PRODUCTIVITY GAINS

The new fertilizer inputs, in combination with better crop rotation schemes (less fallowing), dramatically improved field outputs. At the beginning of the 19th century a good western European harvest returned about 200 times more food energy than it consumed in form of muscle work during production. A century later the energy return was commonly above 500 times the input in muscle energy. The Netherlands achieved the highest crop productivity gains during the Coal Age. Dutch yields averaged one ton of wheat per hectare nationwide by 1800, but they were nearly twice as high in 1900. English yields were already above 1.4 tons per hectare in 1800 and above 2.1 tons per hectare in 1900. France wheat yields showed only a mild upward trend, staying below 1.3 tons per hectare, and U.S. agricultural output per acre actually decreased during the 19th century, sinking well below one ton per hectare.[52]

Americans could afford this sort of inefficiency because they took a lot of new land under cultivation, which greatly increased the overall agricultural output. They also could afford to use a lot of horses. In the early 19th century, the ratio of human to animal power in farming was 1:15 in Europe, while it was 1:100 in the United States in the 1890s. This depended on the availability of a lot of surplus grain. A good Western horse in the late Coal Age worked at a rate equal to that of at least six men, but it ate about nine times as much grain as a single man. The United States set aside about one quarter of the cultivated land to feed horses (typically with maize or oats). This was possible because the United States by 1910 had almost 1.5 hectares of farmland available per capita, 10 times as much as China.

The large number of horses employed in Western agriculture led to impressive productivity gains in terms of output per farmworker. Some 160 hours of human labor in grain farming in New England in 1800 were reduced to nine hours of labor in California in 1900, based on a gradual mechanization of (horse-powered) farm work. The first harvester was patented in 1858: it required only two men to bind the stalks after they were cut. Twenty years later, in 1878, American John Appleby introduced the twine knotter. This innovation allowed for the construction of fully mechanical grain harvesters that could discharge tied grain sheaves ready for stacking. In combination with gang-plowing, twine-binding harvesters opened up huge expanses of grassland for settlers in North America, Argentina, and Australia. However, even the best of these machines were soon outperformed when the first horse-drawn combines (harvester-threshers) were introduced in California during the 1880s. The largest of these machines needed up to 40 horses to operate and could harvest 1 hectare of wheat in less than 40 minutes. (More traditional stationary threshing was otherwise an operation in which steam engines had gradually replaced farm horses. Such machines to remove the husks from grain were developed in the 18th century in Scotland. They were initially water- or horse-powered, but coal-fueled steam threshing had spread by the mid-19th century.)

FOOD CANNING AND REFRIGERATION

As urban centers grew and agricultural produce had to be shipped from further afield to nourish city dwellers, methods of food preservation were improved over pickling and other traditional techniques. A major stimulus came from the military. Motivated by a French government grant (seeking better ways to supply troops during the Napoleonic Wars), Nicolas Appert in 1810 published a new method for preserving food by heating it and sealing it in bottles. (This kills harmful bacteria already present in food.) The British figured that robust tin cans are easier to transport than bottles, and adopted them for their armies. Other nations followed suit, and soon commercial food canning was widely introduced in Western countries.[53]

Justus von Liebig, the influential German Coal Age chemist, invented the beef stock cube. He published a paper warning that the boiling of meat will diminish its nutritional benefits and in 1840 developed a concentrated beef extract that should provide an affordable and nutritious meat substitute for everyone, including those unable to afford real meat. (To be sure, German meat consumption increased rapidly during the Coal Age anyway. It averaged 20 kg per person annually in 1820 and 50 kg per person annually in 1900.) A young engineer, George Christian Giebert, approached Liebig, suggesting setting up a factory in Uruguay (South America), where cattle was slaughtered on a large scale, mainly for the hides industry. From 1866 the Liebig Extract of Meat Company produced a bottled black spread, which became a hugely successful kitchen item in Europe and the US. About 500 tonnes of the extract, which in 1899 became known as Oxo, were produced annually in Uruguay by 1875, and from 1873 on the Liebig company also produced corned beef in tin cans.

Another way of preserving food is to cool or freeze it. This will keep harmful bacteria from proliferating. Francis Bacon, a British scientist and essayist, died in 1626 after eating some chicken that he had stuffed with snow as part of an early refrigeration experiment.[54] Despite such drawbacks, the natural ice industry eventually grew into a considerable business in the United States and Europe. Ice was being cut from lakes in winter and stored during the summer. However, Coal Age scientists also attempted to create ice from liquid water. The principal approach of cooling technologists was based on the observation that liquids absorb heat when they evaporate, and thus cool their environment. Some substances, such as ethers and ammonia (NH_3) are gases under normal temperatures and atmospheric pressure, but turn liquid when compressed. Hence coal energy (or electricity from hydropower stations) could be used to compress such gases until they liquefy, and to allow them to re-evaporate inside the pipelines of a cooling box. German engineer Carl von Linde developed the first efficient and reliable (and portable) compressed-ammonia refrigerators between 1871 and 1877, initially for the beer industry of southern Germany. His company sold internationally to slaughterhouses and all sorts of factories that required cooling, as well as to those who simply produced ice to sell it from ice wagons on the street. (Coal Age people filled this ice in insulated food-storage boxes: These were the days before electric household refrigerators.) In 1877 the first shipload of frozen beef was carried from Argentina to France, while the first cooling ship left Australia to deliver meat and butter to London in 1879. Steam trains and steam ships with cooling equipment changed the global agricultural system and opened up agricultural frontiers around the world. Meat from the American West, South America, Australia, and New Zealand was now increasingly consumed in the population centers of Europe and the American East.

NOTES

52. Vaclav Smil, *Essays in World History: Energy in World History* (Boulder, CO: Westview Press, 1994). (This account served as a general source of historical energy data including the figures on agricultural efficencies provided in this chapter.)
53. "Survey: Food: Make it cheaper, and cheaper," *The Economist*, December 11, 2003, http://www.economist.com/displayStory.cfm?Story_id=2261831.
54. Ibid.

CHAPTER 23

INCREASED MOBILITY AND TRADE: GLOBAL ECONOMIC INTEGRATION

Advances in transportation systems were generally at the core of many changes occurring during the Coal Age. Railroads allowed for inexpensive land transport of people and cargo for the first time in history, while steam ships improved mobility on oceans and rivers. Throughout the Agricultural Age the lack of inexpensive (land) transportation systems had discouraged mass-production, because the cost of transportation would have outweighed any advantage gained from economies of scale in centralized manufacturing. Now, raw materials were shipped from far afield (overseas or the continents' interiors) to coal-fueled factories, and mass produce was cheaply delivered to consumers on global markets. U.S. railway mileage increased from 9,021 to 249,902 miles, and German from 3,637 to 36,152 miles between 1850 and 1910. Africa had 12,000 miles of railroads in 1900, South America had 26,000 miles and Asia 38,000 miles.[55] Ocean transport had long been extensive and inexpensive, but became even cheaper due to steam propulsion and large iron hulls. Thus, world shipping tonnage grew from roughly 4 million to 30 million tons a year during the 19th century, while the value of global trade expanded fivefold between 1750 and 1914.[56]

NAVIGATION AT SEA

One critical prerequisite for the international trade volume and mobility at sea to expand during the Coal Age were breakthroughs in navigation techniques. A certain standard had been reached by the 15th century with the Portuguese techniques of latitude determination by means of solar or stellar observation. Around the same time the measurement of speed and distance was improved. (The speed of a ship was determined by measuring the distance a ship traveled while a sandglass emptied. This was achieved by dropping overboard a log attached to a reel of line knotted at regular intervals. The number of uniformly spaced knots exposed before the sandglass was empty allowed for calculation of the speed in knots.) The quadrant and astrolabe were replaced in the 17th century by the simple backstaff English navigator John Davis had invented in 1594 to measure solar altitude without sighting the sun directly. It was superseded from 1731 by the more precise reflecting octant of English astronomer John Hadley, which was further improved in 1757 by a sextant developed by the British navy for quick and accurate readings of any celestial object against the horizon. Calculating the course of a ship became a lot easier after Scottish mathematician John Napier (Neper) in 1614 developed the concept of the logarithm. English mathematician Henry Briggs popularized the concept and in 1631 advanced logarithms in the form of decimal tables that soon became available to mariners along with logarithmic slide-rules.

The determination of longitude (that is, how far east or west a ship is positioned) remained a problem. The breakthrough came when the longitude problem was linked to the rotation of the Earth, which completes a full turn in 24 hours, or 15 degrees in one hour. If it were possible to have onboard an accurate clock that showed the time at a fixed location (meridian), then the difference between the time on the chronometer and the local time at noon (when the sun was highest in the sky) at any point on the voyage, would give the angular distance traveled, that is longitude. However, even if a chronometer lost no more than a fraction of a second per week, this would still allow for an error margin of a couple of kilometers when positioning a ship after a journey of two to three months. Already in 1598 Philip III of Spain had offered large financial rewards for a method of accurate measurement of longitude, and similar incentives were later offered in France and Holland. In England, the Greenwich Observatory was set up in 1676 to find the data necessary to determine longitude, and following a series of devastating and expensive shipwrecks the British government in 1714 created the Board of Longitude, offering a £20,000 prize for an invention accurate within narrow specifications.[57] Paris-based Swiss clockmaker Ferdinand Berthoud was a leading figure in modern clock making and produced chronometers for French ships, but his 1748 model was outperformed in 1762 by the marine chronometer of John Harrison. Another contribution came from German astronomer Johann Tobias Mayer, who developed lunar tables to assist navigators in determining longitude at sea. Captain Cook used the lunar distance method for determining longitude in his first Pacific voyage, but the observations and calculations required for this method took many hours. When he set out to circumnavigate the world in 1772, Cook took a Harrison-type clock with him. When he returned to Plymouth three years later, the cumulative error in longitude was less than eight miles.[58] A generation later, marine chronometers could be made at low enough prices and in large enough quantities for general Coal Age navigators to use them.

CANAL SHORTCUTS BETWEEN OCEANS

One of the major Coal Age achievements that were motivated by the increasing trade volume (and thereafter facilitated its further growth) was the construction of canals that bridged oceans and provided for extreme shortcuts. The 70 miles (113 kilometers) long Suez Canal, which connected the Indian Ocean and the Mediterranean through the Red Sea, was opened in 1869. This canal project was planned by Austrian engineer Alois Negrelli, who sent a team of engineers to Egypt for measuring work, while the construction itself was in turn overseen by Frenchman Ferdinand de Lesseps. The Suez Canal was a great triumph and inspired attempts to connect the Atlantic and Pacific Ocean in central America, at the isthmus of Panama. This narrow but mountainous stretch of land had received a lot of attention since 1848, when gold was struck in California and a wave of prospectors from the eastern United States and Europe were seeking quick passage. The success of the Panama Railroad, built to serve this traffic, soon gave rise to the idea to construct the 50-mile-long Panama Canal. However, the project turned into a nightmare. Lesseps, now of Suez fame, led the disastrous French efforts of the 1880s. Thousands of workers labored in oppressive heat, threatened by floods and mudslides. Malaria, yellow fever, and other tropical diseases took an enormous toll in lives, and in the end, the project was abandoned. When it was picked up again, this time under American supervision, technology had entirely changed. It took ten years, from 1904 to 1914, to build the canal with the help of 102 steam shovels. Coal energy was in turn also used on the water: As in the Suez canal, purpose-built steamers towed ships through the Panama Canal.

WHO WAS TRADING WHAT?

Coal technology entirely changed the global trade patterns: at the start of the Coal Age Europeans imported luxury articles such as spices, tea, tobacco, all with a high value per weight; plus particular (strategic) bulk commodities such as cotton and raw sugar (which would not grow in Europe), as well as timber (which had become scarce in many regions). From the 1840s it was possible to transport goods cheaply and quickly across the oceans in large screw-propelled iron-hulled steamers. Hence, it became cost-effective to ship even inexpensive commodities of low value per weight, and western Europe began importing agricultural goods from the temperate land-rich countries of white settlement such as the United States, Canada, Argentina, Chile, Uruguay, Australia, New Zealand, and South Africa. In addition western Europe received grain from eastern Europe by rail: German farmers faced an entirely new competition in the 1870s as Russian grain arrived in western Europe from Odessa at the Black Sea. All this further underpinned Europe's urbanization and industrialization.[59]

Europeans (and their offspring in overseas temperate regions) also imported tropical agricultural goods such as rubber, coffee, sugar, vegetable oil, and cotton from such countries as Brazil, Colombia, Cuba, Malaysia, and Ghana (and to some degree the U.S. South). The European Coal Powers made sure their colonies would grow whatever was in demand. Tea cultivation was introduced in India and Ceylon, for instance, and coffee cultivation in Kenya. During the last quarter of the 19th century the rubber tree (formerly exclusively grown in Brazil's Amazon region) was brought to Malaysia, Indonesia and Indochina.[60]

With the railroad and river-going steamboats penetrating Africa's interiors, the European powers to some extent also started to extract and import African mineral resources. However, the industrialization of the West was practically self-sufficient in terms of raw materials.[61] Industrial input materials such as coal, iron ore, glass, clay, cement, and so on were almost exclusively supplied from within the industrialized world even after World War I.

In addition to various mineral raw materials Europe (and the eastern United States) exported mainly manufactured goods, produced cheaply by those able to utilize the energy stored in coal. The value of goods and services exported by Britain in the years prior to World War I amounted to more than a quarter of its national product. About 75 percent of the exports were manufactured goods, and half of these manufactures were textiles. The United States, however, had grown into an enormous empire that stretched from the Atlantic to the Pacific and therefore enjoyed a huge and largely closed internal economy. Nevertheless the United States exported roughly five percent of its national product in the years before World War I, which was similar to Germany or Italy.

MASS MIGRATION

Falling prices for oceanic transport affected not only the movement of goods, but also of people. In the entire 17th century, about 400,000 people came from the British Isles to the Americas. In comparison, emigration from Britain to the United States and overseas colonies amounted to some 250,000 people per year in the 1850s (following the introduction of large screw-propelled iron steamers). The British government encouraged large-scale emigration to counter the rapid population growth within Britain. Some emigration was even enforced. Following U.S. independence the British were no longer able to ship convicts to America and instead sent 150,000 of them to Australia between 1788 and 1867. Others followed voluntarily, enjoying New South Wales' temperate climate and cheap convict labor. People from other European countries migrated to overseas destinations in search for a better life as well. Between 1875 and 1925 some 50 million left (largely from eastern and southern Europe) for Australia and the Americas. Meanwhile another 50 million or so left China, India, and other Asian countries for the Americas, East Africa, and regions surrounding the South China Sea.[62] To be sure, much of the migration of Asians was under deceiving circumstances and only partly voluntary: the Europeans promised illiterate Asian coolies a better life while having them sign contracts agreeing to heavy workloads and inadequate pay in distant countries. The British invented this sort of soft slavery to deal with the labor shortage on colonial plantations after they had finally abolished traditional slavery in 1833. Earlier black slavery had significantly contributed to the redistribution of the world's population as well. British ships alone carried over 3.4 million enslaved Africans across the Atlantic between 1662 and 1807, accounting for more than half of all slaves shipped from Africa to the Americas during that period.

To be sure, the Coal Age Powers managed migrations the way they preferred them. By and large, the Western Coal Powers kept Asian migrants out of the temperate zones of European colonial settlement and directed them to tropical colonies instead: work on tropical plantations paid higher wages than most other work in Asia, but a lot less than work in the West. The stream of immigrating Chinese and Indian workers then kept wages low in such European-controlled regions as Malaysia, Indonesia, the Caribbean, and East Africa.

EFFECTS ON THE GLOBAL ECONOMY

Coal-fueled production and transport, with their effect on international trade, stimulated unprecedented global economic growth. The output of traditional agrarian societies typically grew by a few percentage points per decade, while coal-fueled societies sustained decadal expansion rates of between 20 and 60 percent. (The British economy's output was ten times larger in 1900 than in 1800, and U.S. output doubled between 1880 and 1900.) Such radical economic growth in coal-fueled economies left the rest of the world behind and created substantial differences between the wealth of agricultural versus coal cultural societies. General consumption, however, increased globally, even in those countries that merely imported manufactures from coal-fueled nations. Textiles, for instance, became a lot more affordable everywhere in the world through factory production and international trade.

Austrian economist Joseph Schumpeter described the business cycles that occurred in industrialized Western countries during the Coal Age (and beyond).[63] The first well-documented upswing (1787–1814) coincided with the spread of coal mining and the spread of stationary steam engines. The second expansion wave (1843–1869) was driven by the diffusion of mobile steam engines and advances in iron metallurgy. The third upswing (1898–1924) was influenced by the rise of commercial electricity generation and the introduction of electric motors in factory production.[64] (Note that all upswings were connected to new energy technology.)

As the world experienced unprecedented levels of global economic integration, upswings (or downswings) in one region immediately influenced other regions. The volume of trade, relative to world income, nearly doubled from 10 percent in 1870 to 18 percent in 1914. Meanwhile ownership of foreign assets (mostly Europeans owning assets in other countries) more than doubled from 7 percent to 18 percent of world income, with much capital flowing to rapidly developing regions in the Americas. The large iron steamers tightly connected North American and European markets. In 1870 wheat cost nearly 60 percent more in Liverpool than in Chicago, but by 1895 the difference had shrunk to only 18 percent. Similarly, meat cost 90 percent more in London than in Cincinnati in 1870, while the price difference had shrunk to 18 percent by 1913. And thanks to the European railway network the prices of commodities west of Chicago even affected prices in eastern Europe.[65]

Falling transportation costs and an international division of labor also promoted economic growth far away from the centers of coal technology. American, Argentine, and Australian farmers achieved considerable wealth. Exports as a share of national product doubled in India and in the East Indies (Indonesia), and more than tripled in China between 1870 and 1913. To be sure, the economic fate of people in overseas (tropical) colonies depended heavily on the actions decided by Coal Age Powers. Colonies served as captive markets, which typically ruined their traditional manufacturing base, and regions as diverse as India, the Ottoman Empire, and Latin America experienced a sort of de-industrialization as they became part of the global trade regime.[66] On the other hand, consumers in these regions had now access to inexpensive goods mass-manufactured in the West. And though the Coal Powers often attempted to restrict technology transfer, their firms were eager to sell production equipment globally. What is more, the Coal Age Powers radically improved the infrastructure in their colonies by building railroads, ports, bridges, and so on. Evidently, the role of colonial markets diminished rather fast. Much of international Coal Age trade in terms of both volume and value was really between coal-fueled countries. Western countries had wealthier consumers, and by the end of the 19th century industrial nations had become each other's principal customers.[67]

A NEW VIEW ON TRADE

Diminishing transportation costs alone did not mean that Coal Age powers would immediately start to trade freely with one another. On the contrary, all nations act in their own interest and European countries had long viewed trade as disadvantageous for their prosperity. From the 16th to the 18th century, western European governments conducted an economic policy known as mercantilism. The ruling monarchs and their ministers measured the wealth of a country in terms of gold and silver hoarded inside its borders. Exports that were paid for in foreign gold were thus promoted, while imports (for which money would have flown out of the country) were largely prevented through tariffs. Grain and food products were often the only imports tolerated, as they helped to feed the city dwellers who produced manufactured goods, which were easy to tax. (This allowed the monarchs to acquire the gold they needed to finance the wars in which European nations frequently engaged.) Agricultural output, on the other hand, was more difficult to tax and often rapidly consumed. Hence, farmers were indirectly taxed by being forced to sell at minimal prices to city dwellers, and mercantilism increasingly weakened the position of the countryside. However, agriculture was then still the main source of energy, and most of the population lived in rural areas. But eventually European farmers could no longer make a living. When more and more of them moved to the urban centers during the 18th century, and famines weakened European economies, the 'laissez-faire' principle was formulated to get rid of the privileges of the cities to allow for a natural equilibrium to be found between industrial and agricultural production. The phrase laissez faire, laissez passer (leave matters alone, let goods pass through) suggests more generally that the state should not intervene in economic affairs, and is usually attributed to Frenchman François Quesnay (1694–1774), who developed the first complete theoretic system of economics. Quesnay was a leading figure among the Physiocrats, a group that believed that an economy's power derived from its agricultural sector. They wanted the French government to reduce taxes on French agriculture so that poorer France could emulate wealthier Britain. As the French government had protected French manufacturers from foreign competition, thus raising the cost of machinery for French farmers, the Physiocrats also wanted to get rid of tolls and other regulations that prevented trade.

Quesnay's work paved the way for classical economics, formulated by his contemporary Adam Smith (1723–1790) in Scotland, and later by Englishman David Ricardo (1772–1823). Smith admired Quesnay's work and particularly picked up on his support of free trade. Opposing mercantilism, the classical economists argued that the wealth of a nation is not defined by its possession of gold but by the work of its people and the conditions that allow people to do their work as efficiently as possible.[68] Wealth, according to this definition, cannot be taken away from a country (in the way gold can), which allowed for a more positive attitude toward other nations in general.

To be sure, trade is just about as old as humankind. Even the earliest societies likely bartered, given that their technological standards were equal enough for one group not to attempt to rob the other. If one group was lacking a commodity and had something to offer another group wanted, both were better off through an exchange. More complex societies, if they lived in peace with one another, supplemented each other's resources through trade. Imports often eliminated a lack of resources that limited overall production. (This was similar to Liebig's Law of the Minimum. If a country was short in iron ore but had plenty of fertile land and oxen for draft, imports of iron or iron plows pushed agricultural production to new heights. If the other country needed oxen or food, everyone was better off.) Another simple argument for free trade is economies of scale: the price of production (per unit) tends to decrease with higher production volumes that may in part be exported. (In addition, imports increase competition, which may decrease domestic prices or even break up domestic monopolies.) In order to address the question what should be produced and what should be traded, classical economists transferred the concept of professional specialization from individuals to nations. Professional specialization (division of labor) has obviously helped societies to advance. A gatherer-hunter society in which each member shares all the knowledge and learns all tasks will be quite slow in accumulating expertise. In agrarian societies, however, specialists worked in their respective professions for a lifetime and exchanged their products or services with other specialists. (A baker will buy his shoes from a shoemaker, for instance, who will not make his own bread.) Similarly, the classical economists argued that it would be better for nations to specialize in certain fields and exchange goods through international trade. The decision on what to specialize should be based on a country's natural resources, climate conditions, and so on. Portugal had a warm climate, for instance, and should focus on producing wine, which could, for instance, be traded for textiles made in Britain.[69]

For Britain it should have been quite easy to take the free-trading advice. It is hardly surprising that Adam Smith published his book *The Wealth of Nations* in 1776, just seven years after his fellow Scotsman James Watt had patented an efficient steam engine design, and just one year after this patent had been extended for another 25 years through the Steam Engine Act of 1775. Britain had numerous waterpowered factories by this stage and had differentiated itself from other nations through its increased productivity. Hence, it made perfect sense for Britain to consider concentrating fully on industrial production and to import agricultural goods and raw materials in exchange for manufactures. Nevertheless, Britain only slowly let go of its long tradition of restricting trade. The British Navigation Acts, initially targeted against the Dutch dominance in intermediary European and colonial trade, held that only English ships could transport goods to England from outside of Europe. If any European goods were imported, they had to be shipped either on English ships or by the country of origin. English colonies were prohibited from directly trading outside the British Empire by the Navigation Acts. (Resentment against the Navigation Acts was one of the causes of the American Revolutionary War.) Only in 1849, when the British shipping and ship construction industries had long been sheltered from foreign competition, and Britain had become the world's principal superpower, did the British government adopt a policy of laissez-faire in terms of international shipping.[70] What is more, Britain, like any other country, was striving for self-sufficiency in terms of food supplies. As early as 1361 export of grain from England was officially prohibited, and in turn numerous and complex Corn Laws restricted exports unless the domestic corn price was low and forbade imports unless it was high. The Corn Laws of 1815 placed taxes on imported grain to protect the interests of landlords, but between 1839 and 1846 the Anti-Corn Law League tried to persuade the parliament to remove price controls and import barriers for wheat, arguing that those controls acted as a hefty tax on basic foods and therefore harmed ordinary workers; dampened demand for other goods; raised labor costs; and caused recurrent bouts of starvation.[71] The Irish famine of 1845, which might have been much less severe if total reliance on potatoes would have been decreased by regular grain imports, finally tipped the balance and persuaded Britain to repeal the Corn Laws. Thereafter a long and successful period of free-trade began that lasted virtually until World War I. But even in this period, free trade schemes were abandoned as soon as economic growth slowed. Schumpeter's second (1843–1869) and third (1898–1924) economic Coal Age upswing was interrupted by what became known as the Long Depression of 1873 to 1896, a prolonged period of price deflation punctuated by severe business downturns that prompted European powers to promote home industry and to abandon free trade. During this time their colonial markets became increasingly more important to European Coal Powers to sustain demand for products from the mother country. The motivation to control large overseas areas was thus revived.

NOTES

55. J. Bradford DeLong, "Slouching Towards Utopia?: The Economic History of the Twentieth Century, Part B: The Path from the Pre-Industrial World, VIII. The First Global Economy: Production and Trade," University of California at Berkeley and NBER (National Bureau of Economic Research), January 1997, http://econ161. berkeley.edu/TCEH/2000/eight/html/Slouching2_8preWWI.html.
56. Ricky W. Griffin and Michael W. Pustay, *International Business: A Managerial Perspective* (Reading: Addison-Wesley Publishing Company, 1996).

57. By this stage longitude had been measured with increasing success, but the weight-and pendulum-driven clocks were too inaccurate at sea because of friction and temperature (and humidity) changes and the motion of ships, especially in heavy seas. Besides, Isaac Newton claimed, based on his gravitational theory, that a standard pendulum would beat faster in latitudes further away from the equator than in those nearer to the equator, which was experimentally confirmed by French mathematicians in the 1730s.

58. One of the main Harrison innovations was a unique escapement, with a very subtle lock-and-release action permitting the pallets to work without oil (and consequently be free of the effects of temperature-induced changes in viscosity).

59. J. Bradford DeLong, "Slouching Towards Utopia?"

60. Ibid.

61. Cotton, the agricultural product that was at the core of early industrialization, was somewhat of an exception in this respect. But then again, Britain imported most of its raw cotton from the United States, another Western industrializing country.

62. J. Bradford DeLong, "Slouching Towards Utopia?"

63. Joseph A. Schumpeter, *Business Cycles: A Theoretical and Statistical Analysis of the Capitalist Processes* (New York: McGraw-Hill, 1939).

64. Vaclav Smil, *Essays in World History—Energy in World History* (Boulder, CO: Westview Press, 1994), 240; "Catch the Wave: The long cycles of industrial innovation are becoming shorter," *The Economist*, February 18, 1999; G. O. Mensch, *Stalemate in Technology: Innovations Overcome the Depression* (Cambridge: Ballinger, 1979).

65. J. Bradford DeLong, "Slouching Towards Utopia?"

66. Find more on the issue of the de-industrialization of India, the Ottoman Empire and Latin America, and the decline of traditional Indian cotton textile production in parallel to increasing British imports during the Coal Age, in Daniel Cohen, *Fehldiagnose Globalisierung* (Frankfurt/New York: Campus Verlag, 1999).

67. Griffin, *International Business.*

68. Adam Smith, *The Wealth of Nations* (Buffalo, NY: Prometheus Books, 1991), Original edition published in 1776.

69. To be sure, it may be difficult to figure out what to specialize in, particularly when countries with similar resources and productivity are considered, or in a case where a country is not internationally competitive in the production of anything at all. For the latter case, classical economics holds that even a nation whose productivity is the absolutely lowest in every sector would be better off in a free trading world, because countries that are absolutely best in many sectors would specialize with respect to the abilities of other countries and leave some sectors vacant. (This is like a heart surgeon who is best in lawn mowing. He will spend his time in the hospital, paying somebody to take care of the garden.) A country with lower productivity will remain poorer than its more productive trading partners, but it will (at least) be less poor with trade than without. Hence, free trade increases the wealth (or rather, the maximal possible consumption) in all countries.

Though the theory that free trade increases wealth is widely accepted, it has a few hang-ups. One is that governments tend to oppose free trade when it comes to products of strategic importance. If possible, they usually avoid to rely too much on imported food (and therefore subsidize domestic agriculture) or on the import of iron and iron products (and therefore protect or subsidize domestic steel production). Another problem is that countries as a whole may well be better off under a free trade regime, but that a lot of damage is done to particular domestic sectors, especially until a trade equilibrium is found between the internationally trading countries. Such transitional periods may repeat themselves over and over again whenever the conditions in one trading country change. (The productivity in a sector may regionally decline, for instance, due to the exhaustion of resources, such as soil degradation or depletion of mines, or it may increase due to the discovery of new resources or invention of new technology.) A trained Portuguese weaver might not like British cloth imports, and a British farmer might dislike grain imports. There will always be domestic winners and losers under a free trade regime, but these distributional effects are usually not accounted for, as governments fail to compensate the losers adequately.

Strong criticism of free trade theories derives from the claim that free trade serves the rich and powerful the most. While less productive countries will likely gain from free trade as well, the more productive countries profit more in absolute terms and the gap between rich and poor countries widens. By the end of it, this should not matter. For a poor country it is most important to be less poor and irrelevant if a rich country is more rich. This suggests that trade is to be chosen by all parties.

Nevertheless, countries of low productivity and technological development may have difficulties changing their competitive position if they are locked into a free trading scheme. Some industries simply provide higher earnings than others and generate more general know-how that benefits other industries. Countries therefore attempt to export more technologically advanced, sophisticated, and value-added products for which there is less competition and that provide higher margins of return. Agricultural products are usually at the lowest end of this scale, followed by mined raw materials, which tend to provide less earnings than manufactured products, unless they are very rare. Attempts to build up more advanced and profitable industries may fail if these industries have to compete against imports from more technologically advanced countries. Hence, it may sometimes be wise for a country (or a particular sector) to abstain from trade, while it will be favorable to trade and compete with international quality and price standards at other times.

Find more information on trade and trade theory in Paul R. Krugman, Maurice Obstfeld, *International Economics: Theory and Policy* (New York: HarperCollins, 1994); Herman E. Daly, "The Perils of Free Trade," *Scientific American* 269 (1993): 50–57.

70. Other Western powers, too, decided against free trade regimes during their early industrialization phase. Observing increasingly wealthy and powerful Britain, the continental European nations and the United States attempted to follow Britain's path of industrialization. They closed their borders for British products and protected their immature industries in order for them to catch up with British technological standards while serving domestic markets in the absence of competition. Only later, after their strategic position had improved and their industries had grown strong enough to compete on international markets, did the United States and other industrial countries increasingly started to abstain from protectionism and to promote free trade regimes.

71. "Richard Cobden: An Anti-poverty Hero," *The Economist*, June 3, 2004, http://www.economist.com/opinion/displaystory.cfm?story_id=E1_NSNVDNN.

CHAPTER 24

THE COAL-POWERED EMPIRES

Like trade, fresh efforts in empire building were facilitated by the new transport energy regimes of the Coal Age. Steam trains and ships were not just used to transport goods and migrants; they were employed to move troops swiftly to every corner of the world. Most critical were steam-powered gunboats, armed ships of shallow draft that allowed for the penetration of rivers, the hitherto secure inland lifelines of agricultural societies. Cannon-steamers tipped the balance in favor of Western nations in China and India, for instance, or wherever the previous threat by European wind-powered, ocean-going, cannon-carrying sailing ships had been insufficient. The Coal Powers often did not even have to exert violence to achieve their objectives. The Coal Age was the era of gunboat diplomacy: if an agrarian country was to negotiate, say, trading rights, it would notice one or more cannon steamers appear off its coast. This threat was usually sufficient for Coal Powers to arrange economically advantageous relationships around the world. Those regions that were fully integrated into Coal Age empires as colonies soon saw a harbor and railroad infrastructure being installed. But this infrastructure did not only facilitate trade with the mother country; it also allowed for troops to be rapidly moved around in colonies, discouraging respective independence movements.

To be sure, the motivation for Coal Powers to build empires was different from those of the Agricultural Age, when control of more land (and labor) directly translated into command of more energy. The Coal Age was actually unique in that its wars were not fought over energy at all. (This sets this era aside from the previous and the subsequent Energy Eras.) There were

Militant Steam Power on the Waters   The Black Hawk War's Bad Axe Massacre of 1832, during which the side-wheeled steamboat *Warrior* helped European Americans to defeat Native Americans and thus to open the areas of Illinois and (present-day) Wisconsin for further settlement, was an early example of steam power on the waters to determine military strength. Ships such as the *Warrior* were flexibly fueled by wood or coal, while later specialized cannon steamers relied on coaling stations. Such steamers mastered the high seas as well as relatively shallow waterways, and were thus capable of terrorizing even the rivers and coastlines of highly developed East Asia. (Illustration published by Henry Louis in 1857. Library of Congress image LC-USZ62-90, edited.)

exceptions, of course. Austria and Prussia, for instance, battled over coal-rich Silesia, and Japan conquered coal-rich Korea. But generally all the major Coal Powers had plenty of coal within their own national borders.

Why, then, did the Coal Age Powers build up large international empires anyway? One answer is markets. Selling overseas justified the creation and maintenance of large production and shipping capacities. But on the receiving end, Western products were not always welcome. They ruined the traditional manufacturing base of importing countries, insulted local vested interests, and caused radical social changes (or rather, hardships). Hence, overseas regions had to be politically or militarily controlled to enforce trade relations. The populous and wealthy commercial centers of Asia were the primary international markets targeted by the Western Coal Powers in this respect.

Another reason behind Coal Age empire-building was access to nonenergy resources. Although the West's Coal Age industrialization was largely self-sufficient, there were certain commodities that were to be found or grown only in (hot climate) overseas regions where Europeans either did not want to live, or where the native population was already too large and advanced to be killed or expelled. Such goods included cotton, rubber, timber, tea, coffee, and cocoa, among others. Western Europe also relied on gradually more imports of Fertile Crescent foods produced in temperate overseas regions. These regions were typically populated by European offspring, but they had to be controlled politically or commercially as well to secure food imports for western Europe's growing population. Besides, grain was needed to fuel horses, which were still critical for land transport and in warfare.

Yet another reason behind the build-up of global empires during the Coal Age was prestige. Especially from about 1870 the European Coal Age Powers engaged in a new kind of imperialism that was to a large extent motivated by a wave of nationalistic pride. Europeans also began to explain their technological lead with concepts of cultural and racial superiority. They formulated doctrines of obligations to "civilize the heathen savages" in overseas colonies. The colonization of Africa during the Coal Age was, by and large, a financial disaster. Nevertheless, the European Coal Powers took great pride in acquiring African colonies and engaged in a scramble for the continent.

As coal technology was complex and difficult to copy outside the Western world, the only serious obstacle faced by Coal Powers attempting to build international empires was competition from other Coal Powers. They, too, had the advantage of knowing how to use coal to mass-produce iron for weapons, ships, and trains. And they, too, were using coal energy to move their troops rapidly over the oceans and continents. The major Coal Powers included Britain, France, Germany, Austria, the United States, Russia, and Japan. Except for the last, all of them were of European stock. Hence, it should not surprise us that by 1900 Europeans and their overseas offshoots controlled some 80 percent of the Earth's land area, while representing less than a third of the global population. (To be sure, the European share in the world population had by then radically increased. In 1500, people of European origin and descent accounted for some 100 million people or roughly one sixth of the world population. In 1900, they accounted for nearly 550 million people or about one third of the global population.)

In a way, societies that did not take the step into the Coal Age got off lightly. Gatherer-hunters had been ruthlessly slaughtered by agrarians during the Agricultural Age, but coalians did not usually massacre agrarians. Those looking for markets had no interest in killing their customers. Coal-powered imperialism meant that those who commanded more productive and destructive energy took control over weaker regions to dominate them militarily, economically, politically, and perhaps culturally. Highly-developed but agrarian China came under Western control, and India was made a British colony. The Muslim World, too, came increasingly under the influence of the Coal Powers. (It entirely lost control of the trade between Europe and South/East Asia: ocean transport around Africa became very cheap, and the Trans-Siberian Railway connected China and Europe in the north.) The

Indian Mughal Empire collapsed, and the Ottoman Empire saw much of its industrial base damaged under the de-industrializing influence of trade with the Coal Powers. (The Ottomans also had to accept substantial territorial losses, including those of Greece and Egypt.)

Japan, it seems, found the right solution to deal with the energetic threat imposed by the West. Forced to open its borders, Japan decided to undergo radical changes. By copying Western technology, organization, and administration, the Japanese were the only non-Europeans to make their country a Coal Age Power. And sure enough Japan also turned imperialistic, becoming an aggressive colonial power that invaded and subdued much of East Asia.

To be sure, the presented picture of Coal Age Powers expanding the territories they controlled, but leaving natives in those territories alive in order to preserve markets, is no more than an attempt to explain what has been observed. It may actually be a quite naive attempt, as it implies that Coal Age people were more humane than can possibly be claimed. As the treatment of native Americans by European Americans (or of Australian aborigines by the British) in the later Coal Age shows, coalians were no less brutal than agrarians when they had the means to be. They, too, slaughtered men, women, and children on the largest possible scale when the given energy realities allowed it. Thus, the lack of more outright killing in the Coal Age may in fact have been due to the lack of weapons of mass destruction at the time. If such weapons had been available to Coal Age Powers, they might well have found an excuse to empty the fertile plains of India and China of their native populations and fill them with their own offspring. But fortunately the killing and displacing of the masses of South and East Asians was beyond the technical capabilities of Europeans. And in Africa, though the continent was split between European powers as soon as railways allowed them to penetrate the continent's interiors, the killing of natives did usually not provide gain for the Coal Age people, either none at all, or compared to the costs involved.

THE WESTERN EUROPEAN COAL EMPIRES

Britain profited handsomely in geopolitical terms from being the first nation to enter the Coal Age. The strategic benefits that derived from early coal technology probably helped Britain to win the Very First World War (1754–1763), which handed French Canada to Britain and removed French competition from India. (Britain at this stage mined millions of tons of coal per year, and was the only European nation that had technology to use coke in iron production. The Darby factory at Coalbrookdale had cast over 100 Newcomen cylinders by 1758.)[72] Thereafter Britain built up a huge empire that included the Indian subcontinent (from Pakistan to India, Bangladesh, and Myanmar), Australia, New Zealand, various Caribbean islands, and much of Africa, including the population centers of Nigeria and South Africa. In addition, Britain controlled various regions (such as Egypt) that were not officially part of the empire, but strategically important. At the end of World War I, in 1918, the British Empire comprised a quarter of the world's land area and population. Other western Coal Powers built overseas empires as well, but they never quite caught up with Britain's head start into the Coal Age when it came to colonial expansion.

France

France had emerged as Europe's leading power from the Thirty Years' War (1618 to 1648), but lost this position within a century and was defeated by Britain (and Prussia) in the Very First World War. The latter ended with the Treaty of Paris 1763, in which France decided to forgo Canada in order to keep the Caribbean sugar islands of Guadeloupe and Martinique. France was allowed to keep its enormous but undeveloped Louisiana colony (which stretched from the Gulf of Mexico to Canada and from the Mississippi River to the Rocky Mountains). However, Napoleon Bonaparte, in 1803, sold Louisiana to the United States of America for the bargain price of $15 million dollars, about three cents per acre.

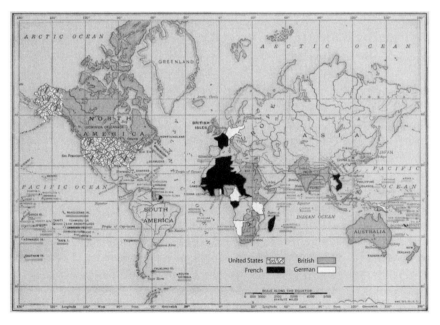

The World in 1910   The British Coal Age empire was the largest, but the other Coal Powers took their share. U.S. overseas possessions included the Philippines, while German colonies existed in Africa and the Pacific, where the widely scattered Caroline Islands became part of German New Guinea. (Based on a map contained in Frank M. McMurry, *The Geography of the Great War*, The Macmillan Co., New York, 1919.)

Napoleon urgently needed the money to finance his many wars. He had initially emerged as a capable military leader from the Revolutionary Wars, which were fought by France against various western European powers following the 1792 deposition of the French king. In 1798 Napoleon attempted to conquer Egypt, but was defeated by the English in the Battle of the Nile. Nevertheless he managed to establish himself as the military dictator of France in a coup d'état in 1799, and crowned himself emperor in 1804. For a while, his many victories during the Napoleonic Wars (1799–1815) gave him control of practically all of continental Europe, but the attempt to conquer Russia in 1812 turned into disaster. In the end, France was once again defeated by Britain and Prussia. Silenced for a while, French imperialism caught up again during the country's rapid coal-based industrialization from the 1860s. Much of it was rooted in a nationalistic urge to rival Britain, with the French taking the lead in constructing the Suez Canal according to the plans of Austrian engineer Alois Negrelli. After its opening in 1869, the Suez Canal assisted the buildup of a French colonial empire that included north and middle Africa as well as Madagascar and Indochina (modern Cambodia, Laos, and Vietnam).

Austria and Germany (Prussia)

The German-speaking area, representing Europe's largest single language, was still not unified as one nation, or even a few nations, at the beginning of the 19th century. Instead, it was an agglomerate of numerous loosely connected and rather independent states, formally under the authority of the Austrian monarch, who was traditionally elected emperor of the Holy Roman Empire of German Nations. In 1806, after the military successes of Napoleon Bonaparte, the last Holy Roman Emperor, Francis II (from 1804: Emperor Francis I of Austria), resigned, and the Holy Roman Empire ceased to exist. However, with peace restored in 1815, the German states formed another loose union, the German Confederation (Deutscher Bund), and it was again Austria that was given special political power, this time with the right to appoint the President of the Diet that was to govern the German Confederation. (Following a wave of consolidation resulting from Napoleon's interference, the German states were now much reduced in number, but they still consisted of 35 monarchies of various kinds plus four free cities.[73])

However, the Austrian Habsburg rulers, residing in Vienna, increasingly focused on their own multi-lingual monarchy (outside the German Confederation), which had grown into the largest empire in Europe except for Russia. The Habsburg monarchy included all of present-day Austria, the Czech Republic and Slovakia, parts of southern Poland (including parts of Silesia and the city of Krakow), parts of the eastern Ukraine (including the city of L'viv or Lemberg), the northern part of Romania, all of Hungary, northern

Serbia, all of Slovenia and Croatia (including the long Mediterranean coastline of Dalmatia), and much of present-day Italy, including the areas around the cities of Venice, Milan, and Florence. (Tuscany was the southernmost province at the time.) In short, the Austrians focused on building a classic contiguous empire. (The Austrian Habsburg Netherlands, comprised of modern Belgium and Luxembourg, were not geographically connected to the rest of the Austrian monarchy, but remained part of it until 1814.) The Austrians operated deep-ocean steamers from their Adriatic ports, planned the Suez Canal, and launched international expeditions. The Austrian Lloyd company, established in Triest on the model of the British Lloyd, traded to the Far East from 1833. And yet Austria hardly engaged in overseas colonialism and had few imperialistic ambitions outside Europe, perhaps because it was preoccupied by domestic problems: The monarchy's various ethnic groups strove for gradually more national independence, which reduced internal coherence. What is more, the German states increasingly viewed the large non-German Austrian Empire as being in conflict with the Austrians' role of being the foremost leaders of the German Confederation of States. The latter problem facilitated the rise of Prussia to German leadership.

Prussia was initially an overwhelmingly rural duchy at the northeastern edge of the German speaking area, stretching along the Baltic Sea shoreline and comprising parts of modern Poland, Russia, and the Baltic countries. Prussia achieved a powerful position after fighting on England's side in the Very First World War (1754–1763) as well as the Napoleonic Wars (1799–1815). After Napoleon's defeat, Prussia gained the industrially advanced German provinces of (Prussian) Saxony, Westphalia, and the Rhine Province, partly as a safeguard against further French aggression against the German-speaking areas. However, the true rise of Prussia is associated with the politics of Otto von Bismarck, Prussian chancellor from 1862, who achieved German unification under Prussian leadership within less than 10 years. In 1866 Prussia concluded an alliance with Italy and started the Seven Weeks' War against Austria, Bavaria, Saxony, Hanover, and several minor German states. Prussia promised Italy the Austrian monarchy's Venetian territory in case of victory, and Italy subsequently declared war on Austria. Austrian troops defeated the Italians at Custozza, but were themselves defeated shortly thereafter by Prussian troops attacking in the north, in Bohemia. Prussian military organization and use of the railway for troop transport were probably most decisive in the Prussian victory at Königgrätz, though the battle went down in history for the Prussians' pioneering breech-loading needle guns. The consequent peace treaty assigned Holstein and other German territories to Prussia, but more importantly, the German Confederation was dissolved and Austria ousted from power in Germany. Instead, a Northern German Confederation under Prussian control was founded.

To be sure, the defeat had major consequences within the Austrian Empire as well. Even though the Austrian navy crushed the Italian fleet just two weeks after the disastrous battle at Königgrätz, Italy gained the Venetian territory when peace was concluded. Internally weakened, the Austrian government also had to hand more power to the Hungarians. The Habsburg ruler was still both Austrian emperor and king of Hungary, but the Hungarians gained control of their internal affairs and the Habsburg Empire was from 1867 referred to as the dual Austro-Hungarian monarchy.

Prussia, on the other hand, continued its ascent towards entire control of all of Germany. France was concerned about this development and (following a diplomatic fall-out) declared war on Prussia in 1870. France was quickly defeated by German troops and had to cede Alsace and East Lorraine to Prussia. What is more, the three independent south German states (Baden, Württemberg, and Bavaria) placed their armies under the command of the Prussian king during the conflict and now agreed to unite with the other states (excluding Austria) into a unified Germany. Prussian King Wilhelm was crowned Emperor of Germany in 1871, and Bismarck became imperial chancellor. To be sure, the German kingdoms, duchies, and principalities were still quite independent, but the Imperial Government was given control over national defense, foreign affairs, and commercial matters. As such, it was able to organize concerted efforts of build an overseas empire. Under Prussian leadership the unified German nation acquired substantial areas of Africa as well as certain Pacific islands (most notably German Samoa) as colonies.

European Empires in Africa

Africa became the principal stage on which western European Coal Powers could put their new technology to work to build empires. Africa had been a source for slaves for centuries, but Europeans generally did not settle and expand there because much of the continent was not suited for Fertile Crescent domesticates. The only substantial European settlements in Africa up until the 1870s were in the very north, in Mediterranean Algeria, and in the very south, at the Cape of Good Hope, both regions situated in temperate climate zones. Otherwise there would actually have been plenty of space in Africa. The continent's only densely populated regions were along the Nile River in the north and along the Niger River in the west.

As soon as Europeans had the means to penetrate the interior of Africa, they began exploring it. Advances in medical science helped to better cope with what had traditionally been a major factor hampering the exploration of this continent: African diseases to which Europeans had no immunity. The first antimalarial drug was a bitter substance called quinine, extracted from the bark of the cinchona tree. (Peruvian Indians taught French missionaries how to use the bark in 1630, but the substance quinine was not isolated until 1820.) And while no treatment was found for yellow fever, it was understood in the 1880s that this disease was mosquito-borne: eradicating mosquito breeding sites and destroying larvae then proved effective in eliminating epidemic transmission of both yellow fever and malaria.

More importantly, Africa's interiors were soon penetrated by European coal-fueled cannon boats steaming up the Nile, Niger, and Congo River systems. Soon European engineers began planning railways to create the infrastructure necessary to exploit such African resources as copper, gold, diamonds, timber, beeswax, ivory, leather, rubber, palm oil, sugar, cacao, tea, and coffee. But this sounds like more than it actually was. By the end of it all, the European colonization of the African continent during the Coal Age was an economic failure. The sparsely settled continent did not provide large markets for European manufactures, and to a large extent it was national pride and prestige associated with owning colonies rather than sound economic analysis that prompted European nations to spend fortunes on developing, administering, and defending empires in Africa.

Britain actually started a new era of imperialism on the African continent when occupying Egypt in 1882 to ensure control of the Suez Canal. Soon thereafter French, Belgian, and Portuguese activity in the lower Congo River region raised concerns that the Scramble for Africa would end in total chaos. Hence Prussian, or rather Imperial German, Chancellor Otto von Bismarck, picking up on a Portuguese suggestion, organized the Conference of Berlin in 1884, to which Britain, France, Germany, Belgium, and Portugal were party. Portugal was not a Coal Power, but had maintained a strong position in Africa ever since its wind-powered colonial head start four centuries earlier. (Portugal had retained the colonies of Mozambique, situated at Africa's southeast coast; Angola, situated at Africa's southwest coast; and the small colony of Portuguese Guinea, or Guinea-Bissau, situated in northwest Africa, opposite the Cape Verde Islands.) The Berlin Conference lined out the rules for empire-building in Africa. The idea was to peacefully partition Africa among the signatory nations, who agreed to notify one another whenever they took possession of an African coast. They would be allowed to control the hinterland to an almost unlimited distance, but they would have to prove their ability to "effectively" occupy the claimed land, that is, to control the indigenous states and peoples. The European Coal Powers in turn brutally put down native African resistance, but at least sometimes the good came with the bad. Germany introduced obligatory elementary schooling for native boys in German East Africa in 1899, for instance, and managed to practically eradicate sleeping disease and smallpox due to an intensive program of vaccination in 1900. The Europeans eventually also managed to close down much of the Arab slave trade in Africa.

Unfortunately, the territorial claims made according to the rules laid out by the Berlin Conference quite arbitrarily determined most of Africa's future internal borders with no regard to traditional tribal boundaries. Ethnic groups were often split apart by colonial borders, while hostile tribes were brought under the same government. This was going to be a major cause for conflicts and civil wars in Africa during the 20th, and even into the 21st century. However, the European Coal Age Powers had the energy to do as they pleased, and by 1914 practically all of Africa was under European colonial rule.

British Africa

Britain alone brought about 30 percent of Africa's population under its control between 1885 and 1914. British troops occupied Egypt in 1882, Uganda in 1894, Kenya in 1895, and Egypt's southern neighbor Sudan in 1896. (In Sudan the Mahdists defeated the British in 1885 to capture Khartoum, but were crushed by Anglo-Egyptian forces under Herbert H. Kitchener in the 1898 Battle of Omdurman that demonstrated the superiority of machine guns: it left 25,000 natives, compared to 48 British, dead.) Thus, Britain controlled the complete stretch from the Mediterranean along the Nile River to Lake Victoria and towards the African east coast. In western Africa, Britain established its influence in densely populated Nigeria through the activities of the National Africa Company (from 1886, the Royal Niger Company), which had acquired Lagos from an African chief in 1861. The two protectorates of North and South Nigeria (proclaimed in 1900) had a population of 15 million people, which was more than all the French (or German) African colonies combined. West of Nigeria Britain maintained Ghana (after defeating the native Ashanti), and Sierra Leone (which had initially been purchased in 1787 by English slavery opponents to found a colony for liberated slaves.)

Britain also acquired the Dutch colony at the horn of Africa. The Dutch East India Company had founded Cape Town in 1652 as a staging post to support its ships sailing between Europe and the Dutch East Indies (Indonesia). The Dutch then expanded the Cape Colony eastward towards the Great Fish River, where they met the first substantial native resistance and fought wars against the Bantu-speaking Xhosa from the 1780s. In 1795, when the Netherlands were occupied by French revolutionary armies, Britain took Cape Town and decided to keep it after the end of the Napoleonic Wars in 1814. (By this time some 20,000 Europeans lived at the Cape.) From 1824 the British government sponsored Britons to settle in the Natal province, and in 1836 some 10,000 Dutch settlers left the Cape region to escape British rule. They founded the republic of Transvaal and the Orange Free State, which are both provinces of the modern Republic of South Africa. All these regions were located substantially north(east) of the Cape, and the Dutch met fierce resistance from the Zulu, a militant south African tribe established by legendary leader Shaka Zulu (circa 1787–1828). Asians were added to South Africa's population mix when British authorities decided to bring in indentured laborers from India to work in Natal's sugar fields (and to some extent in coal mines). Between 1860 and 1911 some 176,000 Indians arrived in Natal, among them, immigrating in 1893, Mahatma Gandhi, the later leader of India's independence movement.

The discovery of diamonds (at Kimberley in 1870) and gold (in the Transvaal in 1886) in the Cape Colony opened a new chapter of African history and exploitation. Prospectors were pouring in from Europe and soon came into conflict with local Dutch farmers, the Boers. British ambitions to expand into the Dutch regions and further towards the north are closely associated with Cecil Rhodes, who arrived as a 17-year-old from Britain and almost immediately made a fortune in diamond and gold mining. In 1880 Rhodes established the De Beers Mining Company (which eventually controlled 90 percent of the world's diamond production), and the year after he gained a seat in the Cape Parliament. Rhodes aspired to establish a Cape-to-Cairo railroad under British control, north to south through the entire African continent, and has been labeled the Coal Age's most ambitious empire-builder.[74] Rhodes died in 1902, at age 49, after having been involved in bringing enormous areas of Africa under British dominion. (What is now Botswana, Zimbabwe, and Zambia were all annexed between 1885 and 1893.) Rhodes was Prime Minister from 1890, but was removed from office six years later for authorizing an aggressive British invasion of the Dutch Boer Republic of Transvaal. A decade later, during the Second Boer War (1899 to 1902), Britain established sovereignty over Transvaal and Orange Free State. These wars were extremely vicious by all accounts, with the Boers employing guerrilla tactics and the British using Maxim machine guns and pioneering concentration camps (that is, prison camps for civilians in wartime). To be sure, Britain by the end of it did not achieve its overall goal until World War I: a British north-south axis through Africa did not become reality, because Germany snatched the regions between Kenya and Zambia.

German Africa

For a while, Germans hardly engaged in any overseas colonialism even after they were united under Prussian leadership in 1871. However, their largely anti-imperialistic attitude changed when Wilhelm II, a grandson of British Queen Victoria, inherited the throne in 1888 and dismissed chancellor Bismarck in 1890. The first German overseas protectorate was established a bit earlier, though, in 1884. It was located in present-day Namibia (northwest of Britain's Cape Colony), where German merchant Adolf Lüderitz had purchased from natives the area around the city still bearing his name. The newly founded Deutsche Kolonialgesellschaft für Südwest-Afrika (German Colonial Society for South West Africa) soon bought the assets of Lüderitz's failing enterprises, and Heinrich Göring, father of the later Reichsmarschall Herman Göring (Hitler's second-in-command), was appointed governor of the colony.

At the time of the Berlin Conference (1884), Germany also established colonies in Cameroon (south of Nigeria) and Togo (east of Ghana), and founded German East Africa, the colony that comprised Tanzania (Tanganyika), Rwanda, and Burundi, hence wrecking the British plans of a Cape-to-Cairo axis. German Africa was the third largest colonial empire in Africa (after Britain and France), but it held merely about 9 percent of the African population.

French Africa

The French possessions in Africa were much larger, with about 15 percent of the African population as of 1914. The French conquest of African areas began in 1830 in Algeria, right across the Mediterranean from France's southern coast. The French were interested in Algeria's fertile coastal regions, but between 1850 and 1870 also occupied the mountainous inland region. In addition France had maintained the relatively small west African colony of Senegal ever since the early slave-trading days.

After the French had been defeated in the Franco-Prussian War of 1871 (which involved the loss of Alsace-Lorraine), a political movement emerged that suggested that France should overcome the humiliation and economic burden of the indemnity payments through prestige and commercial gain derived from the expansion of the African colonies. Algeria's eastern neighbor Tunisia was annexed in 1883, and soon France controlled most of the Sahara desert and hence the caravan trade between the Mediterranean and the semiarid Sahel zone, which was to be colonized as well. In 1895, France founded its colony French West Africa, which included Senegal, Mali, and the two coastal colonies of French Guinea (south of Senegal) and Ivory Coast (neighboring British Ghana and sharing its northern border with Mali). France also gained the relatively small kingdom of Dahomey (Benin, 1892), as well as Upper Volta (Burkina Faso, 1896), and continued the expansion through the Sahel zone, through Niger and Chad, until it clashed with the British who won out in Chad's neighboring region Sudan. France also incorporated Morocco (1912), where Spanish authority had gradually faded away, and Morocco's southern neighbor Mauretania (1892) into French West Africa. Thus, France controlled a huge continuous area that covered North and West Africa, and extended southward to include the Central African Republic, the Republic of Congo, and Gabon. What is more, France managed to gain and maintain formerly Portuguese Madagascar, the large island off the coast of Mozambique.

Belgian Congo

The small country of Belgium played a significant role in the Coal Age colonization of Africa as well. While the independence of the seven northern United Provinces of the Netherlands was generally recognized from 1596, the Catholic provinces covering present-day Belgium and Luxembourg remained Habsburg land. Belgium passed from Spanish Habsburg to Austrian Habsburg in 1713, but was united with the Netherlands at the end of the Napoleonic Wars in 1814. However, Catholic, coal-rich, and partly French-speaking Belgium rejected the union with the Protestant, coal-less, trade-focused Netherlands, and declared its independence in 1830. The subsequent war of independence with the Dutch continued for nine years, even though the major Coal Powers accepted Belgian sovereignty already in 1831, provided that Belgium would remain perpetually neutral. Leopold I, of the German House of Saxe-Coburg and Gotha, was made King of Belgium. (Leopold's sister was the mother of Britain's Queen Victoria, and he was the widower of the daughter of King George IV, who was Victoria's uncle.)

Around 1880 Leopold's son, Leopold II of Belgium, used his private fortune to acquire a huge area of central African land, nearly 80 times larger than Belgium itself. Stretching along the Congo River, from its mouth at the African southwest coast deep into the very heart of Africa, this colony became known as the Congo Free State. Only in 1908, a year before his death, did Leopold II sell the colony to the state of Belgium, which controlled it until 1960. (Known as Zaire from 1971, the country was renamed Democratic Republic of the Congo in 1997.)

Under the rule of Belgian King Leopold II the Congo became the most notorious of all European colonies in Africa. He financed emissaries to establish trade with natives in the 1870s and in turn used his private army to exploit the region. During the 1880s most revenue came from ivory, but soon Leopold earned a fortune from the colony's wild rubber vine. (Rubber was in huge demand in Europe at the time.) Apparently millions of Africans were mutilated or died through murder, starvation, exhaustion, or disease. Whole villages were taken hostage to ensure that the men went into the jungles to tap trees and collect the sap required for rubber production. Those who refused had their hands hacked off and their fellow villagers massacred. Figures such as Guillaume Van Kerckhoven and Leon Rum enjoyed mass executions and had the victims' heads chopped off and exhibited on poles. African soldiers hired by the Belgians were paid five brass rods per human head they returned during the course of military operations. The Belgians also press-ganged tens of thousands of natives into railway construction.[75]

Italian Africa

Italy played a minor role in Africa's colonization. In fact, it was the only European power to face a disastrous defeat by an African nation during the Coal Age. Italy joined the European Scramble for Africa in the 1880s by conquering Italian Somaliland, a coastal strip north of Kenya; and Eritrea, a small coast at the Red Sea, east of Sudan. Between these impoverished regions lay the rich kingdom of Ethiopia, which the Italians attempted to annex. Unexpectedly, native Ethiopian forces crushed the Italian forces in the town of Adowa in 1896. Italy in turn signed a treaty that recognized Ethiopia as an independent nation (though fascist Italy attempted to conquer Ethiopia once again in 1935). In 1911, Italy fought a war against the Ottoman Empire and annexed Libya from the Ottomans.

At the dawn of World War I, the only African areas independent of direct European rule were Ethiopia and Liberia. Similar to Sierra Leone, Liberia was an area in West Africa bought by the American Colonization Society, a philanthropic organization, in order to establish a settlement for liberated black slaves from the southern United States. (The first settlers arrived in 1822, and by 1870 Liberia had attracted 13,000 U.S. immigrants. But in the end Americo-Liberians turned out to be a minority in this independent republic.)

British Australia: Gatherer-Hunter Meets Steam Engine

Unlike the situation in Africa, the Coal Age colonization of Australia was almost exclusively a British affair. However, it was the Dutch who had first sailed to Australia in 1606 (from their Indonesian colony). In 1642 Dutchman Abel Tasman discovered Van Diemens Land, the large island south of Australia that was later named Tasmania in his honor. But that was just about it in terms of Dutch activities on the continent that was known as New Holland until the British renamed it Australia in the Coal Age: The area simply did not seem to offer any resources that would compare to those of the Dutch East Indies.

The English first set foot on New Holland's northwest coast in 1688, but little interest was aroused until James Cook arrived at the continent's fertile eastern coast in 1770, claiming it for England. Around this time Britain had serious domestic problems. The population had soared, land reforms pushed hordes of people into the growing cities, and the oversupply in workers depressed wages and left many unemployed. Consequently the crime rate increased, and the prisons were overflowing. Britain had been shipping convicts to America throughout the 17th century, but stepped up this practice after 1717 to deal even with more petty crimes. America kept on serving as an outlet for British convicts until 1776, but with the U.S. Declaration of Independence the shipments were stopped. Right then Australia came into the picture. In 1788 Britain founded a colony in what is now New South Wales: eleven vessels transporting 736 convicted criminals and some officials to oversee them arrived at the site where the city of Sydney is now located. Thereafter more and more British convicts were shipped to Australia, and free settlers followed from 1819.

Subduing all of Australia was more or less a walk in the park for the British. Australia had been entirely isolated from Eurasia and the rest of the world, and was inhabited by perhaps 500,000 aboriginal gatherer-hunters by the time the Europeans showed up. The native population had probably arrived between 30,000 to 40,000 years ago from Southeast Asia, but Australia was not rich in species that lend themselves to domestication. Thus agriculture never emerged, and people kept living in the Stone Age.[76] (No Australian plant or animal species has been fully domesticated even after European arrival. The macadamia nut is the one exception.)

Australia's people were as diverse as its landscape. Savanna woodlands in the north, a harsh desert outback, tropical rainforests, and temperate woodlands in the south were inhabited by aboriginals who spoke 250 distinct languages, and whose stature varied from the very short (pygmy-like Negritos and Barrineans of north-east Queensland) to the very tall (such as a man from north Queensland who astonished Europeans by his height of 223.5 cm—over 7 feet[77]). Generally, the surviving Australian aborigines tend to have dark skin, eyes, and hair, but some are blond. Tasmanian aborigines, who counted about 4,000 upon European arrival, looked quite different from mainlanders and had a special history. They had become isolated even from the Australian mainland some 10,000 years ago due to rising sea levels. Unlike the mainlanders, Tasmanians did not know how to start a fire, which put them at the lowest possible energetic level. They also had no boomerangs, spear throwers, specialized stone tools, compound tools, or bone tools that would have enabled them to fabricate needles and sew clothing. And even though they lived mostly at the coast, Tasmanians did not catch fish.

The technology and energy gap between Europeans and Australians could hardly have been bigger. Though the first British settlers arrived on wind-powered ships, James Watt's advanced steam engine had already started to spread into commerce when Sydney was founded. Most importantly, the Europeans were equipped with Fertile Crescent domesticates. Despite the aspects of coal energy in British society at the time, this was mainly a classic agricultural expansion. Temperate South Australia is situated at the same distance from the equator as the Fertile Crescent. Europeans simply introduced wheat, barley, cow, sheep, goat, pig, chicken, and horse, plus the American crops potato and maize, and chased the natives out of all fertile areas. What is more, the aboriginal population lacked immunity against Eurasian diseases, and many rapidly succumbed to smallpox, typhoid, venereal disease, influenza, and measles. Perhaps surprisingly, human conduct had not improved much since the Spanish had reached the Americas 300 years earlier: The British ruthlessly massacred native men, women, and children whenever they had an excuse to do so. Tasmania's native population disappeared (almost) entirely, while that of the mainland shrank to about 31,000 by 1911. (Later in the 20th century, the aboriginal population was growing again and was expected to reach the level of 1788 in the early 21st century.)

Australia's new European population had full access to British coal and technology, and *the fifth continent* soon turned out to be itself rich in coal and iron ore. The Australian gold rushes of the 1850s and 1880s promoted more immigration, economic growth, and the construction of railways. Over 20,000 km of track had been laid by 1901, when all states except Western Australia linked their individual rail networks together: This was the year the federal commonwealth of Australia, an independent dominion under the British Crown, was created by an act of Parliament. To be sure, Australia was integrated into the global economy as an agrarian rather than industrialized region, delivering wheat, meat, and wool to international markets. Australia is enormously big (only slightly smaller than the U.S. continuous 48 states), but its total population remained small, growing to around 15 million people in the 20th century. This population was settling in the coastal regions, while the continent's vast arid interior remained practically empty. Australia's Western population, it seems, remained too small to contribute significantly to Coal Age technology, but it pioneered certain aspects of modern agriculture.

British New Zealand: Guns and Sweet Potatoes

Compared to Australia, the European conquest of New Zealand was a totally different affair. When Abel Tasman in 1642 anchored off the coast of New Zealand, a well-organized crowd of natives refused the Dutch explorers access to their land and killed four of them when attacking a Dutch cockboat with canoes. New Zealand was actually one of the world's last fertile areas to be occupied by humans. The native population, the Maori, probably arrived only few centuries before the Europeans did. They were Austronesian farmers who presumably left the Polynesian islands of Tonga and Hawaii for New Zealand from around 950 C.E. The Maori introduced but two animal species in New Zealand, the small Polynesian rat and the dog, both of which served as a food source. They also introduced taro, yam, gourd, and perhaps the sweet potato, which eventually became the staple diet of the Maori living on New Zealand's warmer North Island.

The Polynesian immigrants generally had the problem that their tropical domesticates were poorly suited for New Zealand's temperate climate. Especially those tribes that migrated to New Zealand's colder south increasingly had to rely on gathering-hunting. New Zealand did not offer any large terrestrial mammals, but many flightless bird species, some of which were amazingly big. The largest one, the Giant Moa, stood up to four meters (12 feet) tall and weighed 200 kg. These birds had never encountered humans before and were hence fearless: it did not take long before the Maori had hunted them into extinction. Fur seals suffered the same fate, though they managed to escape total extinction on the South Island. Fish was initially a less important food source than would be expected for people arriving from Polynesia, but it gained importance over time, and eventually the Maori even took up whaling. Whale, dog, and human bones were all used for tools, as the Maori did not know any metals. Otherwise the Maori societies (of the North Island) did exhibit many features typical for Eurasian agricultural communities. They centrally organized the storage of the sweet potato harvest, for instance, and they constructed fortified villages as a response to frequent skirmishes between neighboring communities. Prisoners of war were made slaves or low-status wives. Besides, ritual cannibalism, practiced on the bodies of slain enemies, was deeply rooted in Maori culture.

As Maori men were robust and tall (averaging 175 cm [5'9"] in height, compared to a European average of 160 cm [5'3"] at the time), equipped with spears and clubs, and trained in all aspects of warfare, it was not easy for Europeans to annex New Zealand. James Cook was the first European to set foot on New Zealand in 1769. He claimed the land for Britain even though his party was immediately ousted. British sealing expeditions targeted southern New Zealand from the 1790s, but the first European sealers and whalers to settle permanently in New Zealand did not arrive before the 1830s. These immigrants had problems defending themselves against Maori attacks, but their diseases weakened the natives. What is more, the Maoris began killing one another on a large scale. They had been acquiring guns from European traders even before the first permanent European settlers arrived. This had a devastating effect as the balance of destructive-energy-command was entirely pushed out of equilibrium. The first Maori tribe adopted fire arms around 1818, and during the next 15 years or so the natives fought the Musket Wars, which continued until all tribes either had adopted muskets or had been subjugated or killed off. (What is more, the Maori had guns when they made their infamous 1835 trip to the Chatham Islands, where they slaughtered the natives and reportedly ate some of them.)

The Maori population decreased from around 85,000 in 1769 to 60,000 in the 1850s, to 30,000 around 1900. Meanwhile the European population of New Zealand grew from about 1,000 in the 1830s to 500,000 in the early

1880s. In 1840 the Maori chieftains signed the Treaty of Waitangi, ceding sovereignty of their country to Britain in exchange for protection and guaranteed ownership of their land. But the British immediately began the organized colonization of New Zealand. The Maori were persuaded to cede vast tracts of land for mere token payments, and they initially tolerated the European settlers as they provided metal goods and new domesticates. However, Maori resentment grew and resulted in open warfare between 1860 and 1872. The Maori repeatedly attacked British settlements, but the fighting eventually died down, and the Europeans claimed victory.

Like southern Australia, New Zealand's North Island displays a mild, temperate climate, allowing European settlers to thrive on Fertile Crescent and New World domesticates. (Also like Australia, New Zealand's natural pastures were extremely well suited for sheep-farming.) Besides, the settlers had full access to coal technology from Europe. New Zealand achieved a degree of self-government from 1853, and in 1893 became the first country in the world to introduce entirely equal voting rights to women, including Maori women. In 1907 New Zealand achieved full self-government, following the path of other temperate-climate regions that had been released by Britain into self-governance during the Coal Age: Newfoundland (1855), Canada (1867), Australia (1901), New Zealand (1907), and the newly created Union of South Africa (1910). Britain accepted the loss of some control over these areas of preliminary European settlement to make sure they would not leave the British Empire altogether to follow the path of the United States of America towards full independence.

THE U.S. EXPANSION

The United States of America is a nation that was born at the very beginning of the Coal Age. It emerged out of the Thirteen Colonies of British America during the American Revolutionary War (1775–81), in which it was supported by England's enemies France (from 1778), Spain (from 1779), and the Netherlands (from 1780). A German officer, Baron Friedrich Wilhelm von Steuben, trained the American troops (Continental Army) and was instrumental in George Washington's military success.

The young nation initially stretched along the Atlantic coast, but it almost immediately began to expand quite radically (in terms of both territory and population).[78] The first step was the Louisiana Purchase of 1803, which more than doubled U.S. territory. This opportunity came up because Napoleon needed money to invade the German states, the Austrian monarchy, Spain, and the Russian Empire. The French had lost Canada as a result of the Very First World War, but they had maintained the truly huge Louisiana territory in central North America, stretching from the Gulf of Mexico to

Rupert's Land and from the Mississippi River to the Rocky Mountains. (The territory covered the current states of Louisiana, Arkansas, Missouri, Iowa, Minnesota, North Dakota, South Dakota, Nebraska, New Mexico, Oklahoma, nearly all of Kansas; the parts of Montana, Wyoming and Colorado east of the Rocky Mountains; and the parts of southern Manitoba, southern Saskatchewan, and southern Alberta that drain into the Missouri River.) Most critically, France still owned New Orleans, located at the mouth of the Mississippi. This city controlled the north-south traffic along the river that was vital for shipments to and from U.S. regions located west of the Appalachian Mountains. Hence the United States attempted to buy the city of New Orleans for $2 million, but was suddenly offered the entire French Louisiana colony for $15 million dollars. Even this bargain price was beyond the immediate purchasing power of the American government, but two London-based banks, one of which had actually fled the Netherlands during the Napoleonic invasion, were willing and able to advance the money to Napoleon's war chest and facilitate the deal.[79]

The Burning of Washington   Long before oil-based superpowerdom. In 1812 the United States was still relatively small, and had yet to enter the Coal Age. After the United States had declared war on them, the British in 1814 took Washington and burned down the White House, the Capitol, and other public buildings. (This drawing was published in 1876 under the title "Capture and Burning of Washington by the British, in 1814." Library of Congress image LC-USZ62-117176, edited.)

During the Napoleonic Wars, in 1812, the United States declared war on Britain and invaded Canada. The United States started this war out of frustration over oppressive British maritime practices (designed to enforce a blockade of French ports). American troops soon met with great losses. The British took Washington, burned down the White House (in 1814), and ultimately maintained the borders of their Canadian colony. (The year before, in 1813, U.S. troops had invaded York, now called Toronto, and burned down the Parliament Buildings of Upper Canada.) However, U.S. territorial expansion gained momentum again in 1819, when Florida was purchased for $5 million from the Spanish, who knew that their power in the area was dwindling. Indeed it was just two years later, in 1821, that Mexico finally achieved its independence from Spain. The young nation of Mexico included what are now the U.S. states of New Mexico, Arizona, California and Texas, and parts of Colorado, Utah, and Nevada. This area alone was larger than the original 13 states, and the United States was going to take it by war.

It all began in Texas, a part of Mexico that was ideally suited for cotton cultivation. The United States urged the Mexican government to allow Americans to settle in this sparsely populated region, and Mexico agreed under the condition that the settlers would be loyal to the Mexican government. In 1828 the United States even signed a border treaty that acknowledged Mexico's sovereignty over Texas. But this paper soon turned out to be worthless.

The American settlers used African slaves to build up a Texan cotton economy and were angered when Mexico in 1831 abolished slavery. They refused to accept the new situation, repealed the prohibition on slavery, and declared Texas independent. The U.S. government immediately sent a military force onto Texan soil under the plea of protecting American-Texan settlements from raids by natives (Indians). After a bit of hesitation, the United States then broke the 1828 treaty with Mexico and annexed Texas in 1844. The United States in turn attempted to purchase California and New Mexico from Mexico. When these efforts failed, U.S. troops in 1846 began invading Mexico several days before Congress officially declared war. In this war, which is called the U.S. Invasion in Mexico and the Mexican War in the United States, several U.S.-produced coal-fired (side-wheeled) steamships were in operation to get troops, weapons, and supplies to strategically important sites. Mexico City was taken in 1847, and the year after Mexico had no choice but to negotiate a settlement that was entirely favorable to the United States. Mexico had to abandon its claim to Texas and to cede an additional 525,000 square miles of land to the United States, which was half of Mexico's territory. (The border between the two nations was fixed at the Rio Grande.) In return for this vast territory, the United States gave $15 million and assumed responsibility for paying $3 million in claims of

American citizens against the Mexican government. Mexico was thrown into despair and instability. Pressed for money it sold another 45,535 square miles of land (situated in what is now southwestern New Mexico and southern Arizona) to the United States in 1852. This so-called Gadsden Purchase, named for the politician whose engineers planned a railroad through this area, was a deal worth $10 million and was viewed as conscience money by many Americans who felt that the United States had treated the Republic of Mexico badly.

In the northeast, both Britain and the United States claimed the Oregon Territory, which included the modern states of Oregon, Washington, Idaho, and portions of Montana, Wyoming, and British Columbia. Following a period of joint occupation of the region, the British position began to weaken in the early 1840s due to the large number of American settlers pouring in via the Oregon Trail. In 1846 the British agreed for the border between Canada and the United States to be set at the 49th parallel, which officially added an area of 286,541 square miles to the United States. Alaska, a Russian colony from 1744, was purchased by the United States in 1867 for $7.2 million.

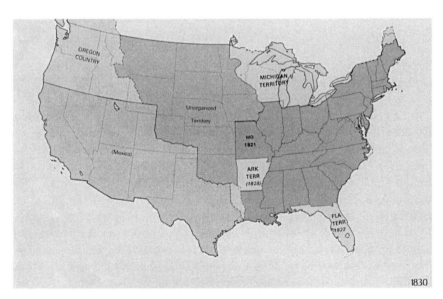

The United States in 1830   The U.S. Expansion was achieved by warfare and purchases. The map shows "Unorganized Territory," which was part of the Louisiana Purchase (1803), while the southwestern regions from Texas to California were conquered in a war against Mexico (1846). Alaska was eventually purchased from Russia (1867). To be sure, all these areas really belonged to Native American populations that were removed by people of European descent. (Reproduced from *The National Atlas of the United States of America*, U.S. Department of the Interior, Geological Survey, 1970. Courtesy of the University of Texas Libraries, The University of Texas at Austin.)

(Russia had been interested in selling Alaska for a number of years as it had problems providing food for its Alaskan settlements.) At a size of 591,000 square miles, twice that of Texas, the purchase of this land increased the now already huge U.S. territory by about 20 percent. But despite the bargain price of less than $1 per acre, the deal was considered a folly by Americans for decades. Only the discovery of gold in Alaska in the 1890s changed the public attitude. (Later on, during the Oil Age, Alaska turned out a highly valuable asset due to its rich oil fields.)

Eradication and Expulsion of Native Americans

No matter how much North American land the US acquired by warfare or purchases from France, Spain, Mexico, Britain and Russia, this was really the land of Native Americans.[80] However, the aboriginal North American population was killed or pushed towards increasingly marginal regions as it stood in the way of the expansion of European Americans. In addition, many succumbed to such Eurasian diseases as small pox, influenza, and cholera. In principle, the U.S. westward expansion was just a continuation of the agricultural expansion that the Spanish had started when they first came to the Americas. People from Europe simply took their Fertile Crescent domesticates and used them in temperate overseas regions where native gatherer-hunters, or less efficient agrarians, sustained lower population densities and enjoyed lower technological standards.

U.S. Coal Age history is full of atrocities against the many different Native American tribes. Those who survived saw their culture being destroyed and found themselves restricted to resource-starved reservations. The Indian Removal Act, passed by Congress in 1830, forced eastern natives to relocate to an Indian Territory west of the Mississippi River. (The term Indian stuck with the natives even though it had long been known that Columbus had not landed in India.) Cherokees contested the act in court, and the U.S. Supreme Court decided in their favor in 1832. However, U.S. President Andrew Jackson ignored the decision. Between 1831 and 1839 the Cherokee, Chickasaw, Creek, Choctaw, and Seminole were forced to relocate. Thousands died during the long Cherokee Trail of Tears towards the west, but this resettlement was by no means the end of the story. North America was the principal outlet to ease population pressures within Europe during the Coal Age, and European-American settlers kept on pressing towards the west. In 1853 the northern portion of the Indian Territory was liquidated and the state of Kansas as well as the Nebraska Territory were established. Between 1853 and 1856 the United States acquired a total of 174 million acres of Indian lands through 52 treaties, all of which were subsequently broken. During the Civil War (1861 to 1865) most of the native tribes remained neutral, while the Five Civilized Tribes supported the southern

Confederacy in exchange for the promise to be allowed to return to their original tribal homelands. By the end of the war they were instead forced to hand the western half of the remaining Indian Territory to 20 tribes from Kansas and Nebraska.

In the aftermath of the Civil War, the full might of the U.S. Army was directed towards the natives. One tribe after another was hunted, cornered, and massacred. In 1869, California was linked by railroad with the eastern states, and in 1871, Congress passed a law that prohibited further negotiations of treaties with native tribes. The phrase "no Indian nation or tribe shall be acknowledged or recognized as an independent nation, tribe, or power with whom the United States may contract by treaty" made clear that the United States would never allow American natives to have their own sovereign state in North America. Congress gave the Army free rein to execute a genocide that would remove the last obstacles to bringing all land between the East coast and the West coast, and between Canada and Mexico, under full U.S. authority. In 1890 the last major battle between American natives and U.S. troops, the Battle of Wounded Knee, destroyed the final band of fighting Sioux. The number of Native Americans in the United States was now some 300,000, down from tens of millions in pre-Columbian times. (For comparison, over nine million Europeans arrived in America between 1880 and 1900.)

Energy Aspects of the Westward Expansion

The U.S. westward expansion from the Atlantic towards the Pacific coast was initially slow and mainly fueled by agricultural energy (muscle power and biomass). The lack of east-west stretching rivers delayed the westward transit, but from 1825 the Erie Canal facilitated the colonization of the mid-West by linking the East coast and the Hudson River to the Great Lakes. In the mid-1840s scores of white settlers used horse-drawn wagons to travel from Missouri along a series of trails to the Columbia River and the Willamette Valley south of Portland. This so-called Oregon Trail was a northern track towards the Pacific coast, while the more southern regions, including California, were still part of Mexico at the time. Mormons (that is, members of a particular Christian sect) traveled for the first time in 1846 from Illinois via Nebraska into Mexican territory to found Salt Lake City in present-day Utah. In California gold was struck in 1848, right after the United States had annexed the region from Mexico. The find was made on the land of Swiss immigrant John Sutter, who had arrived in 1840, taking on Mexican citizenship to qualify for a land grant. At this stage no more than 5,000 Europeans lived in the region, next to some 30,000 natives. But the subsequent Gold Rush greatly accelerated the westward migration of settlers of European descent.

Buffalo and cattle dung were an essential fuel during the early continental crossings and the colonization of the Great Plains. Copying the customs of the native population, European-American travelers on the Oregon and Mormon Trails collected buffalo wood, stacking winter supplies in igloo shapes or against house walls. Also known as cow wood or Nebraska oak, this fuel burned evenly with little smoke or odor. In addition, millions of small windmills (with a large number of fairly narrow blades) were used on the windy Great Plains in the second half of the 19th century. Placed on top of lattice towers these mills pumped water for households, cattle, and steam locomotives. Coal energy gained importance even when the railroad network started to expand: early American steam locomotives were often fueled by wood, but when railroads drove west to the plains and the mountains, coal became more attractive, both because deposits were often found near the new railroad rights of way and because its higher energy content increased the range and load of steam trains. Demand for coal also rose because the railroads were laying thousands of miles of new track and the metals industry needed an economical source of coke to make iron and steel for the rails and spikes.[81] U.S. coal consumption thus slowly caught up with fuel wood consumption to finally provide the basis for the full-fledged U.S. industrialization during the 1880s and 1890s.

U.S. Overseas Coal Age Imperialism

In concert with its increasing level of industrialization the United States also followed the European Coal Powers in terms of overseas imperialistic ambitions. The United States expanded its sphere of influence far beyond the North American mainland and even acquired real overseas colonies. To be sure, the United States was not initially dependent on captive overseas markets. Due to its radical territorial and population expansion the United States enjoyed a huge domestic market on which some 90 percent of all American products were sold still in the 1890s. However, as the United States emerged as the most populous Western Coal Age Power, even the remaining 10 percent represented a major production volume and export value. Besides, American overseas investments between 1875 and 1895 amounted to around one billion dollars. The U.S. government and military assumed the task of protecting these investments. For one reason or another, the United States conducted 103 interventions on foreign territory between 1798 and 1898, which accounts to more than one per year.[82]

The story of U.S. overseas imperialism began long before the mainland expansion towards the Pacific coast had been completed. The Napoleonic Wars, which had opened the opportunity for the United States to purchase the huge Louisiana territory, also created the circumstances under which the United States could replace Spain as the leading foreign power in Latin

America. With the notable exception of Portuguese Brazil, nearly all of
South America was a Spanish colony at the beginning of the Coal Age. Then,
in 1808, French troops invaded Spain, and Napoleon instituted his brother
as the new king in Madrid. This absence of a truly Spanish government in
the mother-country gave the colonies a chance to set themselves free. Spanish South America disintegrated into eight sovereign republics: Paraguay
achieved independence in 1811, Argentina in 1816, Chile in 1818, Colombia
in 1819, Ecuador in 1822, Peru in 1824, Bolivia in 1825, and Venezuela in
1830. Brazil gained independence from Portugal in 1822, and in a disastrous
war with Argentina in 1828 lost the region that emerged as the independent
republic of Uruguay.

Amidst the Latin American independence movement, the United States
of America issued the Monroe Doctrine of 1823. It stated that any further
European colonial ambitions in Latin America (the Western hemisphere)
would be regarded a threat to U.S. peace and security. This was quite a con-
fident statement, given that the British had burned the White House to the
ground less than 10 years earlier. And the fact is that the United States
had by no means the military strength to take on Spain and Portugal at
this time. Much rather the impetus for, and the power behind, the doctrine
came from the British, who shared the (political and commercial) interest
to release Latin America from the Spanish-Portuguese clutch. (Britain did
indeed become the dominant trading partner of Latin American countries.)
In addition Britain had reasons to support the United States: slave-produced
American cotton fiber was critically important for the booming British textile industry.

For the United States, the international acceptance of the Monroe Doctrine set the precedent to operate in the new South American republics
without European interference. The first true victim of these conditions
was Mexico. The United States annexed Texas in 1844 and invaded all of
Mexico in 1846–48. The U.S. government now had state-of-the-art military
equipment (including cannon steamers) as well as trained and experienced
soldiers at its disposal. And before long, the might of the U.S. forces was
directed towards other foreign targets. U.S. marines were landed and maintained in Argentina in 1852–53, for instance, to protect American interests
during a revolution. In 1853 the United States intervened in Nicaragua to
protect American interests during political disturbances. In 1854, Greytown
was destroyed to avenge an insult to the American Minister to Nicaragua,
and in 1855 U.S. (and European) naval forces landed in Uruguay to protect
American interests during an attempted revolution.[83]

But U.S. interventions in foreign countries were not restricted to Latin
America alone: The United States also had plans to control the northern
Pacific. The American aggression against Japan in 1853 enforced access to
Japanese coal, which was crucial for operations of American cannon steamers

in the region. While waiting for a reply to his ultimatum to Japan in 1853–54, Commodore Perry demonstrated U.S. power by landing marines twice on Ryukyu and Bonin Islands, securing a coaling concession from the ruler of Naha on Okinawa (Ryukyu Islands). And in 1859, during the Second Opium War, the United States intervened in Shanghai for the protection of American interests. During the following years the Civil War (1861–1865) was the main stage for U.S. militarism. By the end of it, the U.S. government had even more armed and experienced forces at its disposal. It directed the might of the army toward Native Americans and the navy towards the North Pacific. The USS *Monocacy*, for instance, a 1,370 ton gunboat equipped with steam-powered side wheels, was commissioned in early 1866 as one of nearly four dozen double-enders the Navy built during the Civil War era. (Double-ender refers to a ship with bow and stern of similar shape. It can easily sail back and forth, which is especially useful on narrow rivers.) The USS *Monocacy* served for nearly four decades, mainly operating in Japanese and Chinese waters. The ship was also part of the U.S. expedition into isolationist Korea, where U.S. troops attacked several fortresses and killed a few hundred Koreans. (In diplomatic terms this campaign was a failure: Korea still refused contacts with the Western Coal Powers.)

In the sovereign Polynesian kingdom of Hawaii the United States in 1893 helped to overthrow the indigenous rulers. The new government proclaimed Hawaii a republic and ceded it to the United States within five years. (A century later, in 1993, U.S. President Clinton signed the Apology Resolution, which noted these facts and apologized to Native Hawaiians on behalf of the people of the United States for the overthrow of the Kingdom of Hawaii and the deprivation of the rights of Native Hawaiians to self-determination.)

Cuba, Puerto Rico, and the Philippines

The Spanish-American War of 1898 opened a yet more radical era of U.S. imperialism. After the U.S. warship *Maine* had been destroyed in Havana harbor under circumstances that still remain unclear, the United States declared war on Spain and invaded Cuba and Puerto Rico, the last New World remnants of Spain's once vast overseas empire. At the same time six U.S. warships were ordered to leave Hong Kong for the Philippines, which had also remained a Spanish colony. Both Spanish fleets, the one that anchored in Manila Bay and the one in Cuba, were destroyed by U.S. cannon steamers. Subsequently Spain was forced to cede the Philippines (and Guam) to the United States, and Cuba and Puerto Rico became de facto American colonies. On the island of Puerto Rico an American-appointed governor was to be the executive officer, and the Jones Act of 1917 made Puerto Rico U.S. territory. Cuba officially achieved independence, but the United States installed a military base (at Guantánamo Bay); asserted a right to intervene in internal affairs; and kept Cuba as an economic protectorate (until 1933).

With the Philippines and Guam (east of the Philippines), the United States acquired two real overseas colonies with a substantial native population. The Filipinos, who counted some nine million, attempted to seize the moment and declared their independence in 1898. But the United States would not allow it. Some 16,000 Filipino (guerrilla) fighters put up a three-year fight for their freedom, but in the end they were defeated by U.S. troops that burned and destroyed entire villages and concentrated civilians into so-called protected zones. Estimates of total civilian casualties are in the hundreds of thousands. Ironically, part of President McKinley's justification to make the Philippines a U.S. colony was to Christianize the natives, overlooking the fact that the overwhelming majority of the islands' population was Catholic anyway. However, English was declared the official language, thousands of American teachers were shipped to the Philippines, and the Catholic Church was constrained. (The Philippines achieved internal self-government in 1935, but remained an American colony until 1946.)

Roosevelt Corollary and Panama Canal

Emerging from the war against Spain (and the rapid industrialization period of the 1890s) as an even mightier coal-fueled militaristic power, the United States stepped up its involvement in Latin America significantly during the presidency of Theodore Roosevelt (1901 to 1909). The United States fended off European activities in Venezuela (1902) and the Dominican Republic (1903), for instance. Then, in December 1904, the president issued the Roosevelt Corollary, an extension to the Monroe Doctrine proclaiming a U.S. right to intervene in the internal affairs of Latin American states. And indeed, the United States intervened in as many as six Latin nations between 1900 and 1920. It established protectorates in Haiti and the Dominican Republic, and periodically stationed marines in Nicaragua.

One critical strategic move was for the United States to build a canal at the Isthmus of Panama to link the Pacific and Atlantic Oceans across Central America. The decision to undertake this major construction project was directly related to U.S. imperialistic activities that required cannon steamers to operate in the Atlantic as well as in the Pacific. It took over two months to get from New York to California around Cape Horn, and the canal would reduce that voyage by 8,000 miles. The commercial advantage was obvious as well: Despite the benefits of the railroad, water-borne bulk transport remained a lot cheaper than transport over land.

The Isthmus of Panama was part of Colombia in the 19th century. The United States thus addressed the Colombian government to secure rights to build and operate a canal. When these efforts failed, the United States turned to rebels of the Panama province, promising them assistance if they would proclaim an independent state of Panama, which they did on November 3, 1903. The USS *Nashville* appeared immediately in Panamanian waters impeding any interference from Colombia. Additional U.S. battleships were stationed off the coast of Panama for so-called training exercises. In exchange for the favor, the now independent Panamanians allowed the United States to acquire the Panama Canal Zone in perpetuity for a payment of $10 million, plus $250,000 annually. The canal was then built with the help of over 100 coal-fueled steam shovels between 1904 and 1914.

THE RUSSIAN EXPANSION

Like the United States, Russia kept on widening its borders during the Coal Age through a series of wars. In the 16th century Russia had brought the huge area of the middle and lower Volga as well as Siberia under its control, and in the 17th century the eastern Ukraine was gained from Poland. Yet at the beginning of the rule of Peter I (the Great), in 1689, Russia was essentially still a landlocked country, with no access to either the Black Sea (controlled by the Ottoman Empire), or the Baltic Sea (which was surrounded by land under the control of Sweden and Poland). Archangel, the northern Russian sea port at the mouth of the Northern Dvina (east of present-day Finland), was blocked by ice for half the year and hence of limited use. This was a severe handicap for Russia's development and trade, and Peter made it his main objective to get access to a seaport for trade with western Europe. He created a modern (conscript) army and fought the Northern War (1700–1721) to expand Russia's territory to the Baltic Sea. After acquiring Estonia as well as parts of Latvia and Finland at the expense of Sweden, Peter founded his new capital, Saint Petersburg, at the head of the Gulf of Finland, symbolizing Russia's ambition to become a leading maritime power.

In the south, Russia was fighting the Ottoman Empire to gain access to the Black Sea. Russia managed to annex the southern Ukraine and the Crimea peninsula with the major city of Sevastopol (1783). However, the Ottomans kept controlling the Bosporus Strait, and hence the trade between the Black Sea and the Mediterranean. In the west, Russia expanded at the expense of Poland, which was partitioned between Austria, Prussia, and Russia towards the end of the 18th century. (In 1771 the rural kingdom of Poland, or rather the Polish-Lithuanian Commonwealth, was actually larger than any other European country except Russia and had a bigger population than any other European country except France.)

For Russia, like the United States, the period of the Napoleonic Wars was a time of extraordinary land gains. Russia was warring the Ottoman Empire and annexed Moldova as well as the Caucasian territories of Georgia (1801) and Armenia. A war against the Persians in 1806 handed the Russians Azerbaijan, which eventually was going to prove highly valuable for its rich oil deposits. And in 1809 Russia took Finland from Sweden. To be sure, Napoleon's troops attacked Russia and took Moscow in 1812, but the French armies were nearly annihilated in the winter of that year. And after Napoleon's final defeat and the peace settlement at the Congress of Vienna, Russia and the Austrian monarchy emerged as the two largest and most powerful empires of continental Europe.

Russia's expansion towards the southeast continued from the mid-19th century. Between 1856 and 1860 Russia took advantage of the weakness of the Chinese Empire and occupied the southern section of the Far Eastern Territory (the Russian Far East), the province that includes Asia's entire Pacific northeast coast. Between 1860 and 1884 Russia also annexed the area of Turkestan (which is now divided between the Central Asian Republics Kazakhstan, Kyrgyzstan, Tajikistan, Turkmenistan, and Uzbekistan). The decision to expand into this region was partly influenced by the U.S. Civil War, as the southern United States were Russia's primary cotton supplier. Russia thus in the 1860s took control of Uzbekistan, as the region held a high potential for cotton cultivation.[84] And in 1875 the Russians also took the large northwest Pacific island of Sakhalin, which had formerly been under joint Russo-Japanese control.

Alaska and California

To be sure, the Russian expansion towards the east did not stop once the Pacific coast was reached. Rather, Russians crossed the Bering Strait to settle on the American continent as well. Russian fur traders had hunted in Alaska from the 1740s, and in 1784 founded the first permanent settlement. The Russian-American Company, chartered in 1799, was granted a monopoly over all Russian enterprises in North America and organized the lucrative trade in furs (destined mainly for China). In their efforts to hunt sea otters Russians sailed south along the American West Coast to reach as far as Alta and Baja California. However, the Alaskan colony itself had problems finding or growing enough food to sustain itself. Hence a Russian delegation in 1806 sailed into the San Francisco Bay, then part of Spanish California, to ask for food supplies. The Spanish knew their position in California was weak and feared establishment of the Russians in North America. But following a six-week delay they loaded the Russian ship with grain after the head of the Russian delegation, Count Rezanov, had proposed to marry the teenage daughter of the Spanish commander at San Francisco.

Upon his return to Alaska Rezanov suggested founding a Russian trading post in California, which was established in 1812 under the name Fort Ross.

It was located about 80 miles north of San Francisco, some 13 miles north-west of the mouth of Russian River. Its purpose was to grow grain and raise animals to supply Russian Alaska; to hunt sea otters; and to establish trade relations with Spain. By 1821 the Czar felt that Russia's position in America had grown strong enough as to issue an *ukase* that closed the Pacific Coast north of San Francisco to all but Russian ships. This Russian attempt to control this region was actually part of the reason why the United States issued the Monroe Doctrine in 1823. However, the aggressive hunting out of Fort Ross had already depleted the sea otter population to such degree that agriculture and livestock raising became the main occupation of the colony. But agricultural production proved disappointing as well, and two decades later, in 1841, the Russians left California after selling Fort Ross to one Swiss Mexican John Sutter, who had himself recently arrived in San Francisco on a ship coming from Russian Alaska. (Yes, this was the same Sutter on whose land at the American River the Californian Gold Rush of 1848 was going to start right after the United States had annexed the region.)

In the following years Russian Alaska relied on provisions supplied from settlements in present-day Washington and Oregon by the British Hudson Bay Company. However, in 1867 the Russians decided to sell Alaska to the United States for $7.2 million. This ended the era of Russian colonies in America. Fort Ross had been the furthest outreach of the Russian eastward expansion, and by the time the Russians met the Spaniards in California, the European agrarian expansion on a West-East axis around the world (based on Fertile Crescent domesticates) had been completed.

Energy and People Mix in the Enlarged Russia

Like the U.S. expansion in North America, most of the Russian territorial expansion during the Coal Age was actually fueled by agricultural energy (muscle power and biomass). Russia had wide steppes to source and fuel horses (just as the United States had open prairies), and much of its military strength derived from an efficient cavalry. Also like the United States, the growing Russian railroad system often relied on timber rather than coal for fuel. In 1880 Russia still produced merely three million tons of coal per year, but in 1905, when the Trans-Siberian Railroad opened at full length, Russia had the world's largest railroad network except for the United States, and mined 19 million tons of coal per year.

Both the United States and Russia used the railroad to colonize the enormous, typically sparsely settled, regions they acquired during the Coal Age. While some of the resources that came with these areas were not immediately visible (think: Louisianan, Texan, Californian, Alaskan, and Azerbaijani oil), it was obvious that much of the land had to be converted into farmland if it was to yield agricultural energy. Worldwide, some 440 million hectares of land, an area larger than modern India, were brought under cultivation between 1860 to 1920, and more than half of it was located in the temperate regions of the Russian Empire and North America.

To achieve this enormous conversion of land the United States and Russia competed for foreign immigrants with agricultural skills. In the end of the 18th century Russia invited anyone, with the exception of Jews, to settle in South Russia (New Russia). The call was answered, for instance, by Hutterites, German-speaking Austrians from Moravia (present-day Czech Republic), and by Mennonites from Prussia. (Both these Christian groups were pacifistic Anabaptists, who believed that individuals should have a free and conscious choice to become Christians and should therefore be baptized as adults, not as children. They were at times persecuted by both Protestant and Catholic Christians.) Many Germans then left for Russia (or America) to escape the dire economic situation prevalent in their home countries following the Napoleonic Wars. Swabians of southern Germany, for instance, left for a valley region in Azerbaijan in 1817 following an invitation by Czar Alexander I.

Despite certain similarities in the immigration policy, the Russian Empire ended up being a lot more ethnically, religiously and linguistically divided than the United States. Russia did not simply expand into sparsely settled or less developed regions. Rather, it annexed Protestant regions from Sweden, Catholic regions from Poland, and Islamic regions from the Ottoman Empire and Persia, while keeping Greek-Orthodox Christianity as the state religion. Hence, the Russian Empire never enjoyed the voluntary, internal cohesion that was experienced in the United States. Following a policy of Russification and militarization in the 1870s, many Anabaptist German speakers left Russia for the United States and Canada, for instance, where they felt they had more cultural and religious freedom.

SUBDUING ASIA

While the Russian Empire expanded into the open regions of northern Asia, the rich and densely populated areas of Asia's southern half came under western European control. Ever since the Portuguese had entered the Indian Ocean in the beginning of the 16th century, it was clear that Europeans would be unable to settle and proliferate in this area in the way they did in the Americas. China and the Indian subcontinent, especially, were too populous and technologically advanced even to be militarily or politically controlled. Hence the Europeans soon focused on establishing colonies in less developed (island) regions of Asia (such as the Philippines), and founded coastal trading posts on mainland Asia with the consensus of regional powers. Such "consensus" was often achieved by Europeans threatening piracy or terrorizing coastal regions, which was possible because they had superior oceangoing ships and naval cannons.

During the Coal Age the balance of power, in parallel with the balance of energy command, tipped further to the advantage of the Europeans. The utilization of coal provided Europeans with much more energy than Asians, manifested in a higher productivity, mobility, and fire power. Most critical were cannon-equipped steam boats capable of penetrating rivers, the lifelines of Asia's agrarian societies. Europeans and Americans could now reach inland and terrorize much more than the coastlines. Hence coal energy enabled them to finally complete the subjugation of India and to make a break into China and Japan.

British India

Prior to the Coal Age India was not a colony. The era of direct British rule started regionally with the East India Company taking command over Bengal in 1765. (This was the same year James Watt was commissioned to repair a steam engine of the Newcomen type.) From this foothold, the British East India Company expanded its influence by leveraging its privileged economic position and superior European weapons technology. However, for the next half century the native Marathas, a people devoted to Hinduism, remained the strongest power in India and maintained the largest territory. The British fought three wars against the Maratha confederacy. They lost the first one (1775 to 1782), but won the second one (1803 to 1805): both were fought over Marathi resistance to install a candidate of British liking to the office of peshwa (chief minister). The third and decisive war between the two powers was waged in 1817–18 after the British had invaded Marathi territory in pursuit of robber bands. Native forces reacted by rising against the Europeans, but they were defeated, and the British annexed the territory of the Maratha confederacy. In 1849, when coal-fueled water mobility had matured, the British also conquered Punjab, the five-river region now split between India and Pakistan, and henceforth controlled most of the Indian subcontinent, including present-day Pakistan and Bangladesh.

The British authorities improved Indian infrastructure, built schools to increase general education levels, and organized the construction of the first railway line as of 1853. However, they also adopted English as the official language of Indian law courts (1835), and promoted missionary activity to spread Christianity. Hence, many Indians viewed the foreign rule as nothing but suppressive.

British rule over India became total following the Sepoy Rebellion (Indian Mutiny) of 1857. Sepoys were native Indian soldiers, drawn mostly from

Muslim units from Bengal, who served under British command in great numbers. As the story goes, problems arose with the rifle cartridges, whose base had to be bitten off before use. Rumors circulated among Sepoys that the cartridges had been greased with beef and pork fat, which angered Muslim Sepoys who were not supposed to consume pork, and Hindu Sepoys who were not supposed to eat beef. In turn the Sepoys revolted against the British army and marched to Delhi to offer their services to the (largely powerless) Mughal emperor. In reality, this uprising was the culmination of mounting Indian resentment toward British rule and economic and social policies. It therefore did not take long before much of northern and central India joined in, and the whole region was plunged into a year-long insurrection against the British.

In 1858, after the Sepoy Rebellion had been put down by force (and lots of coal energy), the British government at last made India a colony. The Mughal Empire was liquidated, the Mughal himself exiled to Burma, and the rule over India transferred from the British East India Company to the Crown. The British modernized India in many respects (railroad mileage increased to nearly 15,000 km by 1880 and nearly 70,000 km by 1935), but British India did not industrialize much under foreign rule. The British promoted the development of cotton, tea, and indigo plantations while selling English manufactures to the subcontinent. India had been the world's prime producer of quality cotton textiles at the beginning of the 19th century, but was flooded with English machine-produced cotton products and imported three quarters of its textiles from Britain by the end of the century.

Some technology transfer did occur, though. India had its own textile mills by 1853, and Jamshetji Nusserwanji Tata, known as the father of Indian industry, established several textile factories from the 1870s. In 1903, one year before his death, Tata set up a blast furnace fed by Bihari coke, and his eldest son Dorabji Tata (who was educated in England) supervised the construction of a iron and steel works by Pittsburgh engineers near Calcutta. These installations would become the core of India's largest private industrial group.

For a while Britain actually allowed India to develop industries in the absence of foreign competition, but the import duties were removed in 1883. This exposed India's emerging industries to unfettered British competition and caused much anger among the natives. In combination with the denial of equal status to Indians, this policy prompted the formation of the Indian National Congress, which was initially loyal to the British Empire but later sought Indian self-governance and outright independence. From 1906 the British supported a distinct Muslim political organization in India, and from the 1920s insisted on separate electorates for religious minorities. This policy contributed to Hindu-Muslim discord and the country's eventual partition.

(When British India achieved independence in 1947 it was split into India and the Muslim nation of West and East Pakistan. In 1971, East Pakistan declared itself independent and was renamed Bangladesh.)

Western-Controlled China

During the Coal Age the Europeans also dared for the first time to openly declare war on China. It was especially bizarre that coal energy command should be behind China's defeat, because the Chinese were using coal in iron production much earlier than Europeans did. Already in Han times (before 220 C.E.) the Chinese mass-produced cast iron in four-meter blast furnaces and began to use coal in addition to charcoal in iron smelting. However, this substitution of coal for charcoal was never as overwhelmingly important in China as it was in Europe, and charcoal remained in use in parallel to coal in Chinese iron smelting. Perhaps the landlocked coal provinces of northwestern China (Hebei and Henan) were simply too far away from the coastal centers of population and production. The coal districts were eventually connected to the canal system, but they were still part of the somewhat distant northwestern region that was frequently threatened by nomad invasions. Besides, China never experienced the kind of timber shortage that pushed England into the Coal Age. China had access to plenty of wood supplies that could be transported from inland regions on the rivers and canals. Hence, the Chinese were never forced to dig deep mines, which might have prompted them, too, to develop a coal-fueled prime mover to remove flood-water from the shafts.[85]

But even in the absence of (much) coal energy, China did extremely well. According to the traditional view, western Europe passed China in terms of technological standards from around 1500 C.E., perhaps even in the centuries before. This is now widely disputed.[86] More accurately, China held, or shared with western Europe, technological leadership in many respects right up until the beginning of the Coal Age, when Europe quite suddenly commanded a lot more energy than China. England and the Yangtze (Chang) delta (that is, the leading economic regions of Europe and China) had similar economic performances in the period 1500 to 1800.[87] China as a whole did well despite the enormous population growth from 100 million in 1644 to 300 million in 1800. In 1793 the Chinese emperor could still afford to categorically dismiss a European request for additional trading privileges and a permanent British embassy to China without having to fear any consequences. But soon the tide was going to change, as the Europeans developed stationary steam engines into mobile steam engines that would propel cannon boats up the hitherto secure Asian rivers.

China also faced domestic problems as the rapid population growth at times began outstretching agricultural production in the beginning of the 19th century. This resulted in famines and civil unrest. In addition, the British pushed opium onto the Chinese market. Opium is a highly addictive drug extracted from unripe poppy seeds, which the British cultivated in India on a large scale. Opium was a legitimate trade product for medicinal purposes, but the Europeans had little interest in shipping large amounts of it to their home countries. Instead, they used it to balance their purchases in China. The Chinese had traditionally restricted overseas trade, but ever since the flow of New World precious metals to China had wound down, the Europeans could not even afford to pay for the now relatively small shipments of tea, porcelain, silk, and other oriental goods that were in high demand in Europe. Europe simply did not have anything to offer that the Chinese wanted, and using an addictive drug seemed to be a profitable way to create increasing demand for something the British could deliver from India.

The Opium Wars

When the Chinese authorities saw thousands of their people waste away in addiction, they attempted to enforce the existing prohibitions on the importation of opium. In 1839 they destroyed a large quantity of opium confiscated from British merchants at Guangzhou (Canton). In response Britain declared war on China and sent coal-fueled gunboats to attack several Chinese cities. China was defenseless against the British steam boats that not only attacked coastal cities but also cruised up rivers. Hence China had to give in, and the British coal technologists easily won this (first) Opium War, dictating their terms of trade to the world's most populous country in 1842. China was forced to cede Hong Kong Island, to open five further treaty ports, and to permit Christian missionary activity. Britain achieved unrestricted trade with opium and all other goods, plus allowance for British residency. Within a few years other European powers signed similar treaties with China, received commercial and residential privileges, and began to dominate China's treaty ports.

But the Chinese authorities had even more problems. In 1849–50 a famine triggered the Taiping Rebellion that took over a decade to suppress and cost more than 20 million lives. This popular uprising was directed against the Manchu Dynasty, in part because of disappointment that the Chinese authorities were unable to restrict the opium imports and get rid of the foreign influence. The weakened Chinese rulers then turned to the Western powers for help in putting down the rebellion. Thereafter the British shipped even more opium to China, and resistance grew again. And after the Chinese had conducted an allegedly illegal search of a British-registered ship, the Second Opium War (1856–60) was waged by Britain, this time in alliance with France.

The immediate cause for the French involvement was that the Chinese in 1856 executed a French Christian missionary in southeastern China. In reality, the French saw this war as an opportunity to strengthen their position in Indochina, the region of modern Cambodia, Laos, and Vietnam, into which the Chinese had expanded in the 13th century. The French had established religious (Catholic Christian) and commercial interests in Indochina in the 17th century, but did not have the resources to expand them after they had lost most their overseas empire to the British in the so-called Very First World War (Seven Years' War). Now the French attempted to find a southern trade route into China through Tonkin, the northern region of Vietnam.

With their superior energy command the Europeans easily won the Second Opium War and in 1860 forced the Chinese to sign a treaty to which Russia and the United States were also party. China had to agree to open eleven more ports, to permit foreign legations in Beijing, to sanction Christian missionary activities, and to legalize the import of opium. In addition, China had to cede the southern section of the Far Eastern Territory to Russia, and Indochina to France.

French Indochina and British Burma

More precisely, only southern Vietnam (Cochin China) became a French colony right away. The other regions of Indochina, including Annam (central Vietnam), Tonkin (northern Vietnam), Cambodia, and Laos became French protectorates of some sort. France attempted to unite all these areas into one large colony and seized Hanoi in northern Vietnam in 1882. This led to direct war with China, which the French won. United French Indochina as of 1914 had a population of over 18 million and was 50 percent larger than the French mother country. It served as a source for cotton, rice, tin, pepper, coal, and rubber.

Meanwhile the British gained Burma (now Myanmar). This area (which now borders India, Bangladesh, China, Thailand, and Laos) had also been annexed by the Chinese in the 13th century, but had since regained independence. The British conquered coastal regions of Burma in 1826, Lower Burma in 1852, and Upper Burma in 1886. All of Burma was then integrated into British India and was soon going to serve Britain as an early source of oil. The Kingdom of Siam (now called Thailand) took the role of a neutral buffer region between British Burma and French Indochina, but was under British influence. What is more, Britain from the 1820s progressively established sovereignty in Malaysia, south of Thailand, from where tin and rubber were exported. Singapore was leased by the British East India Company as a trading post from the sultan of Johore in 1819. This free port was of major importance for Britain to get access to the riches of the Indonesian archipelago, which was largely controlled by the Dutch. Singapore turned into Southeast Asia's leading trading center and served as a British military base.

Chinese Expansion into Central Asia (Tibet)

While China lost some of its southern and northern territories to the Coal Powers, the Chinese government realized its own imperialistic goals in such Central Asian areas as Xinjiang and Tibet. Xinjiang is the name the Chinese gave to Eastern Turkestan (or Chinese Turkestan), a large province of about 637,000 square miles (1,650,000 km^2) located between the Kunlun and Altai mountains. The Chinese occupied Eastern Turkestan from 1759, but parts of the region achieved independence in 1864, when China was weakened following the Second Opium War. By 1876 China had regained strength and sent troops that reoccupied the entire region. As of 1888 Eastern Turkestan was formally annexed to China and renamed Xinjiang.

Tibet was freed from Mongolian occupation by China in 1750. The Chinese then installed the 7th Dalai as the Tibetan ruler and declared Tibet their protectorate. Chinese forces intervened in 1788, for instance, when the Gurkha of Nepal invaded Tibet, but thereafter Chinese influence in Tibet began to recede. Things changed dramatically when Britain, in 1903, invaded Tibet out of fear that the Russians would further expand their influence in central Asia. (Tibet was located right at the interface of the British expansion from India and the Russian expansion into central Asia.) The Dalai Lama fled to Mongolia with a Russian adviser, while Britain went ahead to sign an accord with the Dalai-appointed interim ruler. This implied that Tibet was a sovereign power with the right to make treaties of its own. The Chinese objected and actually defeated British troops in this arena. In 1906 the British signed another accord that recognized China's suzerainty over Tibet, which preferred loose nominal Chinese rule over control by Britain or Russia. However, China in 1910 fully occupied Tibet, and the 13th Dalai fled to India under protection of the British.

New Armies and Boxer Rebellion

The reason why China could realize its imperialistic goals in Central Asia, and even defeat British troops in Tibet, was that it radically modernized its troops. The New Armies were trained and equipped according to Western standards. The first of them, founded in 1895, was equipped with German weapons to oppose the Japanese, who aggressively expanded to the Asian mainland. (Germany in 1897 established its Kiautschou colony in the Chinese province of Shandong.) Thereafter the might of the New Armies was directed towards Central Asia, and in 1910 these troops helped to suppress the Boxer Rebellion.

Boxers were religious antiforeign societies that attempted to rid China of Western influence. Several European nations had followed Britain in forcing China to sign unequal treaties, and eventually these powers established spheres of influence within China, which guaranteed specific trading privileges to each Coal Power within its respective sphere. When the United States appeared on the imperialistic scene, it demanded the same trading rights as the European powers, but rather than carving out its own sphere of influence in China, it announced an Open Door Policy in 1899, which stated that all nations should have equal trading rights in China regardless of the existing spheres of influence. This was viewed as yet another insult by Chinese nationalists. They murdered thousands of Chinese Christian converts and besieged the European and U.S. legations in Beijing. Initially, the Boxer Rebellion was directed against the Qing dynasty, but the movement later reconciled itself to the court, which issued edicts in defense of the insurgents. In fact, the rebels were joined by elements of the Imperial army.

The Western Coal Powers reacted by dispatching an international punitive force that eventually numbered 45,000 Austrian, British, French, German, Italian, Russian, American, Japanese, and anti-Boxer Chinese troops. After the uprising had been put down, China was forced to pay a huge indemnity and to grant additional concessions to foreign nations active in China. (By this stage western European energy consumption was at least four times the Chinese mean.)

End of Dynastic Rule

In 1911 the Qing dynasty was overthrown, and 5,000 years of dynastic rule in China came to an end. The revolt also spread to Tibet, where Chinese troops were ousted and the 13th Dalai was reinstalled. (Tibet enjoyed independence during the following three decades and was in turn reoccupied by China.) In China itself, a republic with presidential leaders was established, but the country did not come to a rest. Instability soon increased as local warlords sought regional control, while nationalist leaders such as Sun Yixian attempted to unify China. Following Sun's death in 1925 a Civil War broke out, as Sun's successor Jiang Jieshi was immediately battled by the forces of Mao Zedong, the leader of the Communist Party. (Mao established a communist government in mainland China in 1949, while Jiang Jieshi fled to Taiwan and established his government there.)

Japan's Coal-Fueled Empire

The Coal Powers also forced Japan to open its borders. The Japanese had been increasingly concerned following the Western success in establishing colonial enclaves in China after the first Opium War. This concern was well-founded. The Japanese were indeed defenseless when U.S. cannon steamers arrived in 1853 to force Japan out of seclusion. But after the Meiji restoration of 1867 Japan decided to learn as much as it could from the West, industrialized rapidly, and soon became itself an aggressive imperialistic Coal Power.

Japan had limited arable land to provide food for its growing population, and was only sparsely endowed with mineral resources. It had modest coal deposits, but iron ore was lacking. The increased demand for resources following the onset of industrialization, and the urge to sell manufactures on foreign markets, soon prompted Japan to turn towards the nearby mainland. The government modernized the army and founded a strong navy. And before long Japan built its own coal-fueled gunboats. Coal technology clearly gave Japan an edge over its neighbors, and the government viewed imperialism as a means to establish Japan as a political power equal to the Western nations. (To be sure, there was a lot of prestige in this. After the Meiji restoration nationalistic feelings ran high. The old myths of imperial and racial divinity were revived, and the sentiment of loyalty to the emperor was actively propagated by the new government.)

The first victim of Japan's aggressive expansionary policy was isolationist Korea. In the 1870s Japanese warships threatened the nearby Korean peninsula and struck at the port city of Pusan and at Kanghwado island. As Japan was demonstrating its military might, Korea in 1876 had no choice but to sign a treaty that granted the Japanese extra-territoriality in Korea, exempted them from tariffs, and recognized Japanese currency at Korean ports of trade. (At this stage Japanese merchants purchased agricultural products from the Koreans, while selling them European and American manufactures.)

The development in Korea offended Chinese interests and led to the Sino-Japanese War of 1894–95. Immensely large China had traditionally been the leading power in the region, but Japan enjoyed superior energy command and strong central governance. The Japanese conquered Dalian (Dairen), Shantung, and Seoul, and henceforth officially safeguarded Korean independence. China had to give up its claims to Korea and had to cede Taiwan (Formosa) as well as the Pescadores Islands to Japan. Japan also received the entire Liaodong peninsula with the important naval base Port Arthur (Lüshunkou), and thus was able control the Yellow Sea. However, France, Germany, and Russia successfully pressured Japan into returning the Liaodong peninsula, which Russia in turn leased from China to build (from 1898) a naval base at the ice-free harbor at Port Arthur and a commercial port at Dalian.

But Japan was still growing stronger. In 1900 the Japanese helped putting down China's Boxer Rebellion alongside Western troops, and in 1902 Japan even entered into an alliance with Britain, which de facto granted Japan recognition as an international power. Japan now felt strong enough to object when Russia expanded its claims to rights in Korea and Manchuria. In 1904 Japan launched a surprise naval attack on Russian Port Arthur, starting the Russo-Japanese War (1904–1905), which was fought with large armies using automatic weapons. Russia rapidly completed the railroad to Vladivostok (located at the Korean border) and brought in troops on the Trans-Siberian Railroad while sending its Baltic and Black Sea fleets around Africa. However, after an eight-month-siege Russia in January 1905 surrendered Port Arthur. Russian troops were pushed deep into Manchuria, and the Russian fleet was (nearly) completely annihilated by Japanese cannon steamers. Russia might have been able to send reinforcements to drive the Japanese off the Asian mainland, but a revolution at home forced the Czar to seek peace. Hence, to the surprise of the world, Japan spectacularly defeated the Russian giant and gained the Liaodong Peninsula with Port Arthur; the southern half of Sachalin; and Korea (which was formally annexed in 1910). Manchuria, the huge province north of Korea, had been under Russian military control between 1900 and 1904 and was now returned to China. However, the Japanese received extensive rights in this coal and iron-rich region. Most notably, the Japanese took ownership of the railroad tracks that the Russians had laid in Manchuria.

Japanese administration entirely changed Korea. The southern part of the Korean peninsula, with somewhat greater rainfall, a warmer climate, and plenty of arable terrain, served as an agricultural center, mainly for rice production. The northern part, endowed with coal, iron, and nonferrous ores, as well as major river systems providing vast hydropower potential, was developed into a center of mining and industry. The Japanese authorities organized the construction of roads, railroads, ports, and an electrical power grid, but the Korean population was suffering. The destruction of the traditional feudal agrarian society brought substantial changes, and many Koreans were forced to move to Japan to work in Japanese mines and factories.

Japan's earlier colony Formosa (Taiwan) was utilized for agricultural production. The subtropical climate made it well suited for rice and tea cultivation, but Formosa was also known as a sugar island. By 1908 Japanese authorities had established fifty new sugar mills using steam-powered machinery, and Formosa's sugar exports subsequently increased 6-fold between 1900 and 1910.

Muslim World

Southwestern Asia (the Middle East) came under severe European pressure during the Coal Age as well. Muslim countries had been at the forefront of science and technology for a while during the Middle Ages, but they had since become a technological backwater. Generally, much of their focus was on agriculture and trade rather than manufacturing. The Middle East had once enjoyed a vast income from intermediary trade of oriental goods to the West, but these revenues began to dwindle as soon as Europeans started to sail around Africa. In addition, European demand for Arab sugar was eliminated by Caribbean sugar production, and "Saharan" gold (from West Africa) was replaced by New World gold.

When western Europe advanced into the Coal Age there were three main Islamic powers: The Indian Mughal (Mogul) Empire, Persia, and the Ottoman Empire.[88] These regions were gradually feeling more pressure due to imperial rivalry between the Coal Powers, but they were also in conflict with one another, competing for supremacy in the Islamic World. The Indian Mughal Empire had been founded in 1526 by Babur, and his dynasty lasted until the British retired the last Mughal in 1858. In Persia (present-day Iran), an empire was established by the Safavid dynasty (1501–1723), which had its origins in Azerbaijan. However, Persia lost Azerbaijan to Russia in 1806 and had every reason to fear a further Russian expansion. Persian claims on Herat, an independent emirate in north Afghanistan, led to war between Persia and Britain in 1856. Persia then managed to retain its sovereignty, but both Russia and Britain began expanding their economic influence in the country, dividing Persia into Russian and British spheres of influence, plus an intervening neutral (free or common) zone.

Britain was actually very concerned about the Russians moving into the Muslim areas of the Caucasus and Central Asia, because the Russian Empire was coming very close to British India. Russia gained Georgia in 1801, a large part of Armenia in 1828, and annexed the independent emirate of Bukhara (present-day Uzbekistan) in 1868. The khanate of Khiva (also in present-day Uzbekistan) became a Russian protectorate in 1873, and all of Turkmenistan was subdued during the late 1870s. Afghanistan was therefore (even more than Tibet) the last buffer between the Russian Empire and British-controlled India, and became strategically very important for Britain. However, the subsequent attempts to stall the Russian south-bound expansion into Afghanistan were going to cost the British dearly.

Buddhism was the dominant religion in Afghanistan from the 3rd to the 8th century C.E.; thereafter Islam became entrenched, spreading to all of Afghanistan in the 10th and 11th centuries. From the 16th until the early 18th century the Mughal Empire of India and the Safavid Empire of Persia fought over Afghanistan and variably divided the area between them. (Usually the Mughals held Kabul and the Persians held Herat, with Kandahar frequently changing hands.) An independent Afghan Empire was established in 1747, but already in 1793 two brothers, Shah Shujah and Shah Mahmud, fought over its succession. Shujah ended up going into exile in India, and Mahmud withdrew to Herat, while a number of small principalities emerged in the region.

Things were heating up again when Sikhs of India expanded westward and gained control of the Punjab and the region up to the Khyber Pass. The emir of Kabul, Dost Muhammad Khan, reacted by defeating the Sikhs in

1837. This worried the British, because Dost happened to have friendly relations with Russia, while the Sikhs were considered allies of Britain. Hence
the British in 1838 ordered a military intervention in Afghanistan, with the
British East India Company dispatching an army of 16,500 soldiers, and a
camp of 38,000 persons accompanying them. These troops took Kandahar
and Kabul (and dethroned Dost), but were in turn annihilated by Russian-
supplied Afghan Pashtun tribesman. Britain then allowed Dost to regain
the throne, but he pursued a policy amiable to the British for the rest of his
life. In 1878, when Dost refused to permit British troops to be stationed
in Kabul, the British invaded Afghanistan again, this time with a force of
35,700 soldiers, and again they were being ousted after suffering tremendous losses. Coal energy simply did not mean much in the mountainous
terrain of Afghanistan, where horses were the best means of transportation. However, the British in 1891 managed to annex what is today Pakistan's Northwest Frontier Province, a region that was Afghan territory
until then.

The Ottoman Empire

Further west the Ottoman Empire prospered for about six centuries, roughly
from 1300 to 1920. Named for founder Osman I, this Turkish dynasty warred
with the Byzantines and conquered Constantinople in 1453, making the city
its capital. During the following centuries the Ottomans gradually enlarged
their empire, and though they lost parts of the Balkans to the Austrian monarchy, and the regions north of the Black Sea plus the Caucasus (Georgia, Armenia) to Russia, they still ruled over the southern Balkans (including Greece
and Turkey), the entire Fertile Crescent (Iraq, Syria, Palestine), Egypt, the
fertile North African shoreline (Libya, Tunisia, Algeria), and coastal Arabia
in the beginning of the Coal Age.

Then things changed rapidly. Five centuries of Turkish rule in Greece
ended after a naval force of British, French, and Russian ships in 1827 destroyed the Ottoman fleet at the Bay of Navarino, in southwestern Greece.
Next the Ottoman Empire lost Egypt. The land at the Nile river had been
invaded by French troops under Napoleon in 1798, and though the French
occupation ended already in 1801, Egypt was thrown into chaos and anarchy.
The Ottomans reacted in 1805 by appointing a military officer, Mehmet Ali,
as pasha. They had initially sent this Muslim of Greek-Macedonian (or perhaps Albanian) descent to fight the French, but Mehmet Ali established a
European-style conscript army and massacred the Mamelukes, the dominating political class of Egypt. In 1820 Mehmet Ali used his army to annex
Sudan, and in the 1830s he turned against his Ottoman overlord. The European Coal Powers stepped in to stop Mehmet Ali after his troops had occupied all of Syria as well as Medina and Mecca in Arabia. However, the new

Egyptian ruler in turn cooperated with the Europeans, who helped him to change Egypt profoundly. Mehmet Ali confiscated nearly all Egyptian land and began the most ambitious rebuilding of irrigation systems at the Nile since the times of the Pharaohs. In effect, his construction of dams and canals ended the traditional farming system based on the annual Nile flooding and thereby added months to the growing season. However, what was now cultivated in Egypt was cotton for the European textile mills, while wheat and other food was imported.

Egypt also gained special strategic importance due to the construction of the Suez Canal, a 10-year effort started in 1859. An Anglo-French commission organized the financing of the project, and in 1875 the British government bought the indebted Egyptian ruler's shareholding in the Suez Canal to secure control of this strategic waterway. Following a nationalist revolt the British occupied Egypt in 1882, and the country was henceforth in the hands of British civilian agents. The Suez Canal was firmly under British control, but open to all nations from 1888. In 1914 all of Egypt was officially declared a British protectorate.

Further west, in North Africa, Ottoman power declined as well. The province of Morocco and the city-states of Algiers, Tunis, and Tripoli were collectively known as the Barbary States. From the 17th century they acted quite independently and were called Pirate States. Algiers, in particular, was a center of slave trade up until the Coal Age, and pirates from the Barbary Coast kept on operating against European shipping, taking hostages for ransom. American ships were protected by British naval power and diplomacy until U.S. independence, and by the French in the years thereafter. However, U.S. President Jefferson sent cannon ships to the Barbary States shortly after his inauguration in 1801 to fight the first overseas war in U.S. history. But the piracy against U.S. commercial ships in the Mediterranean resumed (and the American government once again started to pay the pirates off) as soon as U.S. troops were busy fighting the War of 1812 against Britain. In 1815 U.S. ships attacked again, and in 1816 an Anglo-Dutch force bombarded Algiers. In 1830 a French army seized Algiers, and by 1847 northern Algeria was firmly under French control. French settlers were then brought in to develop Algeria into the granary that was to underpin the industrialization of France.

The Fertile Crescent and Mesopotamia (Iraq) remained fully under Ottoman rule. A German company built a railroad from Constantinople to Baghdad and further to the Persian Gulf. Germany attempted to gain economic control of the region and then move on to Persia (Iran) and India, but this was met with bitter resistance by Britain, Russia, and France. (During World War I, the Ottomans joined the alliance of Germany and the Austrian-Hungarian monarchy. At the end of the war the Ottoman Empire was partitioned by the British and French.)

NOTES

72. Coke was widely used in English blast furnaces after 1750, according to Harris. John Raymond Harris, *The British Iron Industry, 1700–1850* (London and Basingstoke: Macmillan Education, 1988), quoted in Vaclav Smil, *Essays in World History: Energy in World History* (Boulder, CO: Westview Press, 1994). According to Hammersley, England's iron industry required coal for further expansion after 1760, while it was not short of affordable fuel (charcoal) in the 1660–1760 period. G. Hammersley, "The Charcoal Iron Industry and its Fuel," *Economic History Review* 2 (26) (1973): 593–613, quoted in Kenneth Pomeranz, *The Great Divergence: Europe, China and the Making of the Modern World Economy* (Princeton: Princeton University Press, 2000), 60.

73. Encyclopædia Britannica, "History of Austria: Conflicts with Napoleonic France," http://www.britannica.com/EBchecked/topic/44183/Austria/33360/Conflicts-with-Napoleonic-France#ref409001.

74. John Haywood, *World Atlas of the Past, Vol. 1–4* (Oxford: Andromeda Oxford/Oxford University Press, 1999). I used the Danish edition: John Haywood, *Historisk Verdensatlas* (Köln: Könemann, 2000).

75. Adam Hochschild, *King Leopold's Ghost* (London: Pan Macmillan, 1998).

76. To be sure, some aborigines apparently practiced a kind of fish farming in irrigated ponds, which is technically a form of agriculture in the wider sense. The term nowadays used for fish farming is aquaculture.

77. M. Gracey, "Nutrition of Australian Aboriginal Infants and Children," *J-Paediatr-Child-Health* 5 (1991): 259–71.

78. "An Outline of American History," Bureau of International Information Programs, U.S. Department of State, http://usinfo.state.gov/products/pubs/histry otln/index.htm; Howard Zinn, *A People's History of the United States: 1492-Present* (New York: HarperCollins, 1999).

79. Glyn Davies, *A History of Money: From Ancient Times to the Present Day* (Cardiff: University of Wales Press, 1994).

80. Vernellia R. Randall, "Indian Redress: How the West Was Stolen," excerpted from: William Bradford, "With a Very Great Blame on Our Hearts: Reparations, Reconciliation, and an American Indian Plea for Peace with Justice," *American Indian Law Review* 27 (1–174) (2002–2003): 19–75, http://academic.udayton.edu/race/02rights/native14.htm.

81. "Energy in the United States: 1635–2000," Energy Information Administration, http://www.eia.doe.gov/emeu/aer/eh/total.html.

82. A State Department list, "Instances of the Use of United States Armed Forces Abroad 1798–1945" (presented by Secretary of State Dean Rusk to a Senate committee in 1962 to cite precedents for the use of armed force against Cuba), shows 103 interventions in the affairs of other countries between 1798 and 1895. Howard Zinn, "The Empire and the People," http://www.historyisaweapon.com/defcon1/zinnempire12.html. Find the complete, updated list in: Richard F. Grimmett, "Instances of Use of United States Armed Forces Abroad, 1798–2007," Updated January 14, 2008, CRS (Congressional Research Service) Report for Congress, http://fas.org/sgp/crs/natsec/RL32170.pdf.

83. Richard F. Grimmett, "Instances of Use of United States Armed Forces Abroad, 1798–2007," Updated January 14, 2008, CRS (Congressional Research Service) Report for Congress, http://fas.org/sgp/crs/natsec/RL32170.pdf.

84. Library of Congress Country Studies, "Uzbekistan—The Russian Conquest," http://lcweb2.loc.gov/frd/cs/uztoc.html. Russian forces managed to take the crucial Samarqand area from the Bukhara Khanate in 1868. A century later, cotton cultivation in Uzbekistan led to enormous environmental disaster. Stalin's plans to produce unrealistic quantities of extremely thirsty cotton in the area resulted in the redirection of two rivers that used to disembogue into the enormous Aral Lake (Sea), which began drying up in the 1960s. At the beginning of the 21st century Uzbekistan was still the world's second-largest cotton exporter.

85. Kenneth L. Pomeranz and R. Bin Wong, "China and Europe: 1500–2000 and Beyond: What is "Modern"?," http://afe.easia.columbia.edu/chinawh/index.html; "China and Europe: 1780–1937, What Happened?," http://afe.easia.columbia.edu/chinawh/web/s6/s6_2.html. There are also plenty of other attempts to explain the divergence between western Europe and China. Mark Elvin, for instance, coined the term high-level equilibrium trap to explain why the Chinese did not experience an indigenous Industrial Revolution. Without being backward, he argues, China was characterized by a kind of technological immobility, because it was unattractive to develop labor-saving machinery under the given circumstances, which included a large population and cheap labor. Mark Elvin, "The High-level Equilibrium Trap: The Causes of the Decline of Invention in the Traditional Chinese Textile Industries" in *Economic Organization in Chinese Society*, ed. W. E. Willmott (Stanford, CA: Stanford University Press, 1972), 137–172.

Other scholars have pointed out that China, unlike Europe, was centrally governed and thus lacked the fragmented political landscape that would unleash the competitive forces that promote progress. There is, however, no good explanation why such difference on the two ends of the Eurasian landmass arose in the first place. Jared Diamond offered a somewhat weak explanation in *Guns, Germs, and Steel* by suggesting that a difference in the occurrence of barriers in the geographical landscape was behind it. If I was to turn to an environmental explanation, I would rather say that differences in the agricultural systems are to be held responsible. China's agriculture, based on a few river systems with canals and much irrigation, called for more central governance than Europe's decentralized, rain-fed agriculture. But this would be a paraphrasing of the *hydraulic civilization* thesis of Wittfogel, which is not all that popular, especially because it involves the term "Oriental Despotism." Karl August Wittfogel, *Oriental Despotism: A comparative Study of Total Power* (New Haven, CT: Yale University Press, 1957).

Various other factors that have been suggested to explain why Europe rather than China experienced an Industrial Revolution (first) are included in the following paper: Peter C. Perdue, "China in the Early Modern World: Shortcuts, Myths and Realities," *Education About Asia* 4, no. 1 (1999): 21–26, http://web.mit.edu/21h.504/www/china_emod.htm, http://ocw.mit.edu/NR/rdonlyres/History/21H-504East-Asia-in-the-WorldSpring2003/70208DF1-1DAB-4B4A-97C8-6800A7D0EE81/0/china_emod.pdf.

86. Find more on this topic in Andre Gunder Frank, *ReOrient: The Global Economy in the Asian Age* (Berkeley: University of California Press, 1998).

87. Kenneth Pomeranz, *The Great Divergence*.

88. Find a short political world history in: John Haywood, *World Atlas of the Past*, *Vol. 1–4* (Oxford: Andromeda Oxford/Oxford University Press, 1999). Detailed histories of many countries can be found in the Library of Congress Country Studies, http://lcweb2.loc.gov/frd/cs/list.html.

BIBLIOGRAPHY TO PART III

Association for Science Education, The. "History of Medicine, 1700–1900: 18th and 19th centuries." http://www.schoolscience.co.uk/content/4/biology/abpi/history/history8.html.

Axon, William E. A., ed. *The Annals of Manchester—A chronological record from the earliest times to the end of 1885.* Manchester: Manchester Central Library, Salford Local History Library, 1886.

Bairoch, Paul. *Economics and World History. Myths and Paradoxes.* Chicago: University Of Chicago Press, 1993.

Bloy, Marjie. "The Anti-Slavery Campaign in Britain." http://www.victorianweb.org/history/antislavery.html.

Bradford, William. "With a Very Great Blame on Our Hearts: Reparations, Reconciliation, and an American Indian Plea for Peace with Justice." *American Indian Law Review* 27 (1–174) (2002–2003): 19–75.

Brimblecombe, Peter, and László Makra. "Selections from the History of Environmental Pollution, with Special Attention to Air Pollution. Part 2: From medieval times to the 19th century." *Int. J. Environment and Pollution* 23 (2005): 354. http://www.sci.u-szeged.hu/eghajlattan/makracikk/Brimblecombe%20Makra%20IJEP.pdf.

Britannica Guide to the Nobel Prizes. "Alfred Nobel—The Man." http://www.britannica.com/nobel/micro/427_33.html.

Bureau of International Information Programs. "An Outline of American History." U.S. Department of State. http://usinfo.state.gov/products/pubs/histryotln/index.htm.

Calvert, J. B. "Hydrostatics." University of Denver. http://www.du.edu/~jcalvert/tech/fluids/hydstat.htm.

Campbell, F. Gregory. "The Struggle for Upper Silesia, 1919–1922." *The Journal of Modern History* 42 (1970): 361–385. http://www.jstor.org/pss/1905870.

Canal Archive: Bridging the Years: "The Bridgewater Canal." http://www.canal archive.org.uk/stories/pages.php?enum=TE133&pnum=0&maxp=7.

Chandler, Alfred D. *The Railroads: The Nation's First Big Business*. New York: Harcourt, Brace & World, 1965.

Cipolla, Carlo M. *The Economic History of World Population*. Hassocks, Sussex: Harvester Press, 1978.

Cohen, Daniel. *Fehldiagnose Globalisierung*. Frankfurt/New York: Campus Verlag, 1999.

Cohen, Joel E. *How Many People Can the Earth Support?* New York: W. W. Norton & Company, 1995.

Cook, Earl. *Man, Energy, Society*. San Francisco: W. H. Freeman, 1976.

Cottrell, Fred. *Energy and Society—The Relation between Energy, Social Change, and Economic Development*. New York: McGraw-Hill Book Company, 1955.

Daly, Herman E. "The Perils of Free Trade." *Scientific American* 269 (1993): 50–57.

Davies, Glyn. *A History of Money: From Ancient Times to the Present Day*. Cardiff: University of Wales Press, 1994.

DeLong, J. Bradford. "Slouching Towards Utopia?: The Economic History of the Twentieth Century, Part B: The Path from the Pre-Industrial World, VIII. The First Global Economy: Production and Trade." University of California at Berkeley and NBER (National Bureau of Economic Research), January 1997. http://econ161.berkeley.edu/TCEH/2000/eight/html/Slouching2_8preWWI.html.

DeLong, J. Bradford. "Slouching Towards Utopia?: The Economic History of the Twentieth Century, VII. The Pre-World War I Economy." University of California at Berkeley and NBER (National Bureau of Economic Research), January 1997. http://econ161.berkeley.edu/TCEH/Slouch_PreWWI7.html.

Diamond, Jared. *Guns, Germs, and Steel: The Fates of Human Societies*. New York: W. W. Norton & Company, 1997.

Digital History. "California Imposes a Tax on Chinese Laborers." http://www.digitalhistory.uh.edu/asian_voices/voices_display.cfm?id=16.

Dukes, J.S. "Burning Buried Sunshine: Human Consumption of Ancient Solar Energy." *Climatic Change* 61 (2003): 31–44. http://globalecology.stanford.edu/DGE/Dukes/Dukes_ClimChange1.pdf.

Economist, The. "Catch the Wave: The Long Cycles of Industrial Innovation Are Becoming Shorter." February 18, 1999.

Economist, The. "Richard Cobden: An Anti-poverty Hero." June 3, 2004. http://www.economist.com/opinion/displaystory.cfm?story_id=E1_NSNVDNN.

Economist, The. "Survey: Food—Make It Cheaper, and Cheaper." December 11, 2003. http://www.economist.com/displayStory.cfm?Story_id=2261831.

Elvin, Mark. "The High-level Equilibrium Trap: The Causes of the Decline of Invention in the Traditional Chinese Textile Industries." In *Economic Organization in Chinese Society*, edited by W. E. Willmott, 137–172. Stanford: Stanford University Press, 1972.

Encyclopædia Britannica, "Cavalry." http://www.britannica.com/EBchecked/topic/100566/cavalry.

Encyclopædia Britannica, "History of Austria: Conflicts with Napoleonic France." http://www.britannica.com/EBchecked/topic/44183/Austria/33360/Conflicts-with-Napoleonic-France#ref409001.

Energy Information Administration. "Energy in the United States: 1635–2000." http://www.eia.doe.gov/emeu/aer/eh/total.html.

Foner, Eric, and John Arthur Garraty. *The Reader's Companion to American History*. New York: Houghton Mifflin Books, 1991.

Frängsmyr, Tore. "Life and Philosophy of Alfred Nobel—A memorial address by The Royal Swedish Academy of Sciences." http://nobelprize.org/nobel/alfred-nobel/biographical/frangsmyr/.

Frank, Andre Gunder. *ReOrient: The Global Economy in the Asian Age*. Berkeley: University of California Press, 1998.

Gracey, M. "Nutrition of Australian Aboriginal Infants and Children." *J-Paediatr-Child-Health* 5 (1991): 259–71.

Griffin, Ricky W., and Michael W. Pustay. *International Business: A Managerial Perspective*. Reading: Addison-Wesley Publishing Company, 1996.

Grimmett, Richard F. "Instances of Use of United States Armed Forces Abroad, 1798–2007." Updated January 14, 2008, CRS (Congressional Research Service) Report for Congress. http://fas.org/sgp/crs/natsec/RL32170.pdf.

Hammersley, G. "The Charcoal Iron Industry and its Fuel." *Economic History Review* 2 (26) (1973): 593–613.

Harris, John Raymond. *The British Iron Industry, 1700–1850*. London and Basingstoke: Macmillan Education, 1988.

Haywood, John. *World Atlas of the Past, Vol. 1–4*. Oxford: Andromeda Oxford/Oxford University Press, 1999. Danish edition: John Haywood, *Historisk Verdensatlas*. Köln: Könemann, 2000.

Headrick, Daniel R. *The Tentacles of Progress: Technology Transfer in the Age of Imperialism, 1850–1940*. New York: Oxford University Press, 1988.

Hochschild, Adam. *King Leopold's Ghost*. London: Pan Macmillan, 1998.

House of Representatives. "107th CONGRESS, 1st session, H. RES. 269, Expressing the sense of the House of Representatives to honor the life and achievements of 19th Century Italian-American inventor Antonio Meucci, and his work in the invention of the telephone." September 25, 2001. http://www.popular-science.net/history/meucci_congress_resolution.html.

Hughes, Stephen. "The International Collieries Study." ICOMOS (International Council on Monuments and Sites) and TICCIH (The International Committee for the Conservation of the Industrial Heritage). International Council on Monuments and Sites, Paris, France. http://www.international.icomos.org/centre_documentation/collieries.pdf.

Jensen, Oliver. *Railroads in America*. New York: Bonanza Books, 1975.

Kirby, Richard Shelton, et al. *Engineering in History*. New York: Courier Dover, 1990.

Koplow, David A. *Smallpox: The Fight to Eradicate a Global Scourge*. Berkeley: University of California Press, 2003. http://www.ucpress.edu/books/pages/9968/9968.ch01.php.

Krugman, Paul R., and Maurice Obstfeld, *International Economics: Theory and Policy*. New York: HarperCollins, 1994.

Library of Congress Country Studies. http://lcweb2.loc.gov/frd/cs/cshome.html. (Alternative site: http://countrystudies.us.) This site presents a comprehensive description and analysis of countries' historical setting, geography, society, economy, political system, and foreign policy. The country studies are on-line versions of books previously published in hard copy by the Federal Research Division of the Library of Congress as part of the Country Studies/Area Handbook Series sponsored by the U.S. Department of the Army between 1986 and 1998.

Library of Congress Country Studies. "Uzbekistan: The Russian Conquest." http://lcweb2.loc.gov/frd/cs/uztoc.html.

Maddison, Angus. *The World Economy:* Volume 1: A Millennial Perspective, Volume 2: Historical Statistics. Paris: OECD Publishing, 2006.

McNeill, John Robert, and William Hardy McNeill, *The Human Web: A Bird's-eye View of World History.* New York: W. W. Norton & Company, 2003.

Mensch, G. O. *Stalemate in Technology: Innovations Overcome the Depression.* Cambridge: Ballinger, 1979.

Nef, John Ulric. *The Rise of the British Coal Industry.* Two volumes. London: Frank Cass, 1966. (Reprint of the original edition from 1932.)

Nuvolari, Alessandro. "Collective Invention during the British Industrial Revolution: The Case of the Cornish Pumping Engine." Eindhoven Centre for Innovation Studies, Faculty of Technology Management, Technische Universiteit Eindhoven, Working Paper 01.04, May 2001. http://www.tm.tue.nl/ecis/Working%20Papers/eciswp37.pdf.

Nye, David E. *Consuming Power: A Social History of American Energies.* Cambridge: MIT Press, 1999.

Online A-Z of 3-Wheeled Cars. "Cugnot." http://www.3wheelers.com/cugnot.html.

Paleontology Portal, The. "The Carboniferous—354 to 290 Million Years Ago." http://www.paleoportal.org/time_space/period.php?period_id=12.

Perdue, Peter C. "China in the Early Modern World: Shortcuts, Myths and Realities." *Education About Asia* 4:1 (1999): 21–26. http://web.mit.edu/21h.504/www/china_emod.htm. http://ocw.mit.edu/NR/rdonlyres/History/21H-504East-Asia-in-the-WorldSpring2003/70208DF1-1DAB-4B4A-97C8-6800A7D0EE81/0/china_emod.pdf.

Pomeranz, Kenneth L., and R. Bin Wong, "China and Europe: 1500–2000 and Beyond: What Is "Modern"?" http://afe.easia.columbia.edu/chinawh/index.html.

Pomeranz, Kenneth L., and R. Bin Wong, "China and Europe: 1780–1937, What Happened?" In "China and Europe: 1500–2000 and Beyond: What is "Modern"?" http://afe.easia.columbia.edu/chinawh/web/s6/s6_2.html.

Pomeranz, Kenneth. *The Great Divergence: Europe, China and the Making of the Modern World Economy.* Princeton: Princeton University Press, 2000.

Public Broadcasting Service (PBS). "Coming to America." *Tesla: Life and Legacy.* http://www.pbs.org/tesla/ll/ll_america.html.

Public Broadcasting Service (PBS). "High Frequency." *Tesla: Life and Legacy.* http://www.pbs.org/tesla/ll/ll_hifreq.html.

Public Broadcasting Service (PBS). "War of the Currents." *Tesla: Life and Legacy.* http://www.pbs.org/tesla/ll/ll_warcur.html.

Public Broadcasting Service (PBS). "Who Invented Radio?" *Tesla: Life and Legacy.* http://www.pbs.org/tesla/ll/ll_whoradio.html.

Randall, Vernellia R. "Indian Redress: How the West Was Stolen." http://academic.udayton.edu/race/02rights/native14.htm.

Russell, Ben. "Powered by Steam: The Steam Engine 1780–1830." Science Museum, London. http://www.fathom.com/course/21701780/session4.html.

Schumpeter, Joseph A. *Business Cycles: A Theoretical and Statistical Analysis of the Capitalist Processes.* New York: McGraw-Hill, 1939.

Science Museum, The. "Centres of Excellence: Engineering pioneers." *Making the Modern World.* http://www.makingthemodernworld.org.uk/stories/manufacture_by_machine/03.ST.01/?scene=5.

Science Museum, The. "Constructing the Railway System." *Making the Modern World.* http://www.makingthemodernworld.org.uk/stories/the_age_of_the_engineer/01.ST.04/.

Science Museum, The. "New Drugs: The industrialisation of pharmaceuticals." *Making the Modern World.* http://www.makingthemodernworld.org.uk/stories/the_second_industrial_revolution/05.ST.01/?scene=4&tv=true.

Science Museum, The. "New Dyes." *Making the Modern World.* http://www.makingthemodernworld.org.uk/stories/the_second_industrial_revolution/05.ST.01/?scene=2&tv=true.

Science Museum, The. "New Power." *Making the Modern World.* http://www.makingthemodernworld.org.uk/stories/the_second_industrial_revolution/05.ST.01/?scene=7.

Science Museum, The. "Power for Production." *Making the Modern World.* http://www.makingthemodernworld.org.uk/stories/the_age_of_the_engineer/03.ST.02/?scene=7.

Science Museum, The. "The Rainhill Trials, October 1829." *Making the Modern World.* http://www.makingthemodernworld.org.uk/stories/the_age_of_the_engineer/01.ST.04/?scene=5.

Science Museum, The. "Stephenson's Rocket Locomotive, 1829." *Making the Modern World.* http://www.makingthemodernworld.org.uk/icons_of_invention/technology/1820-1880/IC.007/.

Smil, Vaclav. *Essays in World History: Energy in World History.* Boulder, CO: Westview Press, 1994.

Smil, Vaclav. *Transforming the Twentieth Century: Technical Innovations and Their Consequences.* New York: Oxford University Press, 2006.

Smith, Adam. *The Wealth of Nations.* Buffalo: Prometheus Books, 1991. Original edition published in 1776.

Smith, Charles H. "The Alfred Russel Wallace Page." Western Kentucky University. http://www.wku.edu/%7Esmithch/index1.htm.

Spartacus Schoolnet. "George Stephenson." http://www.spartacus.schoolnet.co.uk/RAstephensonG.htm.

State Department. "Instances of the Use of United States Armed Forces Abroad 1798–1945" (presented by Secretary of State Dean Rusk to a Senate committee in 1962 to cite precedents for the use of armed force against Cuba). http://www.historyisaweapon.com/defcon1/zinnempire12.html.

Stover, John F. *The Life and Decline of the American Railroad.* Oxford: Oxford University Press, 1970.

Tappenden, Roslyn. "Birmingham's Canal Network—In Brindley's Footsteps." 24 Hour Museum. http://www.24hourmuseum.org.uk/trlout_txo_en/TRA23386.html.